Joseph M. Renes

Quantum Information Theory

Also of Interest

Data Science
Time Complexity, Inferential Uncertainty, and Spacekime Analytics
Ivo D. Dinov, Milen Velchev Velev, 2021
ISBN 978-3-11-069780-3, e-ISBN 978-3-11-069782-7

Quantum Systems, Channels, Information
A Mathematical Introduction
Alexander S. Holevo, 2019
ISBN 978-3-11-064224-7-, e-ISBN 978-3-11-064249-0

Coherent Quantum Physics
A Reinterpretation of the Tradition
Neumaier, 2021
ISBN 978-3-11-066729-5, e-ISBN 978-3-11-066738-7

Electrons in Solids
Mesoscopics, Photonics, Quantum Computing, Correlations, Topology
Bluhm, Brückel, Morgenstern, von Plessen, Stampfer, 2019
ISBN 978-3-11-043831-4, e-ISBN 978-3-11-043832-1

Joseph M. Renes

Quantum Information Theory

Concepts and Methods

DE GRUYTER

Author
Dr. Joseph M. Renes
Quantum Information Theory Group
Insitute of Theoretical Physics
ETH Zurich
Wolfgang-Pauli-Straße 27
8093 Zürich
Switzerland

ISBN 978-3-11-057024-3
e-ISBN (PDF) 978-3-11-057025-0
e-ISBN (EPUB) 978-3-11-057032-8

Library of Congress Control Number: 2022934628

Bibliographic information published by the Deutsche Nationalbibliothek
The Deutsche Nationalbibliothek lists this publication in the Deutsche Nationalbibliografie; detailed bibliographic data are available on the Internet at http://dnb.dnb.de.

Cover image: sakkmesterke / iStock / Getty Images Plus
Typesetting: VTeX UAB, Lithuania
Printing and binding: CPI books GmbH, Leck

www.degruyter.com

To my family

The composition of this book has been for the author a long struggle of escape, and so must the reading of it be for most readers if the author's assault upon them is to be successful,—a struggle of escape from habitual modes of thought and expression. The ideas which are here expressed so laboriously are extremely simple and should be obvious. The difficulty lies, not in the new ideas, but in escaping from the old ones, which ramify, for those brought up as most of us have been, into every corner of our minds.

John Maynard Keynes, The General Theory of Employment, Interest and Money, 1936.

The great thing about physical intuition is that it can be adjusted to fit the facts.

Roger Penrose, quoted in *Quantum Field Theory* by Mark Srednicki.

Preface

Like many such endeavors, this book evolved out of lecture notes for a master level course on quantum information theory that I have given several times at ETH Zürich and at the Technical University of Darmstadt. The main goal of the book, as with the course, is to understand in detail some of the fundamental limitations and possibilities of information processing with quantum-mechanical information carriers. The particular focus is on communication and cryptographic tasks, leaving computational issues for another day.

In outlook and aims, this book is very much inspired by Asher Peres's *Quantum Theory: Concepts and Methods* and John Baez and Javier Muniain's *Gauge Fields, Knots and Gravity*. Peres's "strictly instrumentalist" focus was formative for my own initial understanding of the field, and I follow an operational approach throughout. This helps avoid imbuing the mathematical quantities one can define in quantum theory with any unwarranted physical meaning. The relentless operational approach leads to a slight clash with many other texts in that standard quantities such as distinguishability or fidelity are defined here in variational terms, having to do with optimal measurements or similar, and only later are the corresponding closed forms derived. Continuing in this operational spirit, I felt, justified borrowing a bit from Peres's title. Baez and Muniain is, in my opinion, a real masterpiece of exposition. I have sought (perhaps in vain) to emulate their logical, elegant, and motivated presentation and development of the topics under consideration, though of course the particular topics here are completely different.

I have also borrowed the three-part structure from both. The first covers the formalism of quantum information theory, or really open quantum systems of finite dimension, developing it by explicit analogy with classical probability theory. The second part deals with the tools useful for analyzing information processing tasks, most of which also find application in other parts of the field. Characteristic of Part II is a heavy reliance, if not overreliance, on semidefinite programming to phrase and derive most of the main results. I have attempted to keep the required background for the reader to a modest level—a thorough understanding of linear algebra—and then to derive everything from there. The third part takes up the information processing protocols themselves, with the particular emphasis on so-called "one-shot" statements of structureless resources. These statements are formulated in terms of a quantity involved in the simpler task of binary hypothesis testing. From there, results for the usual setting of i. i. d. resources can be recovered by Stein's lemma relating hypothesis testing to entropy. This is also quite at odds with most texts on information theory, which places entropy front and center, even treating it axiomatically. Again, I follow the strictly operational approach. My aim here has also been to keep the number of different tools developed in Part II and techniques employed in Part III to a minimum. The statements found in the latest literature use a bewildering variety of quantities and methods, and we have to make a cut somewhere. I also thought it useful to make a

https://doi.org/10.1515/9783110570250-201

somewhat different choice of techniques than other texts, if only for variety. Moreover, the relationship between privacy and error correction in the quantum realm, mediated by concrete formulations of the uncertainty principle, is too grand not to explore in more detail.

The material contained in these pages owes much inspiration to the lecture notes by Carl Caves, John Preskill, and Renato Renner, as well as John Watrous's *Theory of Quantum Information* and Mark Wilde's *Quantum Information Theory*. But much more than that it is the result of wrestling with the material contained therein, scrutinizing and rehashing it, adding some bits and subtracting others, and finally molding it to fit my own sense of the landscape of the theory. The two quotes in the epigraph speak beautifully to this point. Readers who really wish to understand the material will have to undertake their own struggle of course, and I hope they (you!) at least gain some inspiration from seeing the particular development here. Working through the exercises will also help.

I have benefited enormously from often lengthy conversations over the past decade with the members of the quantum information theory group at ETH Zürich, and the book would not be what it is were it not for the very stimulating atmosphere of the group and the Institute of Theoretical Physics as a whole. I especially thank Volkher Scholz, Michael Walter, David Sutter, Christophe Piveteau, Ernest Tan, Henrik Wilming, Fred Dupuis, Marco Tomamichel, Mario Berta, and Renato Renner. The material on quantum error correction has benefited greatly from my earlier collaborations with Graeme Smith and especially with Jean-Christian Boileau. The overall presentation has been refined by the very helpful feedback from students and teaching assistants in the quantum information theory course over the years. I am grateful to the assistants Sandra Stupar, Jinzhao Wang, Lisa Hänggli, Philipp Kammerlander, Raban Iten, Alessandro Tarantola, Imre Májer, and Ernest Tan, and students too numerous to list here. To be sure, there is about twice as much material here as for our one semester course, but more than a few rough edges have been polished, and several gaps in the proofs fixed through their feedback, particularly in the earlier material.

Finally, I am immensely grateful to my wonderful wife Andrea. Without her encouragement, love, and support this project could never have been started, let alone finished.

Zürich, May 2022 Joseph M. Renes

Contents

Part III: **Information processing protocols**

List of Figures

https://doi.org/10.1515/9783110570250-202

1 Introduction

"Information is physical" claimed the physicist Rolf Landauer,[1] writing

> [Information processing] is inevitably done with real physical degrees of freedom, obeying the laws of physics, and using parts available in our actual physical universe. How does that restrict the process? [182]

The field of quantum information theory is concerned with what sorts of information processing tasks can and cannot be performed if the underlying information carriers are governed by the laws of *quantum* mechanics as opposed to *classical* mechanics. For example, we might use the spin of a single electron to store information, rather than the magnetization of a small region of magnetic material as in a hard disk drive or even ink marks on a piece of paper as in this book; or we might use the spin of the electron to detect a magnetic field. We might transmit information using just single photons rather than a number so large that we can use the classical wave equation.

The overriding question is what can quantum information processing do that classical information processing cannot, and vice versa? Famously, it is not even possible to copy quantum information, due to the no-cloning theorem. While this fact might make quantum information seem quite useless, it is not so! Quantum computers are, also famously, capable of factoring large integers very efficiently. (To be fair, computer scientists are not certain that standard classical computers are not capable of similar speeds.)

Moreover, information processing by quantum devices makes possible some very counterintuitive protocols. Perhaps most striking is quantum key distribution, which allows two parties separated by a large distance to create a secret key, a random binary string known only to them, by using only *insecure* means of communication. This is surely impossible using classical information carriers, for how would the parties ever know if an eavesdropper spied on all their communication? In contrast to the laws of classical mechanics, quantum mechanics places limitations on the accessibility of information, as exemplified by no-cloning or the uncertainty principle. A would-be eavesdropper cannot spy on quantum communication signals without leaving some evidence of having done so. However, while this is helpful for quantum key distribution, it raises the question of whether, without the possibility of copying, it is possible to protect quantum information from the inevitable noise in actual quantum information processing devices. The simplest method of protecting classical information is, after all, just to repeat it.

The goal of this book is to provide a solid understanding of the mathematical formalism of quantum information theory, with which we can then examine some of the

1 Rolf Wilhelm Landauer, 1927–1999.

https://doi.org/10.1515/9783110570250-001

counterintuitive phenomena in more detail. The focus is on communication and cryptographic tasks, and ultimately we will examine the tasks of protecting quantum information and of quantum key distribution just mentioned. Nonetheless, before embarking on a mathematical treatment, we should first step back and come to a clear understanding of just what information is and why physical law should place any restrictions on information processing at all. Then we can ask what quantum information is and consider what additional physical restrictions or possibilities exist for quantum information processors.

1.1 What is information?

To understand what is meant by information, we look back to its original use in communication engineering. Consider a sender and a receiver who would like to communicate by some means, some *communication channel*, for instance, an electrical telegraph. In this setting, transmitting information refers to the ability to transmit a specific selected message from a set of possible messages. The amount of information is then the number of possible messages the sender *could* reliably transmit to the receiver. Usually, we describe the amount of information logarithmically, in *bits*: n bits refers to the 2^n messages that could be expressed in a binary sequence of length n. The notion of information is less concerned with which one was *actually* sent and is completely unrelated to whether the particular message is meaningful to the recipient. All that is required in the communication scenario is that the sender and receiver agree what the possible messages are in advance, i. e., how to recognize which message is which. This means there is only one kind of information. It does not matter if the message to be conveyed is prose or poetry, an image or a sound, or some combination of all of these.

As a concrete example of a communication device, a very early telegraph from around 1810 used separate wires (!) for different letters and numerals. Current in a wire was detected by passing the wire through a glass tube filled with acid and observing hydrogen bubbles created by electrolysis. Here the messages are plain language, encoded into current in the appropriate wire one letter or numeral at time and read back similarly by the receiver. The subsequent invention of the galvanometer[2] to measure current enabled much simpler, commercially viable designs with fewer wires, such as telegraphs using Morse[3] code.

Information has to do not only with the actual state of affairs, but also with the possible states of affairs. This makes it a very different kind of quantity than energy or momentum, which are inherent properties of a physical system. A physical system,

2 Luigi Galvani, 1737–1798.

3 Samuel Finley Breese Morse, 1791–1872.

i. e., some particular degrees of freedom, has a certain amount of energy and momentum that we can in principle just directly measure; the values of these quantities do not depend on what the system *could* be doing, only on what it is *actually* doing. Information, in contrast, has to take into account all the possible, but indeed counterfactual states of affairs. For this reason, we cannot just measure the information content of a system. Furthermore, the correspondence between messages and physical configurations of the information carrier is not inherent to the system, but relative to the sender and receiver.

Norbert Wiener,[4] one of the pioneers of the field of information theory, put it succinctly:

> Because information depends, not merely on what is actually said, but on what might have been said, its measure is a property of a set of possible messages, or of what is called an ensemble in statistical mechanics. [302]

Wiener's mention of statistical mechanics tips us off to the fact that actually there are physical quantities whose definition relies on an ensemble: the Boltzmann[5] or Gibbs[6] entropies of statistical mechanics. The Boltzmann entropy of a gas, for instance, is proportional to the logarithm of the number of microstates, the possible positions and velocities of the gas molecules themselves, which are consistent with a fixed macrostate, i. e., fixed macroscopic properties such as volume or pressure. The entropy thus depends on which positions and momenta the molecules could have, not merely on which actual configuration they do have. The precise formula, famously inscribed on his tombstone, is $S = k \log W$, where S is the Boltzmann entropy, W is the number of microstates, and k is the proportionality constant, which today we would call the Boltzmann factor. Note that in thermodynamics, we do not directly measure entropy either, but rather infer its value from other directly measurable quantities.

Of course, the difficulty of being able to reliably transmit any one of a large set of messages lies in the inherent noise of the communication channel. This, too, has a counterfactual aspect in that the communication system must be able to contend with different possible noise patterns. Only one noise pattern will actually occur, but which one will not be known beforehand.

The field of information theory did not really take off until the work of Wiener and, particularly, Claude Shannon[7] in the late 1940s, which treated the "counterfactual" aspect of information and noise by using probability theory. In this framework

---

4 Norbert Wiener, 1894–1964.

5 Ludwig Eduard Boltzmann, 1844–1906.

6 Josiah Willard Gibbs, 1839–1903.

7 Claude Elwood Shannon, 1916–2001.

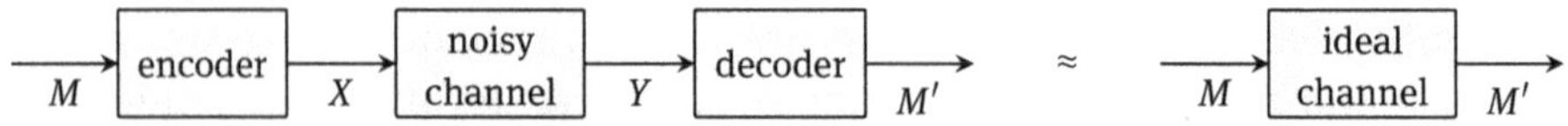

Figure 1.1: The abstract communication scenario. The goal is to transmit messages M from sender to receiver. Ideally, the message M' understood at the receiver is identical to that sent by the sender. In the actual device the noisy communication channel takes input X from the sender and outputs Y to the receiver. The sender and receiver employ some means of encoding and decoding to combat the noise.

the output of an information source or the pattern of noise is treated as a random variable. Shannon was able to give a meaningful measure of the information content of a random variable, nowadays called the Shannon entropy, in his landmark 1948 paper "A Mathematical Theory of Communication". Moreover, he showed that essentially error-free communication over repeated uses of a noisy channel is possible, provided that the rate of communication (the ratio of message bits to channel uses) remains below the *capacity* of the channel, which has an expression in terms of entropy. By giving an abstract mathematical description of the communication scenario, shown in Figure 1.1, and developing useful mathematical tools like entropy for its analysis, Shannon, Wiener, and the other pioneers of information theory dramatically increased the scope of how one could engineer reliable means of communication. Instead of focusing on improving the communication channel itself, their work made it possible to analyze very different encoding and decoding methods used by the sender and receiver.

A striking example of this difference is the development of different modulation techniques of electromagnetic signals. The more straightforward approach to faithfully transmitting, say, voice or music, is to modulate electromagnetic waves using the sound waves in an analog fashion, the method used in AM and FM radio. However, this method does not make very efficient use of the electromagnetic spectrum. Modern communication standards, such as used in digital audio broadcasting or mobile communication, rely on much more intricate means of digital modulation and are able to reliably transmit substantially more information in the same bandwidth.

The above discussion is concerned entirely with classical information, but the setup makes it easy to jump to the quantum world. If instead of transmitting or storing an arbitrary message, we can instead transmit or store different messages in *quantum superposition*, then the channel is capable of sending quantum information. For instance, we may use the two sides of a coin to store a single classical bit: the coin showing heads corresponding to 0 and tails to 1. These two configurations are eminently distinguishable from each other and make for a good encoding. Quantum-mechanically, these two configurations would be represented by quantum states (wavefunctions or wavevectors) $|\text{heads}\rangle$ and $|\text{tails}\rangle$, both elements of a quantum-mechanical state space $\mathcal{H}$. Here we are using Dirac[8] notation to denote the vector states. The state space is a vector space, meaning we can also consider superpositions such as $a|\text{heads}\rangle + b|\text{tails}\rangle$ for any coefficients $a, b \in \mathbb{C}$. The dimension of the vector space spanned by the superpositions, here two, is akin to the number of messages. Quantifying the amount of information logarithmically, as in the classical case, this "quantum coin" example corresponds to a single *qubit* of quantum information. Also, just as in the classical case, when transmitting or storing quantum information, the encoder and decoder will have to overcome the quantum noise that plagues the communication channel.

That quantum mechanics plays an important role in communication and has implications for Shannon's theory of information was discussed by Gabor[9] already in 1950. In particular, quantum effects play an important role in communication at optical and infrared frequencies, whereas the limitations on communication in microwaves and radio were understood to be limited by thermal effects. The ability to treat the most general quantum noise (a term introduced by Gabor) as well as general encoding and decoding operations required further development of the formalism of quantum theory, especially the measurement process. In fact, the ultimate limits on classical communication in the presence of realistic quantum electromagnetic noise were only established in the previous decade. In the early development, quantum effects were seen as a nuisance for the most part, a source of noise mathematically more complicated to handle. That quantum effects could actually be useful for something was first realized by Wiesner[10] in the early 1970s, whose notion of "conjugate coding" anticipated quantum key distribution. The idea of quantum computation was proposed not long thereafter, in the early 1980s, but it was not until Shor's[11] discovery of an efficient factoring algorithm in 1994 that he field of quantum information theory really came into its own.

1.2 Why are there physical limits to information processing?

Now that we have a firmer understanding of the concept of information, we can ask if there really "ought" to be any physical limits on information processing at all. For instance, does computation require energy? How much information can we store in a given volume of space? Again the issue is the counterfactual nature of information. Any single execution of a given information processing task is only concerned with its particular inputs and particular outputs. In each such case, we could imagine that

8 Paul Adrien Maurice Dirac, 1902–1984.

9 Dennis Gabor (Hungarian: Gábor Dénes), 1900–1979.

10 Stephen Wiesner, 1942–2021.

11 Peter Williston Shor, born 1959.

it is possible to design a dynamical system to transform the particular input to the corresponding output without, say, requiring any energy. But for a device to be useful for information processing, it has to be able to appropriately transform *every* possible input to its corresponding output by a *single* mechanism. Its inner workings cannot depend on receiving a particular input or a particular subset of inputs. The physical constraints therefore come from trying to design a single dynamical system that simultaneously satisfies all the requirements implied by the protocol.

1.2.1 Erasure

These issues are superbly illustrated by the very simple task of erasure. This was Landauer's prototypical example of the fact that executing logically irreversible operations requires free energy. Let us recount his argument that erasing one bit of information requires at least $kT \ln 2$ units of work when operating at background temperature T. Again, k is the Boltzmann factor. This requirement is now referred to as *Landauer's principle*.

The bit value is recorded in some physical system, and each of the two possible values must correspond to different values of some degrees of freedom. For instance, we might record the value using the position of a particle in a double-well potential, with the left well corresponding to 0 and the right well to 1. Generally, each value corresponds to some region in phase space, and to have a reliable encoding, these regions must not overlap significantly. The goal of the erasure procedure is to output 0, regardless of which value was input. Therefore, although only one region is occupied for either individual input, the two phase space regions must be merged in the dynamical description of the procedure to produce the desired output.

This is incompatible with Hamiltonian[12] dynamics, however, since by Liouville's[13] theorem phase space volume is constant under such dynamics. Hence dissipation will be required. The entropy of the system will be reduced by $\Delta S = k \ln 2$ according to the Boltzmann formula, since the number of available states is reduced by half. By the second law the entropy will have to be exhausted into the environment, just as in a refrigerator, and this will require free energy. From the relation of entropy to heat, $\Delta Q = T\Delta S$, the amount of free energy required will be at least $kT \ln 2$. Note that, while nonzero, this value is extremely small. At room temperature (20 °C) the required free energy is about 0.02 eV, or about 3×10^{-21} Joules. Erasure is not a practical problem for computers, at least not yet.

In any individual case the transformation that we need to perform, either $0 \to 0$ or $1 \to 0$, does not appear to require any energy. In the former case, we would simply

12 William Rowan Hamilton, 1805–1865.

13 Joseph Liouville, 1809–1882.

do nothing, while in the latter, we would simply interchange the two phase space regions. The free energy constraint only arises because we require a single process that performs the correct action in both cases.

This suggests a possible way around the argument: Why not just look at the bit to determine its value and then take the appropriate action? Since we are interested in information processing devices, we can outsource the "looking" to the device as well. But suppose that we want the device to be reset to its initial state at the end of the process. Now it is susceptible to the same argument as before, for the memory that stores the observation will also have to be reset. The conditional action will have to be represented internally by traversing different, essentially disjoint paths through phase space, but these paths will have to converge for the process to finish.

These considerations were used by Bennett[14] to resolve the paradox of Maxwell's[15] demon. The demon of the paradox, "a being whose faculties are so sharpened that he can follow every molecule in its course" (Maxwell [202]), controls a small door between two chambers of gas. One chamber is initially empty, and when a fast-moving molecule approaches the door from the other chamber, the demon opens the door and lets it through. Slow-moving molecules are left in the original chamber. Thus after some time the two chambers of gas are no longer in equilibrium, an apparent violation of the second law.

Clearly, any resolution will involve treating the demon as a physical system. Initial analyses, most prominently by Szilárd,[16] then later Brillouin[17] and Gabor, concentrated on the demon's observation step, believing that any such process would require free energy and thus restore the second law. However, this is not the case, as Bennett showed in 1982: Measurement of a physical system can in principle be done reversibly. The only truly irreversible step by Landauer's principle is the need for the demon to reset its own memory in order to repeat the sorting process. This cost can be shown to precisely balance the work that would be available from the nonequilibrium state of the gas, restoring the second law. In retrospect, we can appreciate that the initial approach of locating the free energy cost in the observation step was not far off. Though the dynamics of the measurement can be made reversible, we require a properly initialized memory system in which to write the result, the creation of which requires free energy by Landauer's principle.

14 Charles Henry Bennett, born 1943.

15 James Clerk Maxwell, 1831–1879.

16 Leó Szilárd, 1898–1964.

17 Léon Nicolas Brillouin, 1889–1969.

1.3 Quantum limits and possibilities

1.3.1 Copying

Now let us turn to a limitation imposed by quantum mechanics, namely that "a single quantum cannot be cloned". This was first stated in 1982, both by Wootters[18] and Zurek[19] and separately by Dieks.[20] Recall the single qubit, the "quantum coin", only now denote the two states simply by $|0\rangle$ and $|1\rangle$. These states are assumed to be orthogonal. A generic qubit state $|\psi\rangle$ is a vector in $\mathbb{C}^2$ given by $|\psi\rangle = a|0\rangle + b|1\rangle$, with $a, b \in \mathbb{C}$ such that $|a|^2 + |b|^2 = 1$. The qubit is generally not definitely in either state $|0\rangle$ or $|1\rangle$; if we perform a measurement to determine whether the coin is heads ($|0\rangle$) or tails ($|1\rangle$), then the probabilities are

$$\Pr[\text{heads}] = \left|\langle 0|\psi\rangle\right|^2 = |a|^2 \quad \text{and} \quad \Pr[\text{tails}] = \left|\langle 1|\psi\rangle\right|^2 = |b|^2 . \tag{1.1}$$

The state of n qubits is a vector in $\mathbb{C}^{2^n}$, and a convenient basis is given by vectors of the form $|0, \dots, 0\rangle = |0\rangle \otimes \cdots \otimes |0\rangle$, $|0, \dots, 1\rangle$, $|0, \dots, 1, 0\rangle$, etc. Then the quantum state of the entire collection is written as $|\psi\rangle = \sum_{s\in\{0,1\}^n} \psi_s |s\rangle$, where s are binary strings of length n, and once again $\psi_s \in \mathbb{C}$ with $\langle\psi|\psi\rangle = 1 = \sum_s |\psi_s|^2$.

By the Schrödinger[21] equation, allowed transformations of a set of qubits come in the form of *unitary* operators, which just transform one basis of $\mathbb{C}^{2^n}$ into another. This follows because the state $|\psi(t)\rangle$ at time t is related to the state $|\psi(0)\rangle$ at time 0 by the equation $|\psi(t)\rangle = e^{-itH/\hbar}|\psi(0)\rangle$ for a time-independent Hamiltonian H, and the operator $e^{-itH/\hbar}$ is unitary since H is Hermitian.[22] (Here $\hbar$ is the reduced Planck[23] constant.) Time-dependent Hamiltonians will similarly lead to unitary operators.

Suppose then that we have a cloning machine, which should perform the transformation

$$|\psi\rangle \otimes |0\rangle \longrightarrow |\psi\rangle \otimes |\psi\rangle \tag{1.2}$$

for any qubit state $|\psi\rangle$. Here the second system is the equivalent of a blank sheet of paper in a usual copy machine, and formally it is just any fixed state independent of $|\psi\rangle$. According to the laws of quantum mechanics, the transformation should be described by a fixed unitary U, so that

$$U|\psi\rangle \otimes |0\rangle = |\psi\rangle \otimes |\psi\rangle \tag{1.3}$$

---

18 William Kent Wootters, born 1951.

19 Wojciech Hubert Żurek, born 1951.

20 Dennis Geert Bernardus Johan Dieks, born 1949.

21 Erwin Rudolf Josef Alexander Schrödinger, 1887–1961.

22 Charles Hermite, 1822–1901.

23 Max Karl Ernst Ludwig Planck, 1858–1947.

for all $|\psi\rangle$. Now notice that the left-hand side is linear in the coefficients a and b, but the right-hand side is quadratic. Specifically, we have

$$|\psi\rangle \otimes |\psi\rangle = a^2|00\rangle + ab|01\rangle + ab|10\rangle + b^2|11\rangle \tag{1.4}$$

in the latter case, while for the former, we have

$$U|\psi\rangle \otimes |0\rangle = aU|00\rangle + bU|11\rangle = \sum_{j,k\in\{0,1\}} (au_{jk} + bu'_{jk})|jk\rangle , \tag{1.5}$$

where $u_{jk} = \langle jk|U|00\rangle$ and $u'_{jk} = \langle jk|U|11\rangle$ are the components of U. Since by assumption the components of U do not depend on $|\psi\rangle$, (1.3) cannot be satisfied in general.

What does work is $a = 0$ and $b = 1$ or vice versa, i. e., the values where linear and quadratic functions coincide. Thus, although no device can copy an arbitrary quantum state, it is possible to copy an arbitrary element of an orthogonal basis. This is fortunate, since if we believe that quantum mechanics is fundamentally correct and supersedes classical mechanics, then we do not want our argument to rule out the possibility of classical copying machines! Instead, we can take this as an indication that classical information theory can be thought of as a particular case of quantum information theory, one in which we are always dealing with orthogonal quantum states.

Suppose that the basis that is properly cloned is the $|0\rangle/|1\rangle$ basis. Then from $|\psi\rangle$ the "cloning" machine produces the state

$$U|\psi\rangle \otimes |0\rangle = a|00\rangle + b|11\rangle , \tag{1.6}$$

in which the superposition between $|0\rangle$ and $|1\rangle$ inherent in $|\psi\rangle$ has been extended to two qubits. This can be accomplished by the CNOT gate, the unitary which has the action $U|j,k\rangle = |j, j+k\rangle$, where addition is modulo two. It flips the second qubit if the value of the first is $|1\rangle$ and does nothing otherwise.

Instead of cloning, the CNOT gate extends the superposition over two systems, producing an *entangled* state. As we will see, the superposition now manifests itself only in the two systems jointly, not in either system individually. Superposition of two states is often called *coherence*, for just as two classical waves are coherent if they have a definite phase relationship, a given superposition with weights a and b also has a definite phase relationship between the two states (namely, $\arg b/a$). It turns out that for a state like (1.6), the coherence of the first system by itself has completely vanished; there is no more detectable phase relationship between the two states $|0\rangle$ and $|1\rangle$. Of course, the coherence is not irrevocably destroyed, since it can be restored by simply applying U^*. It is now caught up in the entanglement of the state.

The interplay between coherence, cloning, and entanglement already gives us an idea of the delicate nature of quantum information processing. Superposition, or coherence, is the hallmark of the quantum nature of an information processing device. The above example shows that mere copying of the state in one basis, which we think of as copying classical information encoded in this basis, is already enough to destroy the coherence possessed by part of the system. Thus a truly quantum information processing device cannot leak any information whatsoever; it must take care not to become entangled with its environment. It must remain completely isolated to prevent unwanted entanglement but still needs to be somewhat accessible to control its operation. This requirement is one of the daunting challenges of constructing quantum information processing devices.

While no cloning appears to mark significant distinction between quantum information and classical information, there is a sense in which classical information also cannot be copied. In the classical setting, "copying" usually refers to the task of duplicating the value of a random variable. But copying could also refer to the task of duplicating the probability distribution of the random variable. For instance, suppose that we have constructed a random number generator (possibly involving a quantum process, like radioactive decay) that can output a single bit. But now we would like to have two random bits, or perhaps a considerable number, for use in a Monte Carlo calculation or the operation of an online casino.

Do we need to build two random number generators or can we somehow transform the single random bit into two random bits? For this to be useful, the transformation itself must be deterministic, else the additional source of randomness is simply hidden in the transformation. In fact, we cannot copy any probability distribution in this manner, except the trivial cases in which the value of the bit is certain to be 0 or certain to be 1.

Suppose X is the random bit whose distribution P_X should be duplicated, ideally producing an additional bit Y. Here $P_X(x)$ is the probability that $X = x$, i. e., the random variable X takes the particular value x. The joint distribution P_{XY} of X and Y ideally satisfies $P_{XY}(x,y) = P_X(x)P_X(y)$, for then we have two independent instances of the distribution P_X. However, a deterministic function f will just result in the manifestly different joint distribution $P_{XY}(x,y) = P_X(x)\,\mathbf{1}[y = f(x)]$. Here we use the indicator function $\mathbf{1}[\ ]$, which returns 1 when its argument is true and zero otherwise, instead of the usual Kronecker[24] delta $\delta_{y,f(x)}$ simply to better emphasize the equality condition.

The distinction between the two cases is precisely the same as in the quantum scenario: The joint distribution in the latter case depends only linearly on the probabilities $P_X(0)$ and $P_X(1)$, whereas the ideal joint distribution depends quadratically on them. The same reasoning leads to the conclusion that even if we consider stochastic instead of deterministic transformations, there is no single process that can duplicate every distribution P_X. All of this suggests that quantum states $|\psi\rangle$ should not be regarded as the analogs of the values of classical random variables, but rather as probability distributions of random variables.

24 Leopold Kronecker, 1832–1891.

1.3.2 Cryptography

We can also use the delicate nature of quantum information to our advantage, to perform the task of quantum key distribution (QKD). Cryptographic protocols often require the use of *secret keys*, information shared between the legitimate parties and (hopefully) unknown to any would-be eavesdroppers. The task of creating secret keys between separated parties is key distribution. It cannot be accomplished with ultimate security using only classical means of communication, because, in principle, there is no prohibition on copying. However, this conclusion no longer holds in the quantum realm.

Consider first the effect of measurement on quantum systems. For a generic qubit $|\psi\rangle = a|0\rangle + b|1\rangle$ not definitely in one of the states $|0\rangle$ or $|1\rangle$, what happens after a measurement of this basis? Surely, if we repeat the measurement, then we should get the same result (provided that nothing much has happened in the meantime). Indeed, this is the case in quantum mechanics. Starting from $|\psi\rangle = a|0\rangle + b|1\rangle$ and making the $|0\rangle$ versus $|1\rangle$ measurement leaves the system in state $|0\rangle$ with probability $|a|^2$ or the state $|1\rangle$ with probability $|b|^2$. In this way a subsequent measurement yields the same result as the first.

We can measure in other bases as well. For instance, consider the basis $|\pm\rangle = \frac{1}{\sqrt{2}}(|0\rangle \pm |1\rangle)$. Now the probabilities for the two outcomes are

$$\Pr[+] = |\langle +|\psi\rangle|^2 = \tfrac{1}{2}|a+b|^2 \quad \text{and} \quad \Pr[-] = |\langle -|\psi\rangle|^2 = \tfrac{1}{2}|a-b|^2 . \tag{1.7}$$

Thus, if $|\psi\rangle = |0\rangle$, then $\Pr[\pm] = 1/2$, meaning the measurement outcome is completely random. After the measurement, the state is in the corresponding state $|+\rangle$ or $|-\rangle$. In this way, measurement disturbs the system by changing its state.

This phenomenon makes QKD possible. Very roughly, a potential eavesdropper attempting to listen in on a quantum transmission by measuring the signals will unavoidably disturb the signals, and this disturbance can be detected by the sender and receiver. We can get a flavor of how this works by examining the original BB84 protocol, formulated by Bennett and Brassard[25] in 1984. The goal, as in any QKD protocol, is to create a secret key between the two parties, which may then be used to encrypt sensitive information using classical encryption methods.

The BB84 protocol proceeds as follows. One party (invariably named Alice) transmits n quantum states to the other (invariably named Bob), where each state is randomly chosen from the set $\{|0\rangle, |1\rangle, |+\rangle, |-\rangle\}$. Physically, these could correspond to various polarization states of a single photon (horizontal, vertical, +45°, −45°), or anything else whose quantum description is given by the states above. When Bob receives each

---

25 Gilles Brassard, born 1955.

quantum signal, he immediately measures it, randomly choosing either the "standard" $|k\rangle$ basis ($k = 0, 1$) or the "conjugate" $|\pm\rangle$ basis.

The term "conjugate basis" goes back to Wiesner and refers to the fact that if the system is in the state $|0\rangle$ or $|1\rangle$, then it is equally likely to be found in $|+\rangle$ or $|-\rangle$, just as a state of well-defined position has a completely uncertain momentum. It is also appropriate to refer to the conjugate basis as the complementary basis, just as position and momentum are complementary observables. We will use the two terms "conjugate" and "complementary" interchangeably in this context. No matter the specific name, when the bases used to prepare and measure the state are different, Bob's measurement outcome is completely random. These cases are not useful in creating a secret key, so they are discarded by Bob announcing which basis he chose in each case and Alice announcing which cases they should keep.

On the other hand, if the quantum states arrive at Bob's end unchanged, then when he measures in the same basis Alice used to prepare the state, he will certainly obtain the corresponding outcome. When Alice prepares $|0\rangle$, Bob is certain to see $|0\rangle$, again provided that there was no noise during the transmission, so they can create one bit of secret key (with value 0). The same is true of the $|\pm\rangle$ basis, but here we specify the protocol to only attempt key generation from the standard basis.

The conjugate basis is used to detect the presence of a would-be eavesdropper (invariably named Eve) spying on the quantum signals. Suppose Eve intercepts each of the signals, measures it randomly in one basis or the other, and then resends the state corresponding to the outcome she observed. This will cause mismatches in the conjugate basis data, which Alice and Bob will notice.

Specifically, in each round, Eve's "intercept-resend" attack causes a mismatch between Alice and Bob's conjugate basis data with probability $1/4$. This can be seen as follows. For concreteness, suppose Alice sends $|+\rangle$. Half the time Eve measures in the conjugate basis and passes $|+\rangle$ to Bob without error. The other half of the time she measures in the standard basis, which produces a random outcome. Each of the two possible states $|0\rangle$ and $|1\rangle$ has a probability of $1/2$ of generating the correct outcome $|+\rangle$ when measured by Bob, so the overall error probability is $1/4$. This attack nets Eve the value of the key (in the standard basis) with probability $1/2$.

By comparing their conjugate basis data publicly, Alice and Bob can determine if Eve has employed the intercept-resend attack against the standard basis. If they observe no disagreements in this data, they can be relatively certain that Eve did not gain any information about the key. If a substantial mismatch is observed, then Alice and Bob conclude that Eve has spied on their communication, and they discard the key.

Although we have not proven that QKD can be secure against *arbitrary* attacks—Eve is by no means restricted to intercept-resend attacks in quantum mechanics—this example illustrates the basic mechanism of security. The crucial point is that the fragility of quantum information implies that the information gained by Eve about the key is linked to the correlation Alice and Bob observe in the conjugate basis data.

Classical information, in contrast, is not so fragile and shows no evidence of it having been copied. Even though in this example Alice and Bob abort the protocol for any nonzero mismatch rate, it is possible to construct QKD protocols that can tolerate a finite amount. Showing how to accomplish this task is indeed one of the goals of the book.

1.4 Overview of the book

The overarching goal of the book is to analyze the fundamental limits and possibilities of information-processing protocols for communication and cryptography. We adopt a "resource simulation" approach to information processing, for which the setup depicted in Figure 1.1 is the prototype. The aim in building a communication system, for instance, is to simulate the ideal channel by using the actual noisy channel and other resources the sender and receiver have at their disposal. In this case the sender can encode the message, e. g., by adding redundancy, such that the receiver will still be able to decode the transmitted message despite the noise. A specific simulation method, an encoder and decoder, is called a *protocol*.

The aim of Part I of the book is to develop the formalism of quantum information theory to fully characterize the form of the possible resources (noisy quantum channels) and protocols (encoders and decoders): the structure of possible quantum channels, the eavesdropping possibilities for Eve, and the encoding and decoding possibilities. This is done by adopting the analogy that quantum states are akin to classical probability distributions, as mentioned above in the discussion of cloning. Therefore the first part begins with a treatment of classical probability theory and the structure of classical channels to set the stage for their quantum counterparts.

The important aspect of resource simulation approach is that the *quality of a protocol is measured by its ability to allow the real resources to simulate the ideal resource.* Although this seems like the most naive approach, in fact, ad hoc measures of protocol quality are common in the study of information theory and cryptography. The great advantage of focusing on simulation is that it ensures *composability* of resources. Given an ideal resource consisting of several parts, we can build a protocol to simulate the whole by constructing protocols to simulate each part. For instance, in the running example of communication over noisy channels, the incoming message could first be compressed so that transmission would require fewer uses of the ideal channel. The compression task can be studied separately from information transmission by just assuming a noiseless channel. Then compression can simply be combined or composed with the transmission protocol because the latter simulates the ideal noiseless channel. The focus on simulation is especially useful in cryptography because it forces us to be very precise in the desired ideal behavior of the cryptosystem.

For composability of resources to be defined formally, we must choose a suitable measure of "simulatability", the ability of one resource or protocol to simulate another. A simple approach is to define simulatability in an operational way and say that one resource simulates another to the extent that no experiment could tell them apart, except with some small probability. Then composability will follow straightforwardly, as we will see in Part II. This part of the book develops the mathematical tools used in the analysis of information-processing tasks. Here we formalize the notion of approximate simulation of one resource by another and develop the properties of the entropy in the quantum setting. We also establish two information-theoretic uncertainty relations, which have important implications for quantum information processing.

Analysis of various information processing tasks is the subject of Part III. Quantities such as the capacity of a noisy channel, as mentioned above, refer to the setting of asymptotically many uses of the resource channel. However, we take a "one-shot" approach and derive upper and lower bounds on the required resources for various tasks for a single use of the resource. Using methods developed in Part II, the one-shot results can be applied to the many-use setting, and the typical results such as capacity, which will necessarily involve entropy, can be recovered.

Furthermore, our approach will be to study the relations between the various information processing tasks and constructively build up to more complicated tasks such as quantum communication from simpler pieces. Instead of repeatedly using the tools from Part II to prove properties of each task separately, the strategy is to *reduce* one task to another so as to recycle its analysis. For instance, the task of quantum compression can be reduced to classical compression because a suitable modification of any classical compression protocol enables it to perform quantum compression. The two aforementioned uncertainty relations feature prominently in Part III, which culminates in the quantum noisy channel coding theorem and a security proof for the BB84 protocol.

1.5 Notes and further reading

Landauer's phrase "Information is physical" is in fact the title of his survey of the subject [182]. The unit of bit for information stretches back to Bush [51], describing the amount of information that can be stored on a punch card. More details of the early telegraph example can be found in [205]. The quote from Wiener is taken from an early look [302] at the new field of information theory, which is generally regarded to have been founded by Shannon's 1948 paper [258]. For more on the early days of information theory, see [57, 165, 222]. Schumacher [251] coined the name "qubit" for quantum bits. Gabor's introduction of "quantum noise" is from [105], and the ultimate limits on classical communication over so-called Gaussian channels were established in [110]. Though Wiesner's realization that quantum effects could actually be useful, rather than just a hindrance, occurred in the 1970s, his paper [303] on conjugate coding was only published in 1983. Shor's factoring algorithm was reported in [261].

Landauer's argument relating logical and physical irreversibility is found in [181], and Bennett's exorcism of Maxwell's demon in [22]. Penrose [218] gave an exorcism along the same lines earlier, though without the argument that measurement does not increase entropy. Maxwell first mentioned the demon argument in a letter to Tait in 1867 and subsequently included it in his textbook on thermodynamics [202] (though it was Kelvin who called the being a "demon", in the sense of a supernatural being working in the background, as in background daemon process in a computer). Szilard's approach is found in [279], and that of Brillouin and Gabor in [47] and [106], respectively. The interested reader is also directed to the review papers [23, 32, 182, 195, 199, 200]. The Landauer argument is a combination of statements from phenomenological thermodynamics and statistical physics, which raises some important and subtle issues; see [178, 179, 199].

The no-cloning theorem is found in [309] and [81], though it appeared earlier as part of a different argument by Park [216] and is also arguably implicit in the work of Wiesner. For more on the history of the no-cloning argument, see [215, 221]. The BB84 protocol was introduced in [26]. Brassard recollects its history in [46].

Part I: **Formalism of probability and quantum theory**

2 Probability theory

> ...the theory of probability is basically just common sense reduced to calculation...[1]
>
> Pierre-Simon Laplace

In 1815, Laplace[2] reported a calculation of the mass of Saturn to the French National Institute of Sciences and Arts: one part in 3512 of that of the mass of the Sun. The importance of the calculation lies not in its accuracy, though the value differs by less that 0.4 % of the current value of roughly 3499, but that the calculation used the formalism of probability. The difficulty in such a calculation was in combining all the different data existing at the time, stretching from antiquity to contemporary observations, all with different reliabilities. Laplace accomplished the task by making use of probability. He described the certainty of his final result in such terms as well: "My formulas of probability show that there are odds of eleven thousand against one that the error of this result is not a hundredth of its value...".[3] In fact, his calculation makes use of what we today call Bayes'[4] rule, whose general form was published by Laplace himself in 1774.

The underlying approach to probability by both Bayes and Laplace is that probabilities represent *degrees of belief*, applying equally well to the truth of logical propositions or to whether events will (or did) take place. This is referred to as the Bayesian approach to probability, and in this approach, it is perfectly sensible to consider the probability that it will rain tomorrow or that the ratio of mass of the Sun to that of Saturn lies in the range $[3477, 3547]$. In contrast, if we think of probabilities as long-run frequencies of repeated experiments, then the mass of Saturn cannot be treated probabilistically at all.

We adopt the Bayesian approach for our purposes of studying information processing. It is very natural in this setting, where, for instance, the decoder of a communication scheme is interested in the probability that the transmitted message had a particular value. Of course, when regarding probability as a degree of belief, it matters whose beliefs we are referring to—clearly, the sender will, initially at least, have a very different belief about the message than the receiver. In the context of analyzing an information processing protocol, the natural choice is that the beliefs refer to an outside observer of the operation of the protocol.

---

1 On voit par cet Essai, que la théorie des probabilités n'est au fond, que le bon sens réduit au calcul: elle fait apprécier avec exactitude, ce que les esprits justes sentent par une sorte d'instinct, sans qu'ils puissent souvent s'en rendre compte. [183, page cv]

2 Pierre-Simon Laplace, 1749–1827.

3 "Mes formules de probabilité font voir qu'il y a onze mille à parier contre un, que l'erreur de ce résultat n'est pas un centième de sa valeur..." [184]

4 Thomas Bayes, c. 1701–1761.

https://doi.org/10.1515/9783110570250-002

The notion of probability is actually a rather delicate philosophical question, and it is not the topic of this book to address this question in any detail, nor to take sides on the issue. Rather, we rely on the Bayesian approach to provide the intuition for the formalism of probability we will develop in this chapter. Given the fairly simple probabilistic settings we will consider, where it is often sufficient to proceed on intuition, our formalism may seem excessive. However, it will allow us to build up the quantum formalism as a generalization.

2.1 Boolean algebras of events

We will be interested in the probabilities of a finite number of logical propositions or events $\mathsf{A}, \mathsf{B}, \mathsf{C}, \ldots$ and their combinations under the usual logical operations AND, OR, and NOT. We denote these by $\mathsf{A} \wedge \mathsf{B}$, $\mathsf{A} \vee \mathsf{B}$, and $\overline{\mathsf{A}}$, respectively. The value of some physical property (the mass of Saturn) or the result of an experiment will be common events in our setting here. To take a simpler but standard example, suppose we throw two dice. Possible events include things like "the sum of the two numbers is four", "one of the dice shows three", or "the dice both show one". Call these events A, B, and C, respectively. Two events or propositions that cannot both be true are called mutually exclusive or *disjoint*. In the example, B and C are disjoint, as are A and C, but A and B are not, since they are both true when the dice show one and three.

A set of propositions or events and all their combinations under the three logical operations above forms a *Boolean*[5] *algebra*, the algebra of logical relations of the propositions. By its nature a Boolean algebra also includes the false statement (self-contradiction) and true statement (tautology), just by taking $\mathsf{A}\wedge\overline{\mathsf{A}}$ and $\mathsf{A}\vee\overline{\mathsf{A}}$ for some A. We denote these by 0 and 1, respectively.

In the finite case, which is our only concern here, we can always find a "basic" set of disjoint propositions, called *atoms*, such that every proposition can be constructed as the OR of a set of atoms. For our dice example, the atoms are just the possible pairs of values, and all other events are just collections of these pairs. In this sense the algebra is generated from the atoms.

Another immediate implication is that any finite Boolean algebra is equivalent to the powerset (the set of all subsets) of the set of atoms. The logical operations AND, OR, and NOT correspond to the set operations of union, intersection, and complement, respectively. The structure of the Boolean algebra is nicely illustrated by means of a *Hasse*[6] *diagram*, shown for the case of three atoms in Figure 2.1.

Selecting an element X in a Boolean algebra and regarding it as a tautology gives a new Boolean algebra, a subalgebra of the original. Formally, we just map all elements

5 George Boole, 1815–1864.

6 Helmut Hasse, 1898–1979.

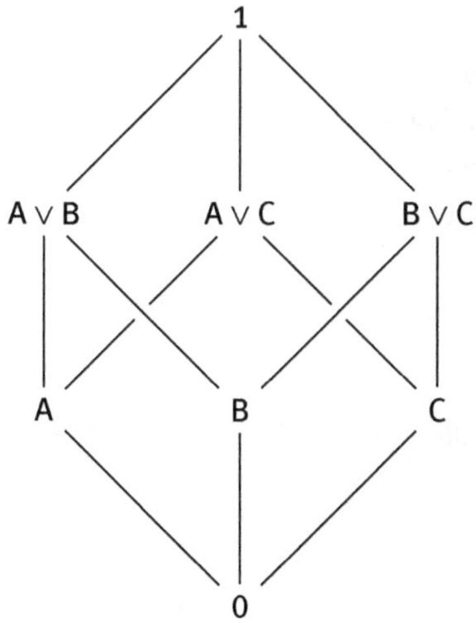

Figure 2.1: Hasse diagram of the Boolean algebra with three atomic elements A, B, and C. Moving upward in the diagram corresponds to OR, and downward to AND, i. e., $\mathsf{B} = (\mathsf{A} \vee \mathsf{B}) \wedge (\mathsf{B} \vee \mathsf{C})$. As A and B are disjoint, $\mathsf{A} \wedge \mathsf{B} = 0$. The three elements are exhaustive, meaning $\mathsf{A} \vee \mathsf{B} \vee \mathsf{C} = 1$.

Y to $\mathsf{Y} \wedge \mathsf{X}$. In the set picture, this amounts to intersecting all sets Y with the given set X. For instance, regarding $\mathsf{A} \vee \mathsf{C}$ as true in our example yields the Boolean algebra generated by A and C.

Most often, we will specify events by the value of some *random variable* or collection of random variables. For instance, M may be the mass of Saturn, or X the value of the first die above and Y the value of the second. In the information-theoretic setting, all inputs and outputs to an information-processing operation are treated as random variables. Moreover, we need not regard these random variables as completely abstract mathematical entities. Following Landauer's dictum, they are encoded into real physical systems, so the random variables refer to the properties of these physical systems. We denote random variables with capital letters, e. g., Z, the alphabet of possible values by calligraphic letters, e. g., $\mathcal{Z}$, and often denote a particular value with lower case letters. Thus the event corresponding to variable Z taking the value $z \in \mathcal{Z}$ is written simply as $Z = z$. The particular alphabets in question are usually not very important in our context, merely that they are finite. We often denote the cardinality of $\mathcal{X}$ for a random variable X by $|X|$ or $|\mathcal{X}|$.

2.2 The rules of probability

2.2.1 Definition

Degrees of belief naturally depend on the background knowledge we already have, so the basic notion in our setting is the conditional probability that a proposition A is true or an event A has occurred or will occur, given background knowledge that proposition B is true. We denote this by $\Pr[A|B]$. The value $\Pr[A|B]$ is a number between zero and one, with larger values representing greater degree of belief. It is important to distinguish $\Pr[A|B]$ from $\Pr[B|A]$, as these are quite distinct. Just because, for instance, most accidents occur near home does not imply that we are safer the farther we are from home!

We encapsulate the notion of probability with the following four basic rules, or axioms:

Definition 2.1 (Probability axioms). The conditional probability $\Pr[A|C]$, defined for all A and $C \neq 0$ on a Boolean algebra, satisfies:

1. Positivity: $\Pr[A|C] \geq 0$,
2. Normalization: $\Pr[A|C] = 1$ if and only if $C \Rightarrow A$ or, equivalently, $A \vee \overline{C} = 1$,
3. Addition rule: $\Pr[A \vee B|C] = \Pr[A|C] + \Pr[B|C]$ for $A \wedge B = 0$,
4. Product rule: $\Pr[A \wedge B|C] = \Pr[A|B \wedge C]\, \Pr[B|C]$.

This axiomatization is due to Rényi.[7] Readers more familiar with the Kolmogorov[8] axioms of unconditional probability and measure theory will note that the first three axioms are such that $\Pr[\cdot|C]$ is a Kolmogorov probability on the Boolean subalgebra resulting from setting $C = 1$.

The fourth axiom ensures that all these probability measures defined on subalgebras are consistent with each other. Suppose we start from the situation in which we condition on C using $\Pr[\cdot|C]$ and want to condition further on B and use $\Pr[\cdot|B \wedge C]$. How should this latter probability be related to the former? The product rule says to simply use $\Pr[A \wedge B|C]$ for the conditional probability of A given B and C, renormalized by the probability $\Pr[B|C]$. (In the Kolmogorov framework, setting $C = 1$ in the fourth axiom gives the definition of conditional probability.)

Note that in all of this, we need to exclude the subalgebra that results from choosing C to be 0. This is sensible as the resulting algebra is trivial: it is the powerset of the empty set and has only one element.

Exercise 2.1. Show that taking $C = 0$ in the probability axioms leads to a contradiction.

There are several ways to motivate these axioms and to confirm that they are consistent with the underlying Boolean structure. We give one known as the "Dutch book" argument later in this section. First, though, let us investigate the structure in a little more detail. Foremost is the very intuitive fact that in the finite setting here the prob-

7 Alfréd Rényi, 1921–1970.

8 Andrey Nikolaevich Kolmogorov, 1903–1987.

ability of any event A is just the sum of the probabilities of its constituent atoms. This follows immediately from the addition rule.

Notice that the addition rule does not discuss the case of nondisjoint events. Nevertheless, it is not too difficult to show that in general

$$\Pr[\mathsf{A} \vee \mathsf{B}|\mathsf{C}] = \Pr[\mathsf{A}|\mathsf{C}] + \Pr[\mathsf{B}|\mathsf{C}] - \Pr[\mathsf{A} \wedge \mathsf{B}|\mathsf{C}]\,. \tag{2.1}$$

By induction, we can extend this to an arbitrary number n of events; using positivity to remove the final term then yields the *union bound*:

$$\Pr\left[\bigvee_{i=1}^{n} \mathsf{A}_i|\mathsf{C}\right] \leq \sum_{i=1}^{n} \Pr[\mathsf{A}_i|\mathsf{C}]\,. \tag{2.2}$$

Exercise 2.2 (Union bound). Show (2.1) and prove the union bound (2.2).

2.2.2 The law of total probability

Missing from the list of axioms, but nonetheless implied by them, is the very intuitive notion that *the unconditional probability is the average of the conditional probability*. This is sometimes called the law of total probability, and it states that for any event A and a set of disjoint events B_i for which $\bigvee_i \mathsf{B}_i = 1$,

$$\Pr[\mathsf{A}] = \sum_i \Pr[\mathsf{A}|\mathsf{B}_i]\ \Pr[\mathsf{B}_i]\,. \tag{2.3}$$

The probability of A is just the average of the conditional probabilities for all the different possible background cases B_i, each case weighted by its probability of occurrence. In light of the aforementioned central importance of conditional probability, we ought to include a background event C in each probability factor. However, it plays no specific role here, so we can assume that $\mathsf{C} = 1$ in this discussion and later insert a nontrivial C if needed.

For example, consider the following urn problem, urn problems being a standard in probability theory since their introduction by Jacob Bernoulli.[9] Suppose there are two urns, the first filled with 50 red and 50 blue balls, and the second with 80 red and 20 blue. Drawing one ball from one of the urns at random, what is the probability of it being red? Intuitively, we just average the probabilities for the two separate cases, so it must be $\frac{1}{2}\frac{1}{2} + \frac{1}{2}\frac{4}{5} = \frac{13}{20}$. This is precisely (2.3). We can also check that this must be the correct answer, since the two urns can simply be combined into one.

To derive (2.3), it is enough to consider the binary case, as the general case is entirely similar. For simplicity, let us abbreviate $\mathsf{A} \wedge \mathsf{B}$ as simply AB. Then, given $\mathsf{B}_1 \wedge \mathsf{B}_2 =$

9 Jacob Bernoulli, 1655–1705.

0, it follows that $AB_1 \wedge AB_2 = 0$. Therefore, by the addition rule, $\Pr[AB_1 \vee AB_2] = \Pr[AB_1] + \Pr[AB_2]$. On the other hand, since $B_1 \vee B_2 = 1$, the proposition $AB_1 \vee AB_2$ is equivalent to A, which completes the derivation.

2.2.3 Bayes' rule

Returning to Bayes and Laplace, the general problem they (and many others) were interested in is that of "inverse probability", which amounts to inverting the order of arguments in the conditional probability. This is in fact the natural problem in most areas of science. We entertain different hypotheses H_i (say, about the mass of Saturn), and we obtain data D_j from different experiments (observations). Then we want to know $\Pr[H_i|D_j]$; ideally, one hypothesis would have most of the probability. What we have, though, are the conditional probabilities $\Pr[D_j|H_i]$.

Bayes' rule addresses precisely this problem. Reverting back to generic events A, B, etc., and using the product rule to decompose $A \wedge B$ in two different ways yields Bayes' rule or Bayes' theorem:

$$\Pr[A|B \wedge C] = \frac{\Pr[B|A \wedge C]}{\Pr[B|C]} \Pr[A|C] . \tag{2.4}$$

The factor $\Pr[A|C]$ is referred to as the *prior probability*, the probability we begin with, whereas $\Pr[A|B \wedge C]$ is the *posterior* probability we are now interested in. The two are related by a factor of the *likelihood*, $\Pr[B|A \wedge C]$, and (inversely) the *evidence* $\Pr[B|C]$. In this setting, it is common to regard the likelihood as a function of A (and therefore *not* as a probability distribution per se). When the prior is uniform, for instance, the most likely value of A is the one with the *maximum likelihood*.

2.2.4 The Dutch book argument

Now let us illustrate the consistency of the probability axioms with the underlying Boolean algebra. We do this with a variant of the so-called "Dutch book" argument, which translates probability statements into betting behavior and shows that probability assignments violating the axioms lead to betting strategies that are certain to lose money. A Dutch book is a collection of bets offered at prices that ensure a loss for the buyer.

We begin by making the notion of degree of belief more concrete, regarding $\Pr[A]$ as the price we are willing to pay for a contract or lottery ticket that states "Collect 1 Swiss franc if A". That is, we are indifferent to holding the contract or $\Pr[A]$ francs, since they are worth the same amount. Conditional probabilities can be handled by contracts that are canceled with refund if the conditioning event does not occur or is found to be false. In particular, consider the ticket that states "Collect 1 Swiss franc if

$\mathsf{A} \wedge \mathsf{B}$; collect p Swiss francs if $\overline{\mathsf{B}}$" for some $p \in [0, 1]$. If we pay p for the ticket, then our money is refunded if B is found to be false. We should be willing to pay $p = \Pr[\mathsf{A}|\mathsf{B}]$, since the situation reduces to the above case when B is true.

The fact that probability is positive is reflected in the fact that either kind of ticket has positive value. We should not pay one franc for the corresponding ticket if A is not certain to occur, lest we possibly face a loss; this also holds in the context of B assumed to be true.

If we value the combination of A- and B-tickets differently than the $\mathsf{A} \vee \mathsf{B}$ ticket when A and B are disjoint, then we are again open to a loss. For instance, when we value the former more than the latter, we buy A- and B-tickets from the bookmaker (who, for whatever reason, is invariably Dutch in this argument) at a higher price than we are happy to sell $\mathsf{A} \vee \mathsf{B}$ tickets to him or her, and ultimately the bookmaker ends up with money no matter what. Note that we can extend this argument to conditional tickets as well. This illustrates the consistency of the addition rule.

The argument for the product rule is similar, based on decomposing the $\mathsf{A}|\mathsf{B} \wedge \mathsf{C}$ ticket into equivalent pieces, which contain the $\mathsf{A} \wedge \mathsf{B}|\mathsf{C}$ and $\mathsf{B}|\mathsf{C}$ tickets. For simplicity, let $p = \Pr[\mathsf{A}|\mathsf{B} \wedge \mathsf{C}]$ and $q = \Pr[\mathsf{A} \wedge \mathsf{B}|\mathsf{C}]$. Consider the ticket for $\mathsf{A}|\mathsf{B} \wedge \mathsf{C}$, which we value at p francs. It has two parts, "Collect 1 if $\mathsf{A} \wedge \mathsf{B} \wedge \mathsf{C}$" and "Collect p if $\overline{\mathsf{B} \wedge \mathsf{C}}$". The second condition can be rewritten as two conditions, so that the entire ticket has three parts:

$$\text{Collect 1 if } \mathsf{A} \wedge \mathsf{B} \wedge \mathsf{C}, \tag{2.5}$$

$$\text{Collect } p, \tag{2.6}$$

$$\text{Pay } p \text{ if } \mathsf{B} \wedge \mathsf{C}. \tag{2.7}$$

The second part pays no matter what, while the third part would require *us* to pay, so it is more sensible to think of it as a contract (or a card from the board game Monopoly). We can add two additional clauses, which cancel each other out:

$$\text{Collect } q \text{ if } \overline{\mathsf{C}}, \tag{2.8}$$

$$\text{Pay } q \text{ if } \overline{\mathsf{C}}. \tag{2.9}$$

Since the combination of (2.5) through (2.9) is equivalent to the original $\mathsf{A}|\mathsf{B} \wedge \mathsf{C}$ ticket, the value of the combination is p. Consistency requires that the total value of all the parts is also p. Notice that (2.5) and (2.8) comprise the $\mathsf{A} \wedge \mathsf{B}|\mathsf{C}$ ticket, meaning their value is q. The value of (2.6) is clearly p. The value of the remaining two, (2.7) and (2.9), is p times the value of "Pay 1 if $\mathsf{B} \wedge \mathsf{C}$" and "Pay q/p if $\overline{\mathsf{C}}$". This is a $\mathsf{B}|\mathsf{C}$ ticket, but now we are holding the other end of the contract, and the implied probability is $\Pr[\mathsf{B}|\mathsf{C}] = q/p$. Observe that this is the product rule. It is then a simple calculation to conform that the value of this decomposition of the original ticket is p as required.

2.3 Random variables

2.3.1 Joint, marginal, and conditional distributions

As we will mostly specify events by the values of random variables, it is convenient to work with the *probability mass function* instead of the full probability function $\Pr[]$. For a random variable Z, we define $P_Z : \mathcal{Z} \to [0,1]$ to be the function $P_Z(z) = \Pr[Z = z]$. Technically, $\Pr[]$ is the *probability distribution*, but we will almost always abuse this convention and refer to the probability mass function as the probability distribution.

For several random variables, say X and Y, we define the *joint probability* $P_{XY} : \mathcal{X} \times \mathcal{Y} \to [0,1]$ as the function $P_{XY}(x,y) = \Pr[X = x \wedge Y = y]$. In physicist's notation, we would drop the label to P and just identify the particular random variables involved by the names of the arguments to P. However, this often leads to confusion in the kinds of calculations we will later perform, so we will be somewhat pedantic and keep the labels. This has the benefit that we can treat the label as part of the name, so that P_X and P_Y refer to distinct probability distributions.

In the context of a joint distribution P_{XY}, the *marginal probability* P_X of just X alone is found from the addition rule:

$$P_X(x) = \sum_{y \in \mathcal{Y}} P_{XY}(x,y)\,. \tag{2.10}$$

Just to emphasize that the particular lower-case variables have no meaning in the definition here, we could just as well write $P_X(y) = \sum_{z \in \mathcal{Y}} P_{XY}(y,z)$. Observe that by regarding the subscript as part of the name of the distribution, P_X and P_{XY} are different distributions, but using the same P for both indicates that the former is the marginal of the latter. We will very often make use of this convention.

With the joint and marginal probabilities, we can use the product rule to write the *conditional probability* $P_{X|Y=y} : \mathcal{X} \to [0,1]$ as

$$P_{X|Y=y}(x) = \frac{P_{XY}(x,y)}{P_Y(y)}\,. \tag{2.11}$$

As for events, the law of total probability says that the *marginal is the average of the conditional*:

$$P_X(x) = \sum_{y \in \mathcal{Y}} P_{X|Y=y}(x)\, P_Y(y)\,. \tag{2.12}$$

In these expressions, we explicitly include the value of the conditioning variable in the subscript. This emphasizes that $P_{X|Y=y}$ is a probability distribution for X, and it is the one conditional on the event $Y = y$. An alternate notation that will also be useful is to write $P_{X|Y} : \mathcal{X} \times \mathcal{Y} \to [0,1]$ for the collection of conditional probability distributions and set $P_{X|Y}(x, y) = P_{X|Y=y}(x)$. Again, the fact that $P_{X|Y}(x|y)$ refers to the conditional probability of X given Y and not the other way around (or some other random variables) is recorded in the subscript, not the arguments.

The conditional and marginal probability distributions make it simple to discuss the probabilities of events in which some of the random variables themselves have specific values but they do not immediately simplify the situation for general events. However, to treat these cases, we can just invent new random variables. The probability of A in the two-dice example, for instance, is just $P_Z(4)$ with $Z = X + Y$, where X and Y represent the values of the two dice, respectively. For B, define the function $f(u, v)$ to be 1 when either argument is three and zero otherwise. Then we have $\Pr[\mathsf{B}] = P_Z(1)$ for $Z = f(X, Y)$. In general, the probability of $Y = f(X)$ taking the value y is just given by adding the probabilities of all x values for which $f(x) = y$, since these are disjoint events:

$$P_Y(y) = \sum_{x \in \mathcal{X}} 1[f(x) = y]\, P_X(x)\,. \tag{2.13}$$

It is easy to see that any event pertaining to a set of random variables can be represented by a suitable function f.

2.3.2 Vector representation

Equation (2.13) exposes the essential linearity of probability, which will be the starting point of our quantum generalization. The equation states that $P_Y(y)$ is the inner product between two vectors, one describing the event $Y = y$ and one describing the probability distribution of the atoms. Choosing some order for the atoms in $\mathcal{X}$ in some arbitrary way, if we let $P = (P_X(x))_{x \in \mathcal{X}}$ and regard the indicator function as the vector $E(\mathsf{A}) = (1[f(x) = y])_{x \in \mathcal{X}}$, then (2.13) is just

$$\Pr[\mathsf{A}] = E(\mathsf{A}) \cdot P\,. \tag{2.14}$$

That is, the atoms of the Boolean algebra are used as basis vectors in the vector representation. This construction relies on the fact that to describe the probability of any event, it is sufficient to work in terms of the atoms.

For a Boolean algebra with n atoms, the function E is a map from the algebra to $\{0, 1\}^n$. Similarly, P is the vector of probabilities of the atoms, so that $P \in [0, 1]^n$. For instance, in the example of Figure 2.1 with three atomic elements, we can set $E(\mathsf{A})$, $E(\mathsf{B})$, and $E(\mathsf{C})$ to be $(1, 0, 0)$, $(0, 1, 0)$, and $(0, 0, 1)$, respectively. Then $E(\mathsf{A} \vee \mathsf{B}) = (1, 1, 0)$. In this way, the vertices of the Hasse diagram correspond to the vertices of the unit cube in $\mathbb{R}^3$. Pointwise multiplication of the vectors corresponds to the logical AND of the propositions, while pointwise application of the function $(x, y) \mapsto x \vee y := x + y - xy$ corresponds to logical OR. The latter rule ensures that the values of the vectors are always either 0 or 1, so that, e. g., $E(\mathsf{A} \vee \mathsf{B}) \vee E(\mathsf{B} \vee \mathsf{C}) = E(1)$. It will be useful to have names for the sets of probability distributions and events:

$$\mathrm{Prob}(n) := \left\{(p_1, p_2, \ldots, p_n) \in \mathbb{R}^n : p_i \geq 0, \sum_{i=1}^{n} p_i = 1\right\}, \tag{2.15}$$

$$\mathrm{Events}(n) := \{(e_1, e_2, \ldots, e_n) \in \mathbb{R}^n : e_i \in \{0, 1\}\}. \tag{2.16}$$

In the context of some collection of random variables, we will slightly overload notation and denote the vector $(P_X(x))_{x\in\mathcal{X}}$ by just P_X and write $\mathrm{Prob}(X)$ for the set of possible distributions for X.

For several random variables, the vector representation formally corresponds to the tensor product of the vector spaces associated with each of the random variables individually. To represent P_{XY}, we need to specify the values of both X and Y, meaning the vector representation has dimension $|X||Y|$. The relevant atomic events for the joint distribution are specified by pairs of X and Y values, so the basis for the joint representation is the product of the individual X and Y representation bases.

Exercise 2.3. Given $P \in \mathrm{Prob}(n)$ and some proposition C, show that the representation $P' \in \mathrm{Prob}(n)$ of the conditional probability $\Pr[\cdot|\mathsf{C}]$ is given by

$$P' = \frac{E(\mathsf{C})P}{E(\mathsf{C}) \cdot P}, \tag{2.17}$$

where juxtaposition of $E(\mathsf{C})$ and P in the numerator denotes the pointwise product.

2.4 Convexity

The average of different possible values is also called a *convex combination* of those values with weights given by the probability. For instance, as we already saw in (2.12), the marginal distribution is a convex combination of the conditional distributions. For a real-valued random variable Z, the expected value $\langle Z\rangle$ is a convex combination of the possible values, $\langle Z\rangle := \sum_{z\in\mathcal{Z}} zP_Z(z)$, and the variance is the convex combination of the squared deviation from the expected value, $\mathrm{Var}(Z) := \sum_{z\in\mathcal{Z}} P_Z(z)(z - \langle z\rangle)^2$.

Convexity plays a pivotal role in both classical and quantum information theory, so let us make a few definitions. A *convex set* is a set closed under convex combinations, i. e., $\mathcal{S}$ is a convex set if there is a meaningful way to add elements of $\mathcal{S}$ and multiply them by real numbers so that, for any $s_1, s_2 \in \mathcal{S}$ and $\lambda \in [0, 1]$, $\lambda s_1 + (1 - \lambda)s_2 \in \mathcal{S}$. Observe that the set $\mathrm{Prob}(n)$ is a convex set in $\mathbb{R}^n$. We will only be interested in convex sets in $\mathbb{R}^n$.

Any set can be extended to a convex set by just taking all possible convex combinations of all the elements; this gives the *convex hull* of the original set. The *extreme points* of a convex set are the elements that cannot be written as a nontrivial convex combination of other elements. Note that the *boundary* of a convex set is not the same as its extreme points, e. g., the extreme points of a triangle are its vertices, but its faces make up its boundary. As we would intuitively expect, any (bounded) convex set is the convex hull of its extreme points, a fact which goes back to Minkowski.[10]

In Prob(n) the extreme points are the deterministic distributions in which one of the components is 1 and the rest are 0. A convex set in $\mathbb{R}^n$ that is a convex combination of a finite number of extreme points is called a *polytope* or *polyhedron*, and evidently Prob(n) is such a polytope. In fact, it is a *simplex*, a convex set for which every point in the set has a unique convex decomposition in terms of the vertices. This follows because Prob(n) has n vertices and can be embedded into $\mathbb{R}^{n-1}$ by discarding the component along the vector whose components are all 1.

A *convex function*, meanwhile, is a function $f : \mathcal{X} \to \mathbb{R}$ from some set $\mathcal{X}$ to the reals such that $f(\lambda x_1 + (1-\lambda)x_2) \leq \lambda f(x_1) + (1-\lambda)f(x_2)$. The function is called *strictly convex* if the inequality is strict. It is almost impossible to remember which way the inequality goes in this definition, but one way to think of it that may help is that the epigraph of a convex function f, the area above the graph of a function, is a convex set. Of course, this is only useful if it is easier to remember that epigraph refers to the area above and not below the function (the area below it is called the hypograph). When the inequality is reversed, the function is called *concave* (its hypograph is a convex set).

For a convex function f on a convex set $\mathcal{X}$, the expectation values of X and $f(X)$ are related by *Jensen's*[11] *inequality*

$$\langle f(X)\rangle \geq f(\langle X\rangle). \tag{2.18}$$

The inequality is essentially a direct consequence of the definition of convexity and is depicted for binary random variables in Figure 2.2. For a strictly convex function f, equality holds if and only if all the possible X values are identical, or only one value of $X = x$ has nonzero probability. Equality also holds if f is affine.

Exercise 2.4. Prove (2.18) and the equality conditions for strictly convex functions.

Note that the set Events(n) is not convex, since the entries are confined to either 0 or 1. However, if we relax this constraint to allow entries in $[0, 1]$, then it is not difficult to see that the resulting set is convex. We can interpret such vectors as a generalization of events in the following sense. First, (2.14) still leads to a *bona fide* probability. Suppose $T \in \mathbb{R}^n$ satisfies $T(j) \in [0, 1]$ for all $j \in \{1, \dots, n\}$ (as with probability vectors, we also use the notation $T(j)$ to refer to the jth component of T). Then, for any $P \in \text{Prob}(n)$, we have $T \cdot P \in [0, 1]$, which we could regard as the probability of T: $\Pr[T] = T \cdot P$.

10 Hermann Minkowski, 1864–1909.

11 Johan Ludwig William Valdemar Jensen, 1859–1925.

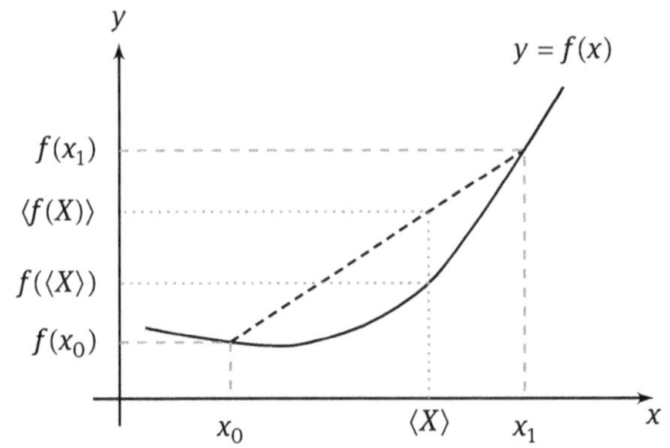

Figure 2.2: Depiction of Jensen's inequality for a convex function of a binary-valued random variable X. The random variable X can take two possible values x_0 and x_1 with corresponding probabilities $P_X(x_0)$ and $P_X(x_1)$. A function f induces a new random variable $Y = f(X)$; for convex f, it follows that $\langle Y \rangle \geq f(\langle X \rangle)$.

Second, one possible class of such vectors is given by "stochastic events" in which we do not perfectly or deterministically check if a random variable takes a certain value or values, but do so only randomly. For instance, in the dice example, suppose that with probability $\lambda_1 \in [0,1]$ we check whether the sum of the dice is even, with probability $\lambda_2 \in [0,1]$ whether one of them shows three, or with probability $1-\lambda_1-\lambda_2 \in [0,1]$ check for "snake eyes" (two ones). The associated vector T is λ_1 times the vector representing the "sum even" event, plus λ_2 times the "at least one three" event vector, plus $1-\lambda_1-\lambda_2$ times the "snake eyes" event vector. The quantity $T \cdot P$ is the probability that the answer to this convex combination of checks is "yes". We can regard the combination as a kind of *test* and $T \cdot P$ as the probability of passing the test. Formally, the set of all possible tests is defined by

$$\mathrm{Tests}(n) := \{(t_1, t_2, \dots, t_n) \in \mathbb{R}^n : 0 \leq t_i \leq 1\}. \tag{2.19}$$

In Chapter 3, we will show that Tests(n) is the convex hull of Events(n), i. e., all tests can be interpreted as convex combinations of events.

2.5 Independence

A similarly sounding but distinct notion to disjointness of events is *independence*. Events A and B are called *independent* when $\Pr[\mathsf{A} \wedge \mathsf{B}] = \Pr[\mathsf{A}] \Pr[\mathsf{B}]$, that is, the joint probability factorizes. By the product rule this is equivalent to $\Pr[\mathsf{B}|\mathsf{A}] = \Pr[\mathsf{B}]$ and $\Pr[\mathsf{A}|\mathsf{B}] = \Pr[\mathsf{A}]$. Disjointness is a statement about the logical relationship of the events, whereas independence involves probability.

Exercise 2.5. Show that A and B are independent iff A and $\overline{\mathsf{B}}$ are independent.

Independence can be extended to conditional probabilities, in keeping with our treating conditional probability as the fundamental concept. Two events A and B are *conditionally independent* when $\Pr[\mathsf{A} \wedge \mathsf{B}|\mathsf{C}] = \Pr[\mathsf{A}|\mathsf{C}]\ \Pr[\mathsf{B}|\mathsf{C}]$.

Exercise 2.6. Suppose that A and B are conditionally independent given C. Show that $\Pr[\mathsf{B}|\mathsf{A}] = \Pr[\mathsf{B}|\mathsf{C}]\ \Pr[\mathsf{C}|\mathsf{A}] + \Pr[\mathsf{B}|\overline{\mathsf{C}}]\ \Pr[\overline{\mathsf{C}}|\mathsf{A}]$.

For three or more events, independence is again defined as having a joint probability that factorizes. Note that pairwise independence is not sufficient, as shown by the following standard example. Take X and Y to be the two dice again and consider the events $X+Y=7$, $X=4$, and $Y=3$. These events are clearly not independent, since $\Pr[X+Y=7|X=4 \wedge Y=3] \neq \Pr[X+Y=7]$. However, they are pairwise independent: $\Pr[X+Y=7|X=4] = \frac{1}{6} = \Pr[X+Y=7]$.

Building on the definition of independence of events, two random variables X and Y are independent when $P_{XY}(x,y) = P_X(x)P_Y(y)$, so that their joint distribution is the product of the marginals. We often write product distributions directly as $P_X \times P_Y$ or just $P_X P_Y$. An important particular case is a collection of independent random variables, each of which has the same underlying distribution. A collection of n random variables $X_1, \ldots, X_n$ each with alphabet $\mathcal{X}$ is said to be *independent and identically distributed (i. i. d.)* if their joint probability distribution has the form

$$P_{X_1 \cdots X_n}(x_1, \ldots, x_n) = P_X(x_1)P_X(x_2)\cdots P_X(x_n) \tag{2.20}$$

for some P_X and all $x_1, \ldots, x_n \in \mathcal{X}$. The i. i. d. property characterizes situations where a certain process is repeated n times independently. In the context of information theory, the i. i. d. property is often used to describe the statistics of noise, for example, in repeated uses of a communication channel. In the context of n random variables X_j, we will sometimes denote the entire sequence by X^n. The corresponding joint distribution can be written $P_X^{\times n}$.

The *law of large numbers* and the *central limit theorem* characterize the "typical behavior" of real-valued i. i. d. random variables $X_1, \ldots, X_n$ in the limit of large n. The law of large numbers states that the sample mean of the X_i tends to the expectation value for large n. It usually comes in two versions, the *weak law* and the *strong law*. As the names suggest, the latter implies the former.

More precisely, let $\mu = \langle X_i \rangle$ be the expectation value of X_i. By the i. i. d. assumption the mean is the same for all $X_1, \ldots, X_n$. Let $Z_n = \frac{1}{n}\sum_{i=1}^{n} X_i$ be the *sample mean*. Then, according to the *weak law of large numbers*, the probability that Z_n is ε-close to μ for any positive ε converges to one as $n \to \infty$:

$$\lim_{n\to\infty} \Pr[|Z_n - \mu| < \varepsilon] = 1 \quad \forall \varepsilon > 0\,. \tag{2.21}$$

The weak law of large numbers will be sufficient for our purposes and is proven in the following sequence of exercises. These also establish two other often useful inequalities.

Exercise 2.7. Show *Markov's*[12] *inequality*: For any random variable X with expectation $\langle X\rangle$,

$$\Pr[X \geq \varepsilon] \leq \frac{\langle X\rangle}{\varepsilon}. \tag{2.22}$$

Exercise 2.8. Using Markov's inequality, prove *Chebyshev's*[13] *inequality*: For any random variable Y with average value μ and variance v,

$$P[(Y-\mu)^2 \geq \varepsilon] \leq \frac{v}{\varepsilon}. \tag{2.23}$$

Exercise 2.9. Use Chebyshev's inequality to prove (2.21).

By contrast, the *strong law of large numbers* says that Z_n converges to μ with probability one,

$$\Pr\left[\lim_{n\to\infty} Z_n = \mu\right] = 1. \tag{2.24}$$

Note that a proper treatment of the strong law is beyond the scope of this book, as the number of random variables is infinite. Here we need the more formal machinery of measure theory. One way to remember the difference between the weak and strong laws is to observe that they are essentially saying the same thing, but using different notions of convergence. The weak law is a statement of *convergence in probability*, whereas the strong law is a statement of *almost-sure convergence*.

While the laws of large numbers tell us about the behavior of the sample mean, the *central limit theorem* gives some insight into the behavior of fluctuations around the mean, at least when the X_i have bounded variance $v = \sigma^2$. In particular, let Φ be the cumulative distribution function of a standard normal distribution, i. e., $\Phi(x) = \frac{1}{\sqrt{2\pi}}\int_{-\infty}^{x} e^{-x^2/2}$, and define the rescaled *fluctuation variable* $Y_n = \sqrt{n}(Z_n - \mu)/\sigma$. Then the central limit theorem states that the cumulative distribution of Y_n converges to that of the normal distribution:

$$\lim_{n\to\infty} \Pr[Y_n \leq y] = \Phi(y). \tag{2.25}$$

This type of convergence is called *convergence in distribution*. It is weaker than either of the other two notions mentioned above.

12 Andrey Andreyevich Markov, 1856–1922.

13 Pafnuty Lvovich Chebyshev, 1821–1894.

The statements above only hold in the limit as $n \to \infty$, but since information processing protocols are finite, it is important to have bounds on the deviation of i. i. d. random variables from their typical behavior for finite n, that is, something that tells us about the rate of convergence to the limit. There is an enormous difference between converging rapidly, say exponentially in n, versus very slowly, say logarithmically in n. In the former case the typical asymptotic behavior is reached for reasonable n, whereas in the latter the convergence is so slow that the asymptotic behavior is hardly relevant for any finite n.

For the deviation from the mean, such a statement can be obtained from the Chebyshev inequality (2.23), which gives $\Pr[|Z_n - \mu| \geq \varepsilon] \leq O(\frac{1}{n\varepsilon^2})$. Much tighter is the *Hoeffding*[14] *bound*. Suppose that the random variables X_j take values in the interval $[a, b]$. Then

$$\Pr[Z_n \geq \mu + \varepsilon] \leq \exp\left(-\frac{2n\varepsilon^2}{(b-a)^2}\right), \tag{2.26}$$

and similarly for $Z_n \leq \mu - \varepsilon$. For fixed ε, i. e., not varying with n, the sample mean converges to the expectation value exponentially fast in n. Observe that the Hoeffding bound is in agreement with the central limit theorem regarding the fluctuations about the mean. From the latter we know that fluctuations are of size proportional to the square root of n. Similarly, if we take $\varepsilon = \frac{c}{\sqrt{n}}$ in the former, then the bound becomes a constant.

2.6 Additional exercises

Exercise 2.10 (Two girls?). Your neighbors have two children. What is the conditional probability of *both children being girls*, given each of the following conditions:

1. You know the first child is a girl.
2. You ask the parents: "Do you have at least one daughter?", and they say yes.
3. You happen to see one of the children in the park, and she is a girl.
4. You ask the parents: "Do you have at least one daughter named Ella?", and they say yes. (Assume that the probability of a girl being named Ella is $p \ll 1$ independently of the name of the other child, even though this includes the possibility of two girls named Ella.)

Assume that the probabilities of any one child being a boy or girl are equal.

Exercise 2.11 (Conditional probabilities: knowing more does not always help). You and your friend are participating in a research project to track the spread of a novel

14 Wassily Hoeffding, 1914–1991.

disease. Your task is to test volunteers randomly selected from the general population for the disease. The tests have a 5 % false positive rate and a 30 % false negative rate. Your friend challenges you to a bet: Correctly predicting whether the next volunteer has the disease. She claims she will not need to know the results of the test to win. You both know that 1 % of the population is currently infected.

1. What are the optimal guessing strategies for you and for your friend?
2. What is the smallest fraction of infected in the population such that the strategy "assume the test is always correct" is better than "claim no one is infected"? The largest such that relying on the test is better than claiming everyone is infected?

Exercise 2.12 (Monty Hall). You are a contestant on a television game show, where you are given the choice of opening one of three doors. Behind one of them is a car, but behind the other two are goats (the doors are soundproof). The host of the show, Monty Hall, knows which door hides the car. After you choose a door, say number 1, the host opens a different door, say number 3, revealing a goat. He then asks if you would like to pick door number 2 instead.

1. Should you (assuming you are interested in winning the car)?
2. Suppose instead that the host did not deliberately open door number 3, but instead slipped on a banana peel and accidentally opened it. Now does it matter if you switch?
3. Suppose that the host does not like to walk and therefore will open the door hiding a goat that has the lowest number. Now what do you do?

2.7 Notes and further reading

Laplace published the general form of Bayes' rule in [185]. For more on the historical development of Bayes' rule, see Dale [67] or Stigler [274]. Boolean algebras were introduced by Boole in 1847 [42]. For more details, consult Givant and Halmos [112] or Whitesitt [301]. Halmos gives a very nice review of the relation of Boolean algebras to probability in [117]. Our approach to formalizing probability follows Rényi [247]. Probability as used in inductive logic, as we have presented it here, is also discussed in more detail by Skyrms [266] and Hacking [115]. Jaynes [155] describes how Bayesian probability "ought" to be applied in science; to say that opinions vary among mathematical and statistical researchers is an understatement. For the contrary view, see Mayo [203]. A nice overview of the issues is given by Gelman and Shalizi [107]. For more on the mathematical structure of probability theory itself, see the introductory text by Ross [246], an intermediate approach by Gut [113], and the more in-depth treatment by Durrett [87]. The Dutch book argument goes back to Ramsey [234] and de Finetti [74]. For much more on convexity, including a proof that bounded convex sets in $\mathbb{R}^n$ are equal to the convex hulls of their extreme points, see the treatments by van Tiel [288], Rockafellar [244], or Barvinok [13]. A nice overview of inequalities such as the Hoeffding bound from [141] is [43]. The two girls problem is from Mlodinow [207], the original Monty Hall problem from Selvin [257], and the two variants (Monty Fall and Monty Crawl) from Rosenthal [245].

3 Classical channels

> I think from where we stand the rain seems random. If we could stand somewhere else, we would see the order in it.
>
> Haskie Jim, from *Coyote Waits* by Tony Hillerman

Channels are the central object of study in information theory, as they are used to describe both the effects of physical noise in a communication medium and the action of the encoder and decoder. Visible light hardly needs to be mentioned as a communication medium, essential as it is to a printed book. Longer distance communication is typically accomplished using longer wavelengths. On Earth, most long-distance communication is via infrared laser in optical fiber. Very long-distance communication, to deep space probes such as Voyager, is via radio frequencies at around 2 GHz. This is essentially the same frequency band used for mobile communication and WiFi. Communication in the broad sense of the term also includes storage. This is also commonly electromagnetic, either magnetic as in tape systems or hard disk drives, or electric as in flash memory.

The difficulty with noisy communication channels, of course, is that the input cannot be unambiguously determined from the output. The breakthrough of Shannon was to treat all channels at an abstract level, as a probabilistic mapping of the input to the output, where both the input and output are described by random variables. That is, there is only one kind of (classical) information, not separate electromagnetic or acoustic information. The encoder and decoder of a communication system are also maps on random variables, though usually deterministic, can be treated formally in the same way.

Perhaps the simplest abstract model of noise is the *binary symmetric channel*, which takes one input bit and reproduces it at the output with some probability, say $1 - p$, and otherwise flips the input value with probability p. We can regard the binary symmetric channel as simply adding a biased noise bit to the input bit modulo two, where the noise bit has the value 1 with probability p. This is depicted in Figure 3.1, which also depicts two other examples, the binary erasure channel and the Z channel.

3.1 Definition

Each particular input, say $X = x \in \mathcal{X}$, to a channel produces a probability distribution over the outputs $\mathcal{Y}$, meaning a channel is specified by a collection of conditional probability distributions $P_{Y|X=x}$. Following the notation for conditional probability distributions, we can then denote a channel by $P_{Y|X}$, though we will more often use $W_{Y|X}$. Now consider the action of an arbitrary channel $W_{Y|X}$ on an arbitrary distribution P_X, which we formally express as

$$P_Y = W_{Y|X} P_X \,. \tag{3.1}$$

https://doi.org/10.1515/9783110570250-003

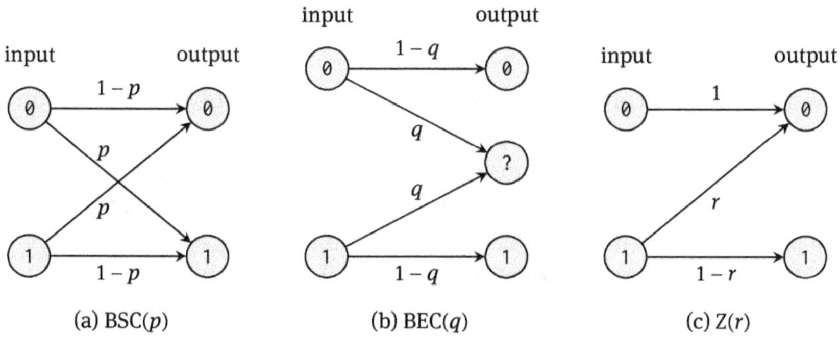

Figure 3.1: The binary symmetric channel (BSC), the binary erasure channel (BEC), and the Z channel. The diagrams indicate various transition probabilities from the channel input to output. BSC(p) flips the input bit with probability p and otherwise leaves the value unchanged. BEC(q) erases the input with probability q, producing the symbol "?" at the output, and otherwise leaves the value unchanged. Z(r) does not change 0, but flips 1 to 0 with probability r.

By the law of total probability we have

$$P_Y(y) = \sum_{x \in \mathcal{X}} W_{Y|X}(y,x)\, P_X(x) \quad \forall y \in \mathcal{Y}. \tag{3.2}$$

Exercise 3.1. Show that the BEC can be transformed into a BSC by acting on the output of the former with another channel. What is the resulting flip probability in terms of the erasure probability? Can a BSC be similarly transformed into a BEC?

Another simple channel is one that just discards the input and outputs a random variable with fixed distribution. That is, the distribution at the output does not depend on the input. For instance, BSC($1/2$) and BEC(1) are such channels.

Exercise 3.2. Show that the BSC is a convex combination of the identity channel and a fixed-output channel.

If we regard P_X as a column vector, then the action of $W_{Y|X}$ corresponds to multiplication from the left by the matrix whose (y,x) entry is $W_{Y|X}(y,x)$, i. e., the matrix formed from the sequence of column vectors $W_{Y|X=x}$. Regarding $W_{Y|X}$ as this matrix, the channel action expressed in (3.1) becomes a statement of matrix multiplication.

Observe that the sum of the entries in every column of any such channel matrix is just 1. Matrices with positive entries whose column sums are all 1 are called *stochastic matrices*. According to this definition, the set of possible channels from $\mathcal{X}$ to $\mathcal{Y}$ is equivalent to the set Stoc(n,m) of $n \times m$ stochastic matrices with $|\mathcal{X}| = m$ and $|\mathcal{Y}| = n$. Note that Stoc(n,m) is the m-fold Cartesian[1] product of Prob(n) with itself.

---

1 René Descartes, 1596–1650.

Exercise 3.3. Show that Stoc(n, m) is a convex set by showing that the Cartesian product of convex sets is convex.

Channels encompass probability distributions and tests: Probability distributions are columns, while tests are rows. By definition the columns of a stochastic matrix in Stoc(n, m) are normalized and positive, meaning they are elements of Prob(n). Put differently, Prob(n) = Stoc$(n, 1)$, as column vectors are matrices acting on a one-dimensional input space. Meanwhile, the rows of Stoc(n, m) must have components no larger than one to meet the column-sum normalization condition, meaning the rows are elements of Tests(m). The set Tests(m) is equivalent to Stoc$(2, m)$. On the one hand, for any $T \in$ Tests(n), we can construct the stochastic matrix with rows T and $1_n - T$, where $1_n \in$ Tests(n) is the vector of all 1s. On the other hand, any $2 \times n$ stochastic matrix specifies a valid test by taking the first (or second) row. Observe that T itself corresponds to the event of passing the test, whereas the associated binary-output channel corresponds to the testing procedure, which results in either pass (the first outcome) or fail (the second). The name "test" is ambiguous in this sense.

Instead of specifying the constraints defining the sets Prob(m), Tests(n), and Stoc(n, m) by components, we will use pointwise inequalities. For instance, $T \le 1_n$ for a test T. Similarly, $P \in$ Prob(n) means $P \ge 0$ and $1_n \cdot P = 1$, equivalently $1_n^T P = 1$, using the transpose. The complete description of tests of size n is any T such that $0 \le T \le 1_n$. For stochastic $n \times m$ matrices M, we write $M \ge 0$ and $1_n^T M = 1_m^T$. In expressions involving matrices, we use the convention that vectors are interpreted as column vectors. We will also adopt the convention that "|" in a subscript separating random variables indicates that we interpret the object as a matrix, mapping the alphabets associated with the random variables on the right to those on the left. For instance, $W_{XY|Z}$ is a matrix taking $\mathbb{R}^{|\mathcal{Z}|}$ to $\mathbb{R}^{|\mathcal{X}||\mathcal{Y}|}$. Fixing values in a subscript allows us to pull out column or row vectors, that is, $K_{Y|X=x}$ is a column vector, whereas $K_{Y=y|X}$ is a row vector. When $K_{Y|X}$ represents a channel, the former is a probability distribution, whereas the latter is a test.

We can just as well regard channels as acting on tests instead of on probability distributions. Given a channel $W_{Y|X}$ and a test T_Y on $\mathcal{Y}$, the probability of T_Y under the distribution $P_Y = W_{Y|X} P_X$ is simply $\sum_{x \in \mathcal{X}, y \in \mathcal{Y}} T_Y(y) W_{Y|X=x}(y) P_X(x)$. It is easy to see that

$$T'_X(x) = \sum_{y \in \mathcal{Y}} T_Y(y) W_{Y|X=x}(y) \tag{3.3}$$

defines a valid test on $\mathcal{X}$, since $W_{Y|X}$ is a stochastic matrix. This can be more compactly written using the transpose as $T'^T_X = T_Y^T W_{Y|X}$ or $T'_X = W_{Y|X}^T T_Y$. We will see later that the two ways of viewing the channel action, as a map either on probabilities or on tests, is completely analogous to the Schrödinger and Heisenberg[2] pictures in quantum mechanics, respectively.

2 Werner Karl Heisenberg, 1901–1976.

Channels also describe physical measurements. Returning to the example of the mass of Saturn, the measurement or measurements in question are observations of Saturn's position in the night sky. Let us describe one such observation by a random variable Y. Given the details of the observation, e. g., which kind of telescope was used, the quality, the observing conditions, etc., we can construct a model for the value of Y given the actual mass $M = m$. The model is a conditional probability distribution $P_{Y|M=m}$, essentially a channel from M to Y. The relevant probability distribution for M then changes from whatever the prior probability P_M was to the conditional probability $P_{M|Y=y}$, since the value of Y is available after the measurement is complete.

Note that if we lose Y or just ignore it, then the relevant probability is again P_M. There are two ways to come to this conclusion. On the one hand, once Y is lost, then we are back to the original state of affairs, so P_M must be the relevant probability. On the other hand, we could just average $P_{M|Y=y}$ over the different possible measurement results $Y = y$. The law of total probability ensures that this also gives P_M.

3.2 Alternate definitions

We should perhaps not be too quick to adopt the definition of channels as stochastic matrices based on the law of total probability, as there are several reasonable alternatives. If we use the "wrong" channel definition, then any result we derive on the physical limits of information processing protocols (which necessarily involves channels) could be invalid. It will also pay off later in the quantum case to have considered this issue here in the simpler setting of probability theory. Two other reasonable definitions of classical channels are the following.

1. (Via convexity). Abstractly, channels should map probability distributions to probability distributions in a way that is compatible with convexity. That is, a channel $W_{Y|X}$ should be a map $W_{Y|X} : \text{Prob}(X) \to \text{Prob}(Y)$ such that, for any random variable Z with distribution P_Z,

$$W_{Y|X}\Big(\sum_{z\in\mathcal{Z}} P_Z(z)P_{X|Z=z}\Big) = \sum_{z\in\mathcal{Z}} P_Z(z)W_{Y|X}P_{X|Z=z}\,. \tag{3.4}$$

   Let us denote by $\text{Cvx}(n, m)$ the set of all possible channels defined in this manner.

   The motivation for this definition is as follows. Here Z is a random variable whose value is correlated to X, but is unaffected by the action of the channel. Given the value $Z = z$, we expect to obtain the channel output to have distribution $W_{Y|X}P_{X|Z=z}$, and if we forget or ignore the value of Z, then we would obtain the average $\sum_z P_Z(z)W_{Y|X}P_{X|Z=z}$. However, Z has nothing to do with $W_{Y|X}$; the channel acts on X, not Z. Therefore averaging over Z should be the same if done before the channel action as after, which is equality (3.4). If the equality did not hold, then the channel would act differently on X depending on whether we know the value of Z.

2. (Via deterministic transformations). Since we want to view the input and output random variables of a channel as the values of physical quantities, we should restrict our attention to channels that in principle perform some deterministic (or even better, reversible) dynamics on the input, but we are not completely certain which. Determinism is in line with the principles of classical mechanics, and our uncertainty is due to the fact that we do not observe all the degrees of freedom involved in the dynamics. (Here we sidestep the issue, raised by the existence of chaotic systems, of whether this is even possible in principle.) Formally, let $\{f_z\}_{z\in\mathcal{Z}}$ be all the deterministic functions from $\mathcal{X}$ to $\mathcal{Y}$ labeled by z. Then we should only consider maps $W_{Y|X}$ that produce P_Y from P_X such that
$$P_Y(y) = \sum_{z\in\mathcal{Z}} P_Z(z) \sum_{x\in\mathcal{X}} 1[f_z(x) = y]P_X(x) \tag{3.5}$$
for some distribution P_Z. This is similar to (2.13), since we generate Y by applying a function to X, but now we average over the particular choice of function. For instance, the BSC is the average of the identity and flipping the value of the bit.
Any function can be represented by a stochastic matrix whose entries are restricted to $\{0, 1\}$. If we set $\mathcal{X} = \{1, \dots, m\}$ and $\mathcal{Y} = \{1, \dots, n\}$, and associate the kth element with the kth unit vector, then the function $f : \mathcal{X} \to \mathcal{Y}$ is represented by the matrix whose xth column has all zeros except for a single 1 in the $f(x)$th row. Let us denote by $F(y_1, \dots, y_m)$ the matrix with a single 1 in the y_xth row of the xth column and call the entire set of such deterministic $n \times m$ matrices $\mathrm{Det}(n, m)$. Then (3.5) amounts to saying that the set of channels is given by $\mathrm{Hull}(\mathrm{Det}(n, m))$, the convex hull of deterministic maps.

From the definitions it is clear that
$$\mathrm{Hull}(\mathrm{Det}(n, m)) \subseteq \mathrm{Stoc}(n, m) \subseteq \mathrm{Cvx}(n, m)\,. \tag{3.6}$$
The first inclusion follows because deterministic matrices are stochastic. The second inclusion follows because any channel defined by a stochastic matrix satisfies (3.4). Hence by adopting the stochastic matrix definition it could be that we either have a too large or too small set to properly describe physical information processing protocols. However, we need not worry: All three channel definitions are equivalent.

Channels as stochastic matrices are linear maps on the input distribution, but this is not explicitly required in the convexity definition. However, linearity is not so far removed from convexity. For any function f defined on a convex domain $\mathcal{S} \subset \mathbb{R}^n$

that is *convex-linear* in that it satisfies $f(\lambda x + (1-\lambda)y) = \lambda f(x) + (1-\lambda)f(y)$ for all $x, y \in \mathcal{S}$ and $\lambda \in [0,1]$, there exists a linear function f^{l} on the span of $\mathcal{S}$ that agrees with f on $\mathcal{S}$. In the present setting, $\mathcal{S}$ is the simplex Prob(n), whose extreme points are the defining orthonormal basis for $\mathbb{R}^n$. Denoting the basis vectors e_j, we can by definition uniquely express any point $y \in \mathbb{R}^n$ as $y = \sum_k c_k e_k$ for some $c_k \in \mathbb{R}$, so the extension must be $f^{\mathsf{l}}(y) = \sum_k c_k f(e_k)$. Therefore all elements of Cvx(n, m) are linear and have a matrix representation.

It remains to show that their matrix representations satisfy the two requirements of stochastic matrices. But both normalization and positivity follow since the inputs are arbitrary elements of Prob(m). Since normalization is preserved for arbitrary inputs from Prob(m), which spans $\mathbb{R}^m$, the columns must each sum to one. Similarly, the output is not guaranteed to be nonnegative for arbitrary inputs unless the matrix components are themselves nonnegative. Hence we have shown the following:

Proposition 3.1. Stoc(n, m) = Cvx(n, m).

The conclusion that convex-linear maps can be extended to linear maps also holds when $\mathcal{S}$ is not a simplex, when there is no unique expansion of points $y \in \mathcal{S}$ with which to define the extension f^{l}. Let us also show this, as it will be useful later.

Proposition 3.2. *For any convex-linear function f whose domain is a convex set $\mathcal{S}$, there exists a linear function f^{l} whose domain is the span of $\mathcal{S}$ and which agrees with f on $\mathcal{S}$.*

The proof relies on the notion of the *convex cone* generated by a convex set $\mathcal{S}$. A convex cone $\mathcal{C}$ is a set for which $ax+by \in \mathcal{C}$ for all $x, y \in \mathcal{C}$ and $a, b \geq 0$. A convex set $\mathcal{S}$ generates the cone $\mathcal{C}$ consisting of all points λx for $x \in \mathcal{S}$ and $\lambda \geq 0$. To ensure the definition indeed generates a convex cone, we need to check that $ax + by$ for $x, y \in \mathcal{S}$ and $a, b \geq 0$ is the scaled version of some element in $\mathcal{S}$. This is indeed the case, since we can write it as $(a+b)(\frac{a}{a+b}x + \frac{b}{a+b}y)$, and the second factor is an element of $\mathcal{S}$ by assumption.

Proof. First, extend f to the cone $\mathcal{C}$ generated from $\mathcal{S}$ by taking $f^{\mathsf{l}}(\lambda x) = \lambda f(x)$ for $\lambda \geq 0$. In particular, $f^{\mathsf{l}}(0) = 0$. Next, consider an arbitrary point $\bar{x}$ in the span of $\mathcal{S}$. By grouping the terms with positive and negative coefficients separately, it is clear that $\bar{x}$ can be expressed as the difference of two points in $\mathcal{C}$, i. e., $\bar{x} = \lambda_1 x_1 - \lambda_2 x_2$ for some $\lambda_j \geq 0$ and $x_j \in S$. We can then define $f^{\mathsf{l}}(\bar{x}) = \lambda_1 f(x_1) - \lambda_2 f(x_2)$.

However, unlike the case of the simplex, the expression for $\bar{x}$ is not unique. We can have $\bar{x} = \lambda_1' x_1' - \lambda_2' x_2'$, and the question arises whether $\lambda_1' f(x_1') - \lambda_2' f(x_2')$ equals $f^{\mathsf{l}}(\bar{x})$ defined from $\lambda_1 x_1 - \lambda_2 x_2$. To see that equality does hold, first define $x' = \lambda_1 x_1 + \lambda_2' x_2'$ and note that it equals $\lambda_1' x_1' + \lambda_2 x_2$ by assumption. Now set $c = \lambda_1 + \lambda_2'$ and observe that x' is c times the convex combination $\frac{\lambda_1}{c} x_1 + \frac{\lambda_2'}{c} x_2'$. Hence $f^{\mathsf{l}}(x') = \lambda_1 f(x_1) + \lambda_2' f(x_2')$. By the same reasoning we have $f^{\mathsf{l}}(x') = \lambda_1' f(x_1') + \lambda_2 f(x_2)$, and therefore the extension is well-defined. □

Now we turn to the remaining equivalence, for which we need to show that the set of stochastic matrices is the convex hull of the deterministic matrices. The deterministic matrices are the extreme points of the set of stochastic matrices, so we have already stated that the conclusion must be true in Section 2.4. However, the proof is simple enough to give here. Due to the equivalence of Stoc$(2, n)$ and Tests(n), this will also imply that Tests(n) = Hull(Events(n)).

Proposition 3.3. Stoc(n, m) = Hull(Det(n, m)).

Proof. In light of (3.6), it remains to show that every $M \in$ Stoc(n, m) is contained in Hull(Det(n, m)). Suppose M has components M_{yx} and define

$$M' = \sum_{y_1,\dots,y_m=1}^{n} M_{y_1,1} \cdots M_{y_m,m} F(y_1, \dots, y_m) . \tag{3.7}$$

The coefficient for a given $F(y_1, \dots, y_m)$ is simply the product of the transition probabilities for $X = 1$ to be mapped to $Y = y_1$ and so forth. We can think of the particular sequence $(y_1, y_2, \dots, y_m)$ as a "path" through the columns of an $n \times m$ matrix, where in each step, we are free to choose a row. The coefficient in the expansion is the product of the probabilities encountered in M on the path $(y_1, y_2, \dots, y_m)$. The coefficients make up a convex combination, since they are positive real numbers whose sum, the product of the column sums of M, is unity. Hence $M' \in$ Hull(Det(n, m)).

Now consider the (j, k) component of M'. This component of $F(y_1, y_2, \dots, y_m)$ is zero unless $y_k = j$, in which case it is one. Thus the only contributions to M'_{jk} from the summation are those with $y_k = j$, i. e., paths through the columns of M that go through (j, k). The (j, k) step of the path contributes a factor of M_{jk}, while the summation over the choice of row in all the other columns just yields one by the same argument as in the normalization statement above. Therefore $M'_{jk} = M_{jk}$. □

For an example of the decomposition used in the proof, consider BSC(p). There are four possible functions from $\{0, 1\}$ to itself: identity, flip, zero, and one, where zero maps everything to zero and one everything to one. Indeed, we have

$$\mathrm{BSC}(p) = (1-p)^2 \begin{pmatrix} 1 & 0 \\ 0 & 1 \end{pmatrix} + p^2 \begin{pmatrix} 0 & 1 \\ 1 & 0 \end{pmatrix} + p(1-p) \begin{pmatrix} 1 & 1 \\ 0 & 0 \end{pmatrix} + (1-p)p \begin{pmatrix} 0 & 0 \\ 1 & 1 \end{pmatrix} . \tag{3.8}$$

Note that, by definition, BSC(p) is $1 - p$ times identity plus p times flip, so that the representation in (3.7) is not unique. The set of stochastic matrices is not a simplex.

In fact, the mismatch between the number of extreme points and the space in which the polytope lives is quite large. Altogether, there are n^m vertices, the number of distinct $n \times m$ deterministic matrices. However, only $m(n-1)$ parameters are needed to specify an arbitrary $n \times m$ stochastic matrix. *Carathéodory's*[3] *theorem* states that any element of a polytope living in an d-dimensional space can be represented as a convex combination of no more than $d + 1$ vertices, so it is always possible to find a decomposition of an $n \times m$ stochastic matrix into no more than $m(n-1)+1$ deterministic matrices. Observe that the binary symmetric channel beats this bound by one.

Exercise 3.4. Determine the decomposition of (3.7) for BEC(q) and Z(r). Are they redundant?

Exercise 3.5. Consider a binary-input channel $W : X \to Y$, where $Y = [-1, 1]$, with the symmetry that $P_{Y|X=1}(y) = P_{Y|X=0}(1 - y)$. That is, the two output distributions are mirror images of each other. Show that W can be regarded as a *heralded* mixture of BSCs in the sense that the output is a mixture of BSCs with different parameters plus an additional piece of information (the herald, so to speak) specifying the parameter of the BSC. Does the BEC fit into this framework?
Hint: Consider the absolute value $|Y|$ *and* $\operatorname{sign}(Y)$.

3.3 Notes and further reading

Proposition 3.3 is adapted from Davis [72]. Readers interested in the proof of Carathéodory's theorem should again consult van Tiel [288], Rockafellar [244], or Barvinok [13].

3 Constantin Carathéodory, 1873–1950.

4 Quantum probability theory

> The important point is that the laws of quantum mechanics can be expressed only in terms of probability connections.
>
> Eugene P. Wigner

In this chapter, we jump right into the probabilistic aspects of quantum theory without yet worrying too much about the physics. Our approach is to obtain the formalism of quantum theory by generalizing that of probability theory. Specifically, we give quantum versions of probability distributions, tests, and channels, as well as the rules by which these are related to each other.

The result does not, on its face, look like it has much to do with the standard, more physical approach to quantum mechanics based on wavefunctions, observables, the Schrödinger equation, and so forth (for a concise formulation, see Appendix A). But we will see that the two approaches are indeed compatible, particularly in the following chapter. By presenting quantum theory as a generalization of probability theory it is much easier to appreciate the relation between the two, and the generalization is quite straightforward besides.

The setting for our generalization is the simple scenario of a generic experiment to measure the value of some physical quantity, such as the energy of a particle. As an example, consider the large hadron collider (LHC). The experiment consists of three basic steps:

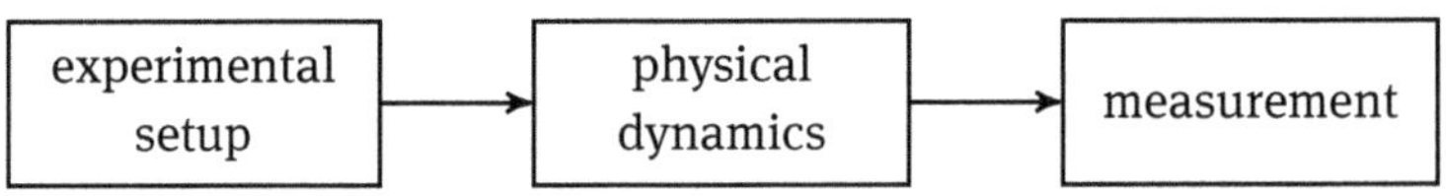

First, the physical system to be investigated is prepared in some way. At the LHC, a great deal of effort goes into generating circulating beams of high-speed protons in two rings. Next, the system is allowed to undergo its own internal dynamics or is subject to an external interaction of some kind, e. g., protons from the two beams are made to collide. Finally, some kind of measurement is made on the result. Abstracting away a colossal amount of detail, a measurement at the LHC can be thought of as having just one of two possible outputs, "particle with energy $E \pm \Delta E$ detected" and "particle with energy $E \pm \Delta E$ not detected".

Indeed, we have abstracted away all of the physics, which consists of knowing which outputs correspond to these two cases. The above is very much the information-theoretic description of the experiment. On this level, there is not that much distinction with an experiment in which two dice are rolled and we check if the output is even. All that really matters is the set of possible outcome events and their probabilities.

In our probabilistic framework the preparation is described by a probability distribution, any possible dynamics is modeled by a classical channel, and the final mea-

https://doi.org/10.1515/9783110570250-004

surement is described by a collection of tests, one for each possible outcome. The system initially prepared is then described by a random variable X with probability distribution P_X, the channel $W_{Y|X}$ may map this to a different random variable Y, and we can denote the final measurement result by the random variable Z. Formally, the final measurement is also a channel, call it $\hat{T}_{Z|Y}$, and the test associated with outcome $Z = z$ is $\hat{T}_{Z=z|Y}$. Let us define outcome $Z = 0$ to be "particle with energy $E \pm \Delta E$ detected", and let $T_Y = \hat{T}_{Z=0|Y}$, treating T_Y as a column vector instead of a row vector. Then the probability to observe this outcome in one run is just $T_Y \cdot W_{Y|X} P_X$.

4.1 States, effects, and measurements

Now that we have the general setup, let us focus on the case in which the channel $W_{Y|X}$ is trivial, i. e., focus just on the probability distributions and tests. We will take up channels in the next chapter. The entire probability structure is given by

$$\begin{aligned} &\Pr_P[T] = T \cdot P, && && (4.1)\\ &P \in \mathbb{R}^n, \quad P \geq 0, \quad 1_n \cdot P = 1,\\ &T \in \mathbb{R}^n, \quad T \geq 0, \quad T \leq 1_n \,. \end{aligned}$$

The formalism of quantum mechanics can be seen as a generalization that retains this structure, but allows the analogs of P and T to reside in a different vector space.

In particular, consider the set $\mathrm{Lin}(\mathcal{H})$ of linear maps or operators on a vector space $\mathcal{H} = \mathbb{C}^d$ for some $d < \infty$. As discussed more thoroughly in Appendix B, $\mathrm{Lin}(\mathcal{H})$ is itself a vector space and can be equipped with the Hilbert[1]–Schmidt[2] inner product $\langle S, T\rangle = \mathrm{Tr}[S^* T]$ for $S, T \in \mathrm{Lin}(\mathcal{H})$ and the trace operation Tr. Furthermore, there is a notion of positivity and a partial ordering of operators by positivity for $\mathrm{Lin}(\mathcal{H})$ as well. Now we are ready for the generalization to the quantum case. We replace P, T, and 1_n by operators ρ, Λ, and the identity operator $\mathbb{1}$ in $\mathrm{Lin}(\mathcal{H})$, but otherwise keep the structure intact. Since positive operators are necessarily Hermitian (see Lemma B.1), this replacement gives

$$\begin{aligned} &\Pr_\rho[\Lambda] = \mathrm{Tr}[\Lambda\rho], && && (4.2)\\ &\rho \in \mathrm{Lin}(\mathcal{H}), \quad \rho \geq 0, \quad \mathrm{Tr}[\rho] = 1,\\ &\Lambda \in \mathrm{Lin}(\mathcal{H}), \quad \Lambda \geq 0, \quad \Lambda \leq \mathbb{1}\,. \end{aligned}$$

Now the preparation of the physical system is described by the operator ρ, called the *density operator*, density matrix, or just the (quantum) *state*. The event associated with

1 David Hilbert, 1862–1943.

2 Erhard Schmidt, 1876–1959.

a particular outcome of the measurement is described by the operator Λ, sometimes called an *effect* or effect operator (the effect produced by the measurement apparatus). The probability of this outcome conditioned on the preparation is given by (4.2), the *Born*[3] *rule*. Indeed, this rule gives a number between zero and one. As both Λ and ρ are positive, by Lemma B.3 $\mathrm{Tr}[\Lambda\rho]$ is, too, and since $\mathbb{1} - \Lambda$ is positive and $\mathrm{Tr}[\rho] = 1$, we have $0 \leq \mathrm{Tr}[(\mathbb{1} - \Lambda)\rho] = 1 - \mathrm{Tr}[\Lambda\rho]$.

The entire quantum measurement is specified by a collection of events such that precisely one of them is bound to occur. In the quantum setting, this collection is called a POVM, short for *positive operator valued measure*,[4] a set $\{\Lambda(x)\}_{x=1}^n$ of effects such that $\sum_{x=1}^n \Lambda(x) = \mathbb{1}$. Often, the $\Lambda(x)$ are called POVM elements instead of effects. The outcome of the measurement is a random variable X with distribution $P_X(x) = \mathrm{Pr}_\rho[\Lambda(x)]$. The probability of any one of the outcomes occurring, i. e., $X = x_1 \vee X = x_2 \vee \cdots \vee X = x_n$, is, by the addition rule of probability and linearity of the Born rule,

$$\sum_{x=1}^{n} P_X(x) = \mathrm{Tr}\left[\sum_{x=1}^{n} \Lambda(x)\rho\right] = \mathrm{Tr}[\rho] = 1 . \tag{4.3}$$

Thus the completeness condition $\sum_{x=1}^n \Lambda(x) = \mathbb{1}$ reflects the fact that the POVM covers every possible outcome.

In this presentation the density operators and POVM elements have no particular physical status; they are only mathematical objects used to generate a probability distribution for the final measurement results. That is not to say that these objects could not have a more physical meaning, just that we are eschewing that question here. Our focus is the statistical structure of quantum theory. Note also that we have only dealt with the (classical) outcomes of the measurement, and the post-measurement quantum state will be treated in the following chapter.

A word on notation: Often subscripts are used to denote elements of a sequence, e. g., Λ_x for the xth element of a POVM. However, just as with probability, this will be unwieldy as we have other uses for the subscript. Instead, we write $\Lambda(x)$ for the xth POVM element. Occasionally, we will violate this rule when expedient.

The wavefunctions and projective measurements described by complete sets of orthogonal projection operators familiar from the usual treatment of quantum mechanics are particular cases of this general formalism. Wavefunctions or wavevectors $|\psi\rangle \in \mathcal{H}$ correspond to density operators $|\psi\rangle\langle\psi|$. Here we make use of Dirac notation, the details of which can be found in Section B.2. Projection operators are a particular

---

3 Max Born, 1882–1970.

4 The mouthful "POVM" comes from the more general setting in which the measurement outcome can take a continuous range of outcomes. This necessitates the use of measure theory. Then for each measurable set of outcomes, we require a positive operator such that the probability that the measurement result is in the set is given by the Born rule. Hence the measure assigned to each measurable set takes values in the positive operators.

kind of effect, and the normalization condition of POVMs becomes the completeness relation for projection operators.

The set of states of a d-dimensional quantum system is given by

$$\mathrm{Stat}(d) := \{\sigma \in \mathrm{Lin}(\mathbb{C}^d) : \sigma \geq 0, \mathrm{Tr}[\sigma] = 1\}, \tag{4.4}$$

and for systems with explicit names, e. g., system A, we also write $\mathrm{Stat}(A)$. We could also name the set of effects and the set of POVMs with a fixed number of outcomes, but we will not need them very often. To learn something about the extreme points of the set of states, observe that by the spectral theorem we can decompose any positive operator $M \in \mathrm{Lin}(\mathcal{H})$ into a summation of projection operators of rank one

$$M = \sum_{j=1}^{d} \lambda_j |\lambda_j\rangle\langle\lambda_j|, \tag{4.5}$$

where $\lambda_j \geq 0$ are the eigenvalues, and $|\lambda_j\rangle \in \mathcal{H}$ are the associated normalized eigenvectors. When $\mathrm{Tr}[M] = 1$ so that M is a density operator, the eigenvalues λ_j form a probability distribution. The projection operators $|\psi\rangle\langle\psi|$ associated with wavefunctions $|\psi\rangle$ are called *pure states*. Thus the extreme points of the set of states are necessarily pure states.

Exercise 4.1. Show that a density operator ρ is a pure state if $\mathrm{Tr}[\rho^2] = 1$.

In fact, all pure states are extreme points, which is to say that if $|\psi\rangle\langle\psi| = p|\varphi\rangle\langle\varphi| + (1-p)|\theta\rangle\langle\theta|$ for normalized $|\varphi\rangle$ and $|\theta\rangle$ and $p \in [0, 1]$, then either $p = 0$ and $|\theta\rangle = |\psi\rangle$, $p = 1$ and $|\varphi\rangle = |\psi\rangle$, or $p \in (0, 1)$ and $|\varphi\rangle = |\theta\rangle = |\psi\rangle$.

Exercise 4.2. Show this. *Hint: One option is to use the fact that $\rho^2 = \rho$ for projection operators ρ. Another is to show that $|\psi\rangle\langle\psi| - p|\varphi\rangle\langle\varphi| \ngeq 0$ for $p > 0$ and $|\psi\rangle \neq |\varphi\rangle$.*

From (4.5) it therefore follows that the set of states is the convex hull of the pure states. Density operators that are not pure are called *mixed states*. Unlike its classical analog, the set of probability distributions, the set of states is not a simplex. In other words, the decomposition of an arbitrary mixed state into a convex combination of pure states is not unique. Consider the case of the maximally mixed state in some arbitrary dimension d, the operator $\frac{1}{d}\mathbb{1}$, which we will always denote by π. It is clearly the convex combination (with equal weights) of the d pure states associated with any orthonormal basis.

This has important consequences for the interpretation of mixed states. Consider the probability of some effect Λ for a mixed state ρ. For $\rho = \sum_j \lambda_j |\lambda_j\rangle\langle\lambda_j|$, the probability is $\mathrm{Tr}[\Lambda\rho] = \sum_{j=1}^{n} \lambda_j \mathrm{Tr}[\Lambda|\lambda_j\rangle\langle\lambda_j|]$. This is the average of conditional probabilities of Λ given the various pure states $|\lambda_j\rangle\langle\lambda_j|$. We might therefore be tempted to view the indeterminacy in the outcome Λ as partly due to the quantum nature of $|\lambda_j\rangle$ and partly due to the ignorance of the value of j. However, it is impossible to uniquely make such a division since the set of states is not a simplex.

We have less to say about the set of effects and the set of POVMs. The set of effects has a nice convex structure as well, and it is not difficult to see that projection operators are the extreme points. Thus general effects are to projection operators what general (classical) tests are to the more basic events. However, the set of POVMs is not as simple. In particular, not all POVMs are mixtures of projective measurements, as we will see later, in Exercise 4.6.

Exercise 4.3. Show that all projections are the extreme points of the set of effects. *Hint: Recycle the methods used in Exercise 4.2 for one direction. For the other, use the spectral decomposition to express an arbitrary effect as a convex combination of projections of all ranks, including zero.*

4.2 Qubits

The simplest quantum system, the qubit, has just two levels, a state space of $\mathcal{H} = \mathbb{C}^2$. We typically denote a "standard basis" for a qubit by the states $|0\rangle$ and $|1\rangle$. A qubit is any system, or more precisely degree of freedom, whose state vector $|\psi\rangle$ can be written as $|\psi\rangle = a|0\rangle + b|1\rangle$ with $a, b \in \mathbb{C}$ and $|a|^2 + |b|^2 = 1$. Table 4.1 lists several examples of qubit systems.

Table 4.1: Examples of qubit systems.

Degree of freedom	**Basis states $\|0\rangle$ and $\|1\rangle$**	
Spin-1/2	$\|J_z = +1/2\rangle$	$\|J_z = -1/2\rangle$
Photon polarization	$\|\text{horizontal}\rangle$	$\|\text{vertical}\rangle$
Electron level in an atom	$\|\text{ground state}\rangle$	$\|\text{excited state}\rangle$
Position in a double well potential	$\|\text{left}\rangle$	$\|\text{right}\rangle$

A useful parameterization of states comes from the spin-1/2 picture. Any state $|\psi\rangle$ can be associated with a point on the unit sphere described by spherical coordinates (θ, φ) via the relation

$$|\psi\rangle = \cos\tfrac{\theta}{2}|0\rangle + e^{i\varphi}\sin\tfrac{\theta}{2}|1\rangle\,. \tag{4.6}$$

This sphere of states is called the *Bloch*[5] *sphere*, as depicted in Figure 4.1.

5 Felix Bloch, 1905–1983.

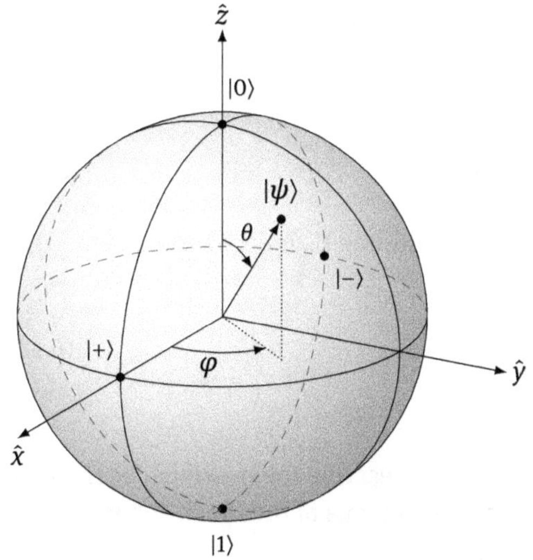

Figure 4.1: The Bloch sphere. Every pure qubit state can be associated with a point on the unit sphere.

Equivalently, we can label states by *Bloch vectors*, unit vectors $\hat{n} = \hat{x}\sin\theta\cos\varphi + \hat{y}\sin\theta\sin\varphi + \hat{z}\cos\theta$. Then it is easy to see that the states $|\hat{n}\rangle$ and $|-\hat{n}\rangle$ are orthogonal. The states along the six cardinal directions ($\pm\hat{x}$, $\pm\hat{y}$, and $\pm\hat{z}$) form three orthogonal bases, and the states $|\pm\hat{x}\rangle = \frac{1}{\sqrt{2}}(|0\rangle \pm |1\rangle)$ are usually just denoted $|\pm\rangle$. These three bases are the eigenbases of the three *Pauli*[6] *operators*:

$$\sigma_x = |0\rangle\langle 1| + |1\rangle\langle 0| \simeq \begin{pmatrix} 0 & 1 \\ 1 & 0 \end{pmatrix}, \tag{4.7}$$

$$\sigma_y = -i|0\rangle\langle 1| + i|1\rangle\langle 0| \simeq \begin{pmatrix} 0 & -i \\ i & 0 \end{pmatrix}, \quad \text{and} \tag{4.8}$$

$$\sigma_z = |0\rangle\langle 0| - |1\rangle\langle 1| \simeq \begin{pmatrix} 1 & 0 \\ 0 & -1 \end{pmatrix}. \tag{4.9}$$

Here the matrices are representations in the basis $\{|0\rangle, |1\rangle\}$. A linear combination of Pauli operators with real coefficients leads to a Hermitian operator.

These three operators, together with the identity operator $\mathbb{1}$, form a very convenient basis for operators on $\mathbb{C}^2$, i. e., a basis for $\mathrm{Lin}(\mathbb{C}^2)$. This follows because we can very easily construct the matrices $\left(\begin{smallmatrix} 1 & 0 \\ 0 & 0 \end{smallmatrix}\right)$, $\left(\begin{smallmatrix} 0 & 1 \\ 0 & 0 \end{smallmatrix}\right)$, etc. from the Pauli operators, and the latter is evidently a basis for $\mathrm{Lin}(\mathbb{C}^2)$. Writing $A = a_0\mathbb{1} + \vec{a}\cdot\vec{\sigma}$ for an operator A with $\vec{\sigma} = \hat{x}\sigma_x + \hat{y}\sigma_y + \hat{z}\sigma_z$, $\vec{a} = \hat{x}a_x + \hat{y}a_y + \hat{z}a_z$, and $a_0, a_x, a_y, a_z \in \mathbb{R}$, it is straightforward to

6 Wolfgang Ernst Pauli, 1900–1958.

verify that $|\pm\hat{a}\rangle$ are the eigenstates of A with eigenvalues $\lambda_{\pm} = a_0 \pm \|\vec{a}\|$. Here $\hat{a}$ is the normalized version of $\vec{a}$. Using this relation, we can immediately infer that the projection operators $\Pi_{\hat{n}} := |\hat{n}\rangle\langle\hat{n}|$ take the form

$$\Pi_{\hat{n}} = \tfrac{1}{2}(\mathbb{1} + \hat{n} \cdot \vec{\sigma}) \,. \tag{4.10}$$

Then it is simple to verify that the probability of obtaining $\Pi_{\hat{n}}$ for the state specified by $|\hat{m}\rangle$ is just

$$\Pr_{\hat{m}}[\Pi_{\hat{n}}] = \tfrac{1}{2}(1 + \hat{n} \cdot \hat{m}) \,. \tag{4.11}$$

Exercise 4.4. Show (4.11).

Exercise 4.5. For a given qubit density operator ρ, describe the possible decompositions of ρ into convex combinations of pure states in terms of the Bloch sphere.

Exercise 4.6. Consider three unit vectors $\hat{n}_k$ that point to the vertices of an equilateral triangle. Show that the operators $\Lambda(k) = \frac{1}{3}(\mathbb{1} + \hat{n}_k \cdot \sigma)$ form a POVM. What changes in $\Lambda(k)$ if the $\hat{n}_k$ point to the vertices of a regular tetrahedron?

Observe that the measurements in Exercise 4.6 imply that not all POVMs are convex combinations of simple projective measurements. The only way a convex combination of effects can give the rank-one $\Lambda(k)$ there is for each one to be equal to $\Lambda(k)$ itself. On the other hand, mixtures of projection measurements are also useful POVMs; for instance, Bob's measurement in the BB84 protocol can be described by a POVM with four outcomes, two aligned or antialigned with $\hat{z}$ and the other two aligned or antialigned with $\hat{x}$.

Our generalization to density operators and POVMs is based on obtaining sensible results from the Born rule. We can turn the setup on its head and ask what form the Born rule can take, given that measurements are associated with projection operators. That is, we are interested in a function Pr that gives the probability of any projector Π_j, and which satisfies the constraints that $\Pr[0] = 0$, $\Pr[\mathbb{1}] = 1$, and $\Pr[\Pi_j + \Pi_k] = \Pr[\Pi_j] + \Pr[\Pi_k]$ for $\Pi_j\Pi_k = 0$. Interestingly, in dimensions three and higher, the Born rule $\Pr[\Pi_j] = \mathrm{Tr}[\Pi_j\rho]$ for some density operator ρ is the only possibility, a statement known as *Gleason's*[7] *theorem*.

Exercise 4.7. Show that the Born rule is not necessary for qubits, i. e., construct a sensible probability function Pr that is not based on a density operator.

7 Andrew Mattei Gleason, 1921–2008. Known for saying that mathematical proofs "really aren't there to convince you that something is true—they're there to show you why it is true."

4.3 Comparison with probability theory

Having established the basics of quantum probability by analogy with standard probability theory (which we will often refer to as "classical" probability theory), let us examine the similarities and differences in more detail. First, it is important to note that the quantum formalism encompasses probability, simply by using commuting operators. For any test T and probability distribution P, we have

$$T \cdot P = \mathrm{Tr}\big[\mathrm{diag}(T)\,\mathrm{diag}(P)\big]\,, \tag{4.12}$$

where $\mathrm{diag}(P)$ is the operator whose matrix representation is diagonal with entries $P(x)$ and similarly for T. Thus the probability rule can be implemented via the Born rule. We will further discuss describing classical distributions in the quantum language in Section 4.4.4.

In our presentation, probability theory was built on the foundation of Boolean algebras of events. As the events are extreme points of the set of tests, the natural analogy in the quantum case are projection operators, the extreme points of the set of effects. However, the crucial distinction between the two formalisms is that the set of projectors does not form a Boolean algebra. Rather, it forms a whole collection of Boolean algebras! Something like this must hold, since, as we have just seen, the quantum formalism encompasses probability theory. More concretely, any complete set of projectors forms a Boolean algebra, which holds because a complete set must consist of commuting or, equivalently, disjoint projectors.

Exercise 4.8. Given a set of $\Pi_k \in \mathrm{Lin}(\mathcal{H})$ such that $\Pi_k^2 = \Pi_k$ and $\sum_k \Pi_k = \mathbb{1}$, show that $\Pi_j \Pi_k = \Pi(j)\delta_{jk}$.

For disjoint projectors, we can define AND, OR, and NOT of the projections to be intersection, span, and orthogonal complement of the underlying subspaces $\Pi \wedge \Pi' = \Pi\Pi'\Pi$, $\Pi \vee \Pi' = \Pi + \Pi'$, and $\neg\Pi = \mathbb{1} - \Pi$. (Here we switch convention, denoting NOT by $\neg$ instead of an overline. We will use overline for complex conjugation of vectors and matrices later on.) Generally, we can say, in the context of the quantum formalism, that

$$\text{classical} = \text{commuting}\,. \tag{4.13}$$

On the other hand, the entire collection of Boolean algebras formed from all sets of complete commuting projectors does not form one larger algebra, because the representation of AND, OR, and NOT cannot be extended to noncommuting projectors. This is a consequence of Gleason's theorem. If it were possible, it would also be possible to consistently assign TRUE or FALSE to all projections. Such an assignment would be a function Pr satisfying the premises of Gleason's theorem. However, there is no density operator ρ for which $\mathrm{Tr}[\Pi\rho] \in \{0,1\}$ for all projections Π.

This leads directly to the indeterminacy of quantum theory. The theory gives the probabilities of the possible outcomes in a given measurement, but does so in a way incompatible with the notion that for every possible outcome of every possible measurement, there is a fact of the matter about whether or not it would occur if actually measured. That is, probabilities in quantum mechanics are not due to ignorance of the "true state of affairs". We will examine this issue in more detail in Chapter 7.

4.4 Composite systems

Let us turn back to the structure of the theory itself. As in the probabilistic case, we use the tensor product to describe multiple quantum systems. We usually give the systems different names or labels, e. g., $A, B, C, \ldots$, and denote the associated vector spaces $\mathcal{H}_A$, $\mathcal{H}_B$, $\mathcal{H}_C$, and so forth. Occasionally, we write $\mathcal{H}_{AB}$ for $\mathcal{H}_A \otimes \mathcal{H}_B$ when there is no possibility of confusion. The advantage of naming the systems is that, just on the level of (multi)linear algebra, we can specify the input and output spaces of operators by their subscripts, e. g., M_A is an element of $\mathrm{Lin}(\mathcal{H}_A)$, while $M_{AB} \in \mathrm{Lin}(\mathcal{H}_{AB})$. Borrowing the notation from probability, $M_{B|A}$ is an element of $\mathrm{Lin}(\mathcal{H}_A, \mathcal{H}_B)$, and there is nothing wrong with $M_{AB|A} \in \mathrm{Lin}(\mathcal{H}_A, \mathcal{H}_{AB})$. We also write $M_{B|A}^*$ for the adjoint of $M_{B|A}$ instead of $(M_{B|A})^*$ to avoid parenthetical clutter. Doing so unfortunately collides with the | notation just introduced in that $M_{B|A}^*$ is an element of $\mathrm{Lin}(\mathcal{H}_A, \mathcal{H}_B)$. Parentheses, instead of subscripts, will be used for function or sequence arguments, so that $\rho_B(x)$ denotes the density operator on $\mathcal{H}_B$ labeled by x, echoing the notation $P_X(x)$ from probability. In the context of composite systems, it pays to be a little more careful with how Dirac notation is interpreted; see Section B.2.

4.4.1 Entangled states

The structure of composite quantum systems is quite different than for classical random variables, due to the existence of *entangled* states. As we will see, in one form or another, entanglement is responsible for the strangeness of quantum mechanics.

The simplest situation for a composite system of two parts A and B is that each is in its own pure state, say $|\varphi\rangle \in \mathcal{H}_A$ and $|\theta\rangle \in \mathcal{H}_B$. We will usually use subscripts to the kets to denote the system, e. g., $|\varphi\rangle_A$ and $|\theta\rangle_B$. The corresponding state in the composite state space $\mathcal{H}_{AB}$ is simply the *product state* $|\varphi\rangle_A \otimes |\theta\rangle_B$.

Since $\mathcal{H}_{AB}$ is a vector space, it also contains all superpositions of product states. Any state that is not a product state is said to be *entangled*. For instance, the two-qubit state $|\Phi\rangle_{AB} = \frac{1}{\sqrt{2}}(|0\rangle_A \otimes |0\rangle_B + |1\rangle_A \otimes |1\rangle_B)$ is an entangled state. As we will see later, there is good reason to think of it as a "maximally" entangled state.

Entangled states such as $|\Phi\rangle_{AB}$ are quite unlike anything we encountered in classical probability theory for the following reason. Since we can create a measurement out of any complete set of projection operators, for every entangled state, there is some measurement whose outcome is certain. In the measurement represented by the projections Π_{AB} and $\mathbb{1}_{AB} - \Pi_{AB}$ for $\Pi_{AB} = |\Phi\rangle\langle\Phi|_{AB}$, the former outcome is certain. Now suppose that instead we measure σ_z on system A, i. e., the observable $(\sigma_z)_A \otimes \mathbb{1}_B$. The measurement has two outcomes, associated with the two projectors $|j\rangle\langle j|_A \otimes \mathbb{1}_B$ for $j \in \mathbb{Z}_2$; here each outcome occurs with equal probability. It can be verified that *every* nontrivial measurement of A results in a uniform distribution of outcomes, and the same holds for B.

Exercise 4.9. Show this. *Hint: note that the difference in outcome probabilities in the example is related to the expectation value of σ_z.*

From the classical point of view, this is a very strange state of affairs. If two random variables X and Y have a deterministic (extremal) joint distribution P_{XY}, then the marginals P_X and P_Y are necessarily also deterministic. Not so in the quantum realm. The joint state of two systems can be pure (extremal) without the marginals also being pure. In some sense the whole is more than the sum of the parts.

Mixed states can be also be entangled. In terms of their density operator, product states take the form $\rho_{AB} = \theta_A \otimes \varphi_B$ for some $\theta_A \in \mathrm{Stat}(\mathcal{H}_A)$ and $\varphi_B \in \mathrm{Stat}(\mathcal{H}_B)$. Such states can be regarded as classical in the sense that there is a well-defined state for each of the constituent systems A and B. This notion continues to hold for mixtures of product states, such as

$$\sigma_{AB} = \sum_{k=1}^{n} P(k)\rho_A(k) \otimes \varphi_B(k) \tag{4.14}$$

with $P \in \mathrm{Prob}(n)$, since then each system again has its own well-defined state conditional on the parameter k of the mixture. Any quantum state of the form (4.14) is called *separable*, and any state that is not separable is said to be entangled.

4.4.2 Bell bases and Weyl–Heisenberg operators

Given orthonormal bases $\{|b_j\rangle\}_{j=1}^{d_A}$ and $\{|b'_k\rangle\}_{k=1}^{d_B}$ for $\mathcal{H}_A$ and $\mathcal{H}_B$, the set of all products $|b_j\rangle_A \otimes |b'_k\rangle_B$ is a basis for $\mathcal{H}_{AB}$. Indeed, this is one way to define the tensor product space. However, $\mathcal{H}_{AB}$ itself is quite indifferent to any possible underlying tensor product structure; it is just a vector space of dimension $d_A d_B$. Thus, whereas we can use the product basis above, we can also find bases of entangled states. For two qubits, for instance, a very useful basis is the *Bell*[8] *basis*:

$$\begin{aligned} |\Phi_{00}\rangle &= \tfrac{1}{\sqrt{2}}(|00\rangle + |11\rangle), \quad &|\Phi_{01}\rangle = \tfrac{1}{\sqrt{2}}(|00\rangle - |11\rangle), \\ |\Phi_{10}\rangle &= \tfrac{1}{\sqrt{2}}(|01\rangle + |10\rangle), \quad &|\Phi_{11}\rangle = \tfrac{1}{\sqrt{2}}(|01\rangle - |10\rangle). \end{aligned} \tag{4.15}$$

The first state is just $|\Phi\rangle$, while the other three differ by the action of a Pauli operator on only one of the qubits. The indices are chosen so that

$$|\Phi_{jk}\rangle = \mathbb{1} \otimes (\sigma_x^j \sigma_z^k)|\Phi\rangle\,, \tag{4.16}$$

as can be verified by direct calculation.

Exercise 4.10. Show that the Bell basis states are eigenstates of the operators $(\sigma_x)_A \otimes (\sigma_x)_B$ and $(\sigma_z)_A \otimes (\sigma_z)_B$. Which eigenvalue combination corresponds to the state $|\Phi_{jk}\rangle$?

Exercise 4.11. Show that $|\Phi\rangle\langle\Phi| = \frac{1}{4}(\mathbb{1} \otimes \mathbb{1} + \sigma_x \otimes \sigma_x - \sigma_y \otimes \sigma_y + \sigma_z \otimes \sigma_z)$. What are the other states of the Bell basis in these terms?

We can generalize the Bell basis to arbitrary dimensions using the Weyl[9]–Heisenberg operators, a generalization of the Pauli operators. First, for a basis $\{|k\rangle\}_{k\in\mathbb{Z}_d}$ of $\mathbb{C}^d$ (which is then referred to as the standard basis), we define the *canonical maximally entangled* state on $\mathcal{H}_A \otimes \mathcal{H}_B$ with $\mathcal{H}_A \simeq \mathcal{H}_B$ and $d = \dim(\mathcal{H}_A)$ by

$$|\Phi\rangle_{AB} := \frac{1}{\sqrt{d}} \sum_{k\in\mathbb{Z}_d} |k\rangle_A \otimes |k\rangle_B\,. \tag{4.17}$$

Here we label the basis states from 0 to $d-1$, i. e., by elements of $\mathbb{Z}_d$, as this is convenient for the following calculations.

Meanwhile, to define the Weyl–Heisenberg operators, start with the *shift* and *clock* operators

$$U = \sum_{k\in\mathbb{Z}_d} |k+1\rangle\langle k|\,, \tag{4.18}$$

$$V = \sum_{k\in\mathbb{Z}_d} \omega^k |k\rangle\langle k|\,, \tag{4.19}$$

where $\omega = e^{2\pi i/d}$. We can regard $|k\rangle$ as the position of a d-dimensional "clock" showing the "time" ω^k. The shift operator advances the time by ω. The Weyl–Heisenberg operators are the d^2 operators $\{U^x V^z : x, z \in \mathbb{Z}_d\}$.

8 John Stewart Bell, 1928–1990.

9 Hermann Klaus Hugo Weyl, 1885–1955.

Exercise 4.12. Show that $UV = \omega VU$ and that the eigenvectors of U are related to the eigenvectors of V by the discrete Fourier[10] transform:

$$|\tilde{x}\rangle = \frac{1}{\sqrt{d}} \sum_{z\in\mathbb{Z}_d} \omega^{xz}|z\rangle\,. \tag{4.20}$$

Just as the Bell states are obtained by the action of the Pauli operators on the maximally entangled state for qubits, the states $|\Phi_{jk}\rangle_{AB} := \mathbb{1}_A \otimes U_B^j V_B^k |\Phi\rangle_{AB}$ for $j,k \in \mathbb{Z}_d$ form an orthonormal basis in dimension d.

Exercise 4.13. Confirm that the $|\Phi_{jk}\rangle_{AB}$ form an orthonormal basis for $\mathcal{H}_{AB}$.

4.4.3 Marginal states and the partial trace

Let us return to the discussion of measuring part of a pure entangled state from Section 4.4.1. In particular, consider an arbitrary measurement performed on one part, system A, of bipartite system described by the pure state $|\Psi\rangle_{AB}$. Here $\mathcal{H}_A$ and $\mathcal{H}_B$ need not be of the same dimension. Call the POVM elements describing the measurement $\Lambda_A(x)$. We can perform the calculation of the probability of outcome x slightly differently, explicitly writing out the trace on system B using the basis $\{|b_k\rangle\}$ to obtain

$$\begin{aligned} P_X(x) &= \mathrm{Tr}_{AB}[\Lambda_A(x)\otimes \mathbb{1}_B|\Psi\rangle\langle\Psi|_{AB}] = \mathrm{Tr}_A[\Lambda_A(x)\ \mathrm{Tr}_B[|\Psi\rangle\langle\Psi|_{AB}]] \\ &= \mathrm{Tr}_A\left[\Lambda_A(x) \sum_{k=0}^{d_B-1} {}_B\langle b_k|(|\Psi\rangle\langle\Psi|_{AB})|b_k\rangle_B\right]. \end{aligned} \tag{4.21}$$

The operation taking $|\Psi\rangle\langle\Psi|_{AB}$ to $\sum_{k=0}^{d_B-1} {}_B\langle b_k|(|\Psi\rangle\langle\Psi|_{AB})|b_k\rangle_B]$ is referred to as the *partial trace* over system B and denoted Tr_B. It is not difficult to show that the result is a valid density operator.

Exercise 4.14. Show that $\mathrm{Tr}_B[|\Psi\rangle\langle\Psi|_{AB}] = \sum_{k=1}^{d_B} \langle b_k|(|\Psi\rangle\langle\Psi|_{AB})|b_k\rangle_B$ is a valid density operator.

Exercise 4.15. Show that $\mathrm{Tr}_B[|\Phi\rangle\langle\Phi|_{AB}] = \frac{1}{2}\mathbb{1}_A$ for the qubit maximally entangled state $|\Phi\rangle_{AB}$.

Thus, if we are only interested in probabilities of measurements on system A, then we can just as well use the density operator given by the partial trace over B. It is often called the *reduced state* or *marginal state*. Marginal states are precisely the quantum analog of marginal probability distributions. We treat them similarly at the level of

10 Joseph Fourier, 1768–1830.

notation. In the context of a joint state ρ_{AB}, we denote by ρ_B the marginal state on system B, i. e., $\rho_B = \mathrm{Tr}_A[\rho_{AB}]$, just as P_X is the marginal of P_{XY}.

The existence of marginal states is an important locality feature of quantum theory. Despite the possibility of entanglement, no action performed on B can, just by itself, affect or influence ρ_A. To see this, first consider a unitary operation on B, such as would arise from time evolution according to the Schrödinger equation. More generally, consider any isometry $V_{B'|B}$. By the cyclic property of the trace, $\mathrm{Tr}_B[|\Psi\rangle\langle\Psi|_{AB}] = \mathrm{Tr}_{B'}[V_{B'|B}|\Psi\rangle\langle\Psi|_{AB}V^*_{B'|B}]$. Next, consider a measurement of B in basis $|b'_k\rangle$ whose result is unknown to A. For outcome x, the conditional state of A is $|\psi_k\rangle_A = {}_B\langle b'_k|\Psi\rangle_{AB}$. This state is not properly normalized, and its norm encodes the probability of outcome k. Averaging over the outcomes by their probabilities gives ρ_A again; this is just the same calculation as in (4.21). Thus neither unitary evolution nor measurement on B has any effect on the description of A alone. We will see in the following chapter that this conclusion extends to arbitrary quantum channels.

4.4.4 Classical-quantum states

As we saw in (4.12), density operators can also be used to represent classical random variables, and the Born rule will faithfully replicate the probability rule (4.1). The key feature is that classical information is encoded in orthogonal quantum states by using diagonal operators. Representing the states of classical values $X = x$ with mutually orthogonal vectors $|b_x\rangle$ on a Hilbert space $\mathcal{H}_X$, the density operator associated with probability distribution P_X is just

$$\rho_X = \sum_{x\in\mathcal{X}} P_X(x)|b_x\rangle\langle b_x|_X\,. \tag{4.22}$$

Composite states with a classical and a quantum part are called classical-quantum (CQ) states. Given some distribution P_X and a collection of states $\varphi_A(x)$, the following is a CQ state:

$$\rho_{XA} = \sum_x P_X(x)|b_x\rangle\langle b_x|_X \otimes \varphi_A(x)\,. \tag{4.23}$$

CQ states are closely related to *ensemble decompositions* of density operators. Consider an arbitrary density operator ρ_A on system A. Suppose we find a decomposition into positive operators, $\rho_A = \sum_{x=1}^n \hat{\varphi}_A(x)$. The $\hat{\varphi}_A(x)$ are necessarily subnormalized, and the set of $P_X(x) = \mathrm{Tr}[\hat{\varphi}_A(x)]$ forms a probability distribution, as can be seen by taking the trace of both sides. Defining $\varphi_A(x) = \hat{\varphi}_A(x)/P_X(x)$, we can then write $\rho_A = \sum_{x=1}^n P_X(x)\varphi_A(x)$. The set of normalized states $\varphi_A(x)$, each one paired together with its associated probability $P_X(x)$, is an ensemble decomposition $\{(P_X(x), \varphi_A(x))\}_{x=1}^n$ of ρ_A. For instance, the spectral decomposition gives an ensemble decomposition into pure states. However, decomposition into orthogonal pure states is not the only kind of pure state ensemble decomposition. In particular, there may be more elements to the ensemble than the dimension of the density operator. This is especially easy to see for qubits using the Bloch representation. Any point in the interior can be decomposed into an average of an arbitrary number of points $n \geq 2$ on the surface in many different ways.

Physically, we could realize state preparation of ρ_A by randomly preparing $\varphi_A(x)$ with probability $P_X(x)$. This ensemble decomposition corresponds to the CQ state in (4.23), so that indeed $\rho_A = \mathrm{Tr}_X[\rho_{XA}]$. Observe that $\rho_{XA} = \sum_x |b_x\rangle\langle b_x|_X \otimes \hat{\varphi}_A(x)$, a more direct link between the CQ state and the associated decomposition into positive operators. The CQ state corresponds to a state preparation device in which $\varphi_A(x)$ is randomly prepared with probability $P_X(x)$ and the particular value of x is also recorded in X at the output. Ignoring X by tracing it out yields the original density operator ρ_A.

4.5 Isomorphism of operators and bipartite vectors

We will frequently make use of the unnormalized version of the canonical maximally entangled state $|\Phi\rangle_{AA'}$,

$$|\Omega\rangle_{AA'} := \sum_{k=0}^{d-1} |b_k\rangle_A \otimes |b_k\rangle_{A'}\,, \tag{4.24}$$

where $\mathcal{H}_{A'} \simeq \mathcal{H}_A$. With it we can convert operators in $\mathrm{Lin}(\mathcal{H}_A, \mathcal{H}_B)$ to elements of $\mathcal{H}_A \otimes \mathcal{H}_B$ and vice versa. In particular, define the map $\mathsf{V} : \mathrm{Lin}(\mathcal{H}_A, \mathcal{H}_B) \to \mathcal{H}_A \otimes \mathcal{H}_B$ by the action

$$\mathsf{V}(M_{B|A}) \mapsto \mathbb{1}_A \otimes M_{B|A'}|\Omega\rangle_{AA'}\,. \tag{4.25}$$

Writing out $M_{B|A}$ in the basis defining $|\Omega\rangle_{AA'}$ and $|\Omega\rangle_{BB'}$, we see that V "vectorizes" the matrix representing M by just stacking its columns into a giant column vector.

Exercise 4.16. Show that $\mathsf{V}(|\varphi\rangle_B\langle\psi|_A) = |\overline{\psi}\rangle_A \otimes |\varphi\rangle_B$, where $|\overline{\psi}\rangle_A$ is the vector whose components in the basis defining $|\Omega\rangle_{AA'}$ are the complex conjugates of those of $|\psi\rangle_A$.

Importantly, V is an isomorphism. The inverse V^{-1} is just

$$\mathsf{V}^{-1} : |\Psi\rangle_{AB} \mapsto {}_{AA'}\langle\Omega|\Psi\rangle_{A'B}\,. \tag{4.26}$$

Here we use the capabilities of Dirac notation to give such a compact expression, but let us write it out more explicitly to confirm that it is correct.

First, we should check that ${}_{AA'}\langle\Omega|\Psi\rangle_{A'B}$ is an element of $\mathrm{Lin}(\mathcal{H}_A, \mathcal{H}_B)$. It is the composition of $|\Psi\rangle \in \mathrm{Lin}(\mathbb{C}, \mathcal{H}_{A'} \otimes \mathcal{H}_B)$ followed by $\langle\Omega| \in \mathrm{Lin}(\mathcal{H}_A \otimes \mathcal{H}_{A'}, \mathbb{C})$, so the first complication we face is that formally the composition is not well-defined. Inserting appropriate identity maps does yield a valid composition. Namely, we regard ${}_{AA'}\langle\Omega|\Psi\rangle_{A'B}$ as the composition $((\langle\Omega|_{AA'}\otimes\mathbb{1}_B)\circ(|\Psi\rangle_{A'B}\otimes\mathbb{1}_A)$, which indeed is an element of $\mathrm{Lin}(\mathcal{H}_A,\mathcal{H}_B)$. Here $\circ$ denotes composition of the linear maps. Henceforth we assume that identity operators needed to make such expressions well-defined are implicitly included.

Now we must determine that ${}_{AA'}\langle\Omega|\Psi\rangle_{A'B}$ is the correct operator for V^{-1}. For any basis $\{|b_k'\rangle\}$ of $\mathcal{H}_B$, we have

$$\begin{aligned}{}_{AA'}\langle\Omega|\Psi\rangle_{A'B} &= \Big(\sum_j \langle b_j|_A\otimes\langle b_j|_{A'}\Big)\Big(\sum_{k\ell}\Psi_{\ell,k}|b_\ell\rangle_{A'}\otimes|b_k'\rangle_B\Big)\\ &= \sum_{jk\ell}\langle b_j|b_\ell\rangle\Psi_{\ell,k}|b_k'\rangle_B\,\langle b_j|_A = \sum_{jk}\Psi_{j,k}|b_k'\rangle_B\,\langle b_j|_A\,.\end{aligned}\tag{4.27}$$

Here we make full use of the possibilities of Dirac notation for composite systems. Let us explain the calculation carefully, as we will very often employ similar sleights of hand. The expression $\langle b_j|_A\circ|b_k'\rangle_B$ is more correctly written as $((\langle b_j|_A\otimes\mathbb{1}_B)\circ(\mathbb{1}_A\otimes|b_k'\rangle_B)$. By the definition of ket in (B.2) and the form of tensor products of operators in (B.13), the first operator is an element of $\mathrm{Lin}(\mathcal{H}_A)\otimes\mathrm{Lin}(\mathbb{C},\mathcal{H}_B)\simeq\mathrm{Lin}(\mathcal{H}_A\otimes\mathbb{C},\mathcal{H}_{AB})\simeq\mathrm{Lin}(\mathcal{H}_A,\mathcal{H}_{AB})$ since $\mathcal{H}_A\otimes\mathbb{C}\simeq\mathcal{H}_A$. In particular, $\mathbb{1}_A\otimes|b_k'\rangle_B$ acting on an arbitrary $|\psi\rangle_A$ produces $|\psi\rangle_A\otimes|b_k'\rangle_B$. Similarly, the second operator is the element of $\mathrm{Lin}(\mathcal{H}_{AB},\mathcal{H}_B)$ that takes an arbitrary $|\xi\rangle_A\otimes|\varphi\rangle_B$ (the collection of which spans $\mathcal{H}_{AB}$) to $\langle b_j|\xi\rangle|\varphi\rangle_B$. Therefore the composition must be the map from $\mathcal{H}_A$ to $\mathcal{H}_B$ that takes $|\psi\rangle_A$ to $\langle b_j|\psi\rangle|b_k'\rangle_B$. This is indeed $|b_k'\rangle_B\langle b_j|_A$.

Observe that this reasoning is consistent with applying the composition to the two tensor factors separately, following (B.12). Performing the calculation this way gives $\langle b_j|_A\circ|b_k'\rangle_B = \langle b_j|_A\otimes|b_k'\rangle_B\in\mathrm{Lin}(\mathcal{H}_A,\mathbb{C})\otimes\mathrm{Lin}(\mathbb{C},\mathcal{H}_B)$. By (B.13) this space is again equivalent to $\mathrm{Lin}(\mathcal{H}_A,\mathcal{H}_B)$. Dirac notation efficiently encapsulates all the tensor product equivalences so that we do not have to deal with them directly.

Now back to the calculation at hand. A particular case of (4.27) is

$${}_{AA'}\langle\Omega|\Omega\rangle_{A'B} = \sum_j |b_j\rangle_B\,\langle b_j|_A \tag{4.28}$$

for $\mathcal{H}_B\simeq\mathcal{H}_A$, which is essentially the "identity operator" taking $\mathcal{H}_A$ to $\mathcal{H}_B$ (for which we of course need the basis used in the definition of $|\Omega\rangle_{AA'}$). This particular case makes it simple to complete the proof that V is an isomorphism. Consider an arbitrary $M_{B|A}$ and compute

$$\mathsf{V}^{-1}\circ\mathsf{V}(M_{B|A}) = {}_{AA'}\langle\Omega|\mathbb{1}_{A'}\otimes M_{B|A''}|\Omega\rangle_{A'A''} = M_{B|A''}\,{}_{AA'}\langle\Omega|\Omega\rangle_{A'A''} = M_{B|A}\,. \tag{4.29}$$

This establishes one half of the isomorphism, and proving other is entirely similar.

Exercise 4.17. Show that $\mathsf{V}\circ\mathsf{V}^{-1}: |\Psi\rangle_{AB}\mapsto|\Psi\rangle_{AB}$ for any $|\Psi\rangle_{AB}\in\mathcal{H}_{AB}$.

The map V is actually an isometry in that it transforms the Hilbert–Schmidt inner product on $\mathrm{Lin}(\mathcal{H}_A, \mathcal{H}_B)$ into the usual inner product on $\mathcal{H}_A \otimes \mathcal{H}_B$:

$$\langle S_{B|A}, T_{B|A} \rangle = \mathrm{Tr}[S_{B|A}^* T_{B|A}] = {}_{AA'}\langle \Omega | S_{B|A}^* T_{B|A} | \Omega \rangle_{AA'} \,. \tag{4.30}$$

Another useful property of $|\Omega\rangle_{AA'}$ is the following:

Exercise 4.18. Show that for any $M_{B|A} \in \mathrm{Lin}(\mathcal{H}_A, \mathcal{H}_B)$,

$$\mathbb{1}_A \otimes M_{B|A'} |\Omega\rangle_{AA'} = (M_{B'|A})^T \otimes \mathbb{1}_B |\Omega\rangle_{BB'} \,, \tag{4.31}$$

where $(M_{B|A})^T \in \mathrm{Lin}(\mathcal{H}_B, \mathcal{H}_A)$ is the operator whose matrix components relative to the bases defining $|\Omega\rangle_{AA'}$ and $|\Omega\rangle_{BB'}$ are just the transpose of those of $M_{B|A}$. Usually, we will write the transpose without the parentheses. As with vectors, $\overline{M}$ denotes the operator whose matrix components are the complex conjugates of M.

Exercise 4.19. Show that $\mathsf{V}(L_{D|C} M_{C|B} R_{B|A}) = (L_{D|C} \otimes R_{B|A}^T) \mathsf{V}(M_{C|B})$.

Exercise 4.20. What are the marginal states on systems A and B of the density operator $\rho_{AB} = |\Psi\rangle\langle\Psi|_{AB}$ in terms of $M_{B|A} = \mathsf{V}^{-1}(|\Psi\rangle_{AB})$?

4.6 Notes and further reading

The quote from Wigner appears in [304]. Density operators were independently introduced by von Neumann [292] and Landau [180]. The name "effects" is due to Ludwig [194], while POVM goes back to mathematical work on functional analysis. The importance of POVMs for the general setting of quantum mechanics was realized by Jauch and Piron [153], Davies and Lewis [71], and Holevo [143]. The discussion in Section 4.3 follows Bub [48]. That commuting projections offer the only possible realization of Boolean algebras in the most general quantum setting was shown by Varadarajan [290], building on earlier results from von Neumann [293, Ch. 2, § 10]. It is not necessary to appeal to Gleason's theorem to infer that sets of nonorthogonal projection operators cannot form a Boolean algebra, and a more direct argument using a finite set of projectors was given by Kochen and Specker [169]. For more on the formalism of quantum systems and measurements, see [70, 145, 171, 220, 299, 307].

5 Quantum channels

I think I can safely say that nobody understands quantum mechanics.

Richard Feynman

Just as with classical channels, quantum channels are meant to describe the change in an experimental setup due to time evolution, external interference, measurement, and so forth. If we prepare a quantum system in a certain way, intending to measure it with a given POVM, then the Born rule gives us the probabilities of the various measurement results. But suppose that we wait some time before completing the measurement and possibly let the system interact with additional degrees of freedom or perform some *other* measurement. Now the probabilities of the original measurement are presumably different. How should we describe the measurement outcome probabilities in the new setup in terms of the old? We can either ascribe this change to a change in the quantum state or to a change in the POVM elements. The Schrödinger picture is the former choice, and the Heisenberg picture the latter. Here we will mostly follow the Schrödinger picture and focus on the states.

For instance, suppose we prepare an electron or some other spin-$1/2$ system so that it points up along the $\hat{z}$ axis, i. e., in the state $|\uparrow\rangle$. If we immediately measure its angular momentum along this axis, say using a magnetic field as in a Stern[1]–Gerlach[2] device, then the measurement result should be $+1/2$, i. e., "up", with high probability. However, if we wait too long, then stray magnetic fields near the electron could change its angular momentum. The probability of "up" will presumably decrease the longer we wait.

We can model this "noise" by saying that the quantum state of the electron is transformed by a *depolarizing channel*, which is a kind of quantum analog to the binary symmetric channel. The output of a depolarizing channel is just the maximally mixed state $\pi = \frac{1}{2}\mathbb{1}$ with probability p, whereas with probability $1-p$ the output is the same as the input. Formally, the depolarizing channel is a map $\mathcal{N} : \rho \mapsto (1-p)\rho + p\,\mathrm{Tr}[\rho]\pi$. Since $\mathrm{Tr}[\rho] = 1$ for quantum states, the factor $\mathrm{Tr}[\rho]$ is not necessary in this case, but it emphasizes that the depolarizing channel is a linear map. In terms of the Bloch sphere, the Bloch vector of the input simply shrinks to $1-p$ times its original length.

Another common kind of noise encountered in real devices is that it is difficult to maintain superpositions of relatively stable states. This kind of noise can be modeled by a *dephasing channel*. Consider the qubit formed by the ground and excited states of a trapped ion, which we denote by $|0\rangle$ and $|1\rangle$, and suppose these are relatively stable for some given amount of time. The depolarizing channel is therefore an inappropriate

1 Otto Stern, 1888–1969.

2 Walther Gerlach, 1889–1979.

https://doi.org/10.1515/9783110570250-005

noise model. The dephasing channel, in contrast, simply removes all the off-diagonal elements of the density operator in the $|0\rangle/|1\rangle$ basis with some probability p and leaves the state untouched with probability $1-p$. This is formally described by a map $\mathcal{N} : \rho \mapsto (1-p)\rho + p\,\mathrm{diag}[\rho]$, where $\mathrm{diag}[\rho]$ removes the off-diagonal terms of ρ (as opposed to "diag" applied to a vector, which creates a diagonal operator). It also has an intuitive action in the Bloch representation.

Exercise 5.1. Describe the action of the dephasing channel in the Bloch representation.

5.1 Definition

5.1.1 First considerations

Now let us turn to the general definition. From the point of view of compatibility with the Born rule, it is clear than any purported quantum channel will have to map density operators to density operators. Concretely, let us consider a map from $\mathrm{Stat}(A)$ to $\mathrm{Stat}(B)$ and denote it by $\mathcal{E}_{B|A}$. For precisely the same reasons involving convexity as described in Section 3.2, quantum channels must also be represented by convex-linear maps. By Proposition 3.2 any such map $\mathcal{E}_{B|A}$ can be uniquely extended to a linear on the span of density operators, which is all of the Hermitian operators. We may further extend $\mathcal{E}_{B|A}$ to a linear map on all operators as follows. Observe that an arbitrary $M_A \in \mathrm{Lin}(\mathcal{H}_A)$ can be decomposed into "real" and "imaginary" parts $H_A^r = \frac{1}{2}(M_A + M_A^*)$ and $H_A^i = \frac{1}{2i}(M_A - M_A^*)$, so that $M_A = H_A^r + iH_A^i$. Since both H_A^r and H_A^i are Hermitian, we can define $\mathcal{E}_{B|A}[M_A] := \mathcal{E}_{B|A}[H_A^r] + i\mathcal{E}_{B|A}[H_A^i]$.

Thus a quantum channel $\mathcal{E}_{B|A}$ ought to be a linear map from $\mathrm{Lin}(\mathcal{H}_A)$ to $\mathrm{Lin}(\mathcal{H}_B)$ (often called a superoperator) satisfying the two conditions

1. (Positivity) $\mathcal{E}_{B|A}[\rho_A] \geq 0$ for $\rho_A \geq 0$ and
2. (Trace preservation) $\mathrm{Tr}[\mathcal{E}_{B|A}[\rho_A]] = 1$ for $\mathrm{Tr}[\rho_A] = 1$.

A trivial example of a map satisfying both conditions is the *identity map* from $\mathrm{Lin}(\mathcal{H})$ to itself, denoted $\mathcal{I}$.

Exercise 5.2. Check that the depolarizing and dephasing channels satisfy both these conditions.

We will denote superoperators using calligraphic capital letters, except for $\mathcal{H}$, which is already reserved for state spaces. Subscripts indicate the input and output spaces using | as above; a superoperator $\mathcal{E}_A$ maps $\mathrm{Lin}(\mathcal{H}_A)$ to itself. The set of maps or superoperators from $\mathrm{Lin}(\mathcal{H}_A)$ to $\mathrm{Lin}(\mathcal{H}_B)$ is denoted $\mathrm{Map}(\mathcal{H}_A, \mathcal{H}_B)$, and $\mathrm{Map}(\mathcal{H})$ denotes the maps from $\mathrm{Lin}(\mathcal{H})$ to itself. We use square brackets to denote application of a map to an operator, in contrast to parentheses, which are used to denote application of a function to a number or an index. Observe that the trace Tr already fits this convention by regarding the output as an operator on a one-dimensional space.

A superoperator that preserves the trace is called *trace-preserving*, and one that preserves positivity is called *positive*. As a linear map on the inner product space of operators, every superoperator $\mathcal{E}$ has an adjoint $\mathcal{E}^*$ defined in the usual way as the unique map such that $\mathrm{Tr}[\Lambda\,\mathcal{E}[\rho]] = \mathrm{Tr}[\mathcal{E}^*[\Lambda]\,\rho]$ for all $\rho \in \mathrm{Lin}(\mathcal{H}_A)$ and $\Lambda \in \mathrm{Lin}(\mathcal{H}_B)$. Another interesting class of superoperators are *unital*, meaning they preserve the unit (identity) operator: $\mathcal{E}[\mathbb{1}] = \mathbb{1}$.

Exercise 5.3. Show that $\mathcal{E}^*$ is unital iff $\mathcal{E}$ is trace-preserving.

However, positivity alone is actually not enough to ensure compatibility with the Born rule. Consider the superoperator $\mathcal{T} : S \mapsto S^T$ resulting from matrix transposition in a fixed basis $\{|b_k\rangle\}$, that is, for $S = \sum_{jk}\langle b_j|S|b_k\rangle|b_j\rangle\langle b_k|$, the output is simply $S^T = \sum_{jk}\langle b_k|S|b_j\rangle|b_j\rangle\langle b_k|$. For more general $S \in \mathrm{Lin}(\mathcal{H}_A, \mathcal{H}_B)$ with bases $|b_j\rangle_A$ and $|b'_k\rangle_B$, the transpose is just $S^T = \sum_{jk}\, {}_B\langle b'_k|S_{B|A}|b_j\rangle_A\, |b_j\rangle_A\langle b'_k|_B$. Clearly, $\mathcal{T}$ is trace-preserving, since the transpose does not affect the diagonal elements of a matrix. Similarly, it does not change the eigenvalues, since the characteristic polynomials of both S and S^T are the same: Using $\det(M^T) = \det(M)$, we have $\det(\lambda 1 - S^T) = \det((\lambda\mathbb{1} - S)^T) = \det(\lambda\mathbb{1} - S)$ for $\lambda \in \mathbb{R}$.

Now suppose $\mathcal{T}$ acts on one part of a bipartite entangled state, say the B system of the maximally entangled state $|\Phi\rangle_{AB}$ from (4.17):

$$\sigma_{AB} = \mathcal{I}_A \otimes \mathcal{T}_B[|\Phi\rangle\langle\Phi|_{AB}] = \frac{1}{d}\sum_{jk}|k\rangle\langle j|_A \otimes |j\rangle\langle k|_B\,. \tag{5.1}$$

Observe that the state $\frac{1}{\sqrt{2}}(|j\rangle|k\rangle - |k\rangle|j\rangle)$ for any $j \neq k$ is an eigenvector of σ_{AB} with eigenvalue $-1/d$. Therefore σ_{AB} is not positive, leading to nonsense probabilities from the Born rule (4.2). It should be emphasized that these are not physical constraints on the allowable form of quantum channels, but rather statistical.

Henceforth we will often abuse notation and write Φ_{AB} for the density operator associated with $|\Phi\rangle_{AB}$, simply for notational convenience. This extends to Ω_{AB} and other contexts in which a pure state is defined and denoted with a Greek letter, i. e., when working with pure state $|\varphi\rangle_A$, the density operator will just be written φ_A.

Exercise 5.4. Show that $\Upsilon_{AB} := \mathcal{I}_A \otimes \mathcal{T}_B[\Omega_{AB}]$ is the *swap operator* (or *flip operator*) on $\mathrm{Lin}(\mathcal{H}_A \otimes \mathcal{H}_B)$ with $\mathcal{H}_A \simeq \mathcal{H}_B$, satisfying $\Upsilon_{AB}|\psi\rangle_A \otimes |\varphi\rangle_B = |\varphi\rangle_A \otimes |\psi\rangle_B$ for arbitrary $|\psi\rangle, |\varphi\rangle \in \mathcal{H}_A$.

Exercise 5.5. What is the action of transposition on qubits in terms of the Bloch vector?

Exercise 5.6. Show that $\mathrm{Tr}[\Upsilon_{AB}\,\sigma_{AB}] \geq 0$ for all separable states. That is to say, the negativity of the swap operator can only be detected by entangled states.

5.1.2 Complete positivity

To avoid this problem, we must demand that quantum operations be *completely positive*, i. e., positive on arbitrary and possibly entangled input states.

Definition 5.1 (Completely positive map). A superoperator $\mathcal{E}_{B|A} \in \mathrm{Map}(\mathcal{H}_A, \mathcal{H}_B)$ is said to be *completely positive* if the map $\mathcal{E}_{B|A} \otimes \mathcal{I}_R$ is positive for all $\mathcal{H}_R$.

Clearly, $\mathcal{I}_A$ is completely positive. Conjugation by a unitary operator, i. e. $\rho \mapsto U\rho U^*$, is also completely positive, and in fact this holds for any linear operator.

Exercise 5.7. Suppose $N_{B|A}$ is an arbitrary linear map from $\mathcal{H}_A$ to $\mathcal{H}_B$. Show that the superoperator $M_A \mapsto N_{B|A}M_A(N_{B|A})^*$ is completely positive, where $M_A \in \mathrm{Lin}(\mathcal{H}_A)$.

Exercise 5.8. Show that the adjoint $\mathcal{E}^*$ of any completely positive superoperator $\mathcal{E}$ is itself completely positive.

Sums and compositions of completely positive maps are also completely positive. For a given channel $\mathcal{E}_{B|A}$, a collection of operators $\{K_{B|A}(j) \in \mathrm{Lin}(\mathcal{H}_A, \mathcal{H}_B)\}_{j=1}^n$ such that

$$\mathcal{E}_{B|A} : \rho_A \mapsto \sum_{j=1}^{n} K_{B|A}(j)\, \rho_A\, K_{B|A}^*(j) \tag{5.2}$$

is called a *Kraus*[3] *representation* (or operator-sum representation) of $\mathcal{E}_{B|A}$, and the $K_{B|A}(j)$ are called *Kraus operators*. Existence of a Kraus representation is sufficient for complete positivity, and we will return to the question of necessity in Section 5.4.

Exercise 5.9. Find a Kraus representation of the partial trace map.

The superoperator "diag" in the dephasing channel is an example of a *pinch map*, a map whose Kraus operators are projections. We will usually denote such maps by $\mathcal{P}$. An important property of the pinch map with rank-one projections is that it creates a CQ state when applied to one part of a bipartite state.

Exercise 5.10. Show that "diag" is a pinch map with projections $|k\rangle\langle k|$ as Kraus operators. Deduce that $\mathcal{P}_A[\rho_{AB}]$ is a CQ state for arbitrary states ρ_{AB}.

Exercise 5.11. Suppose $\mathcal{P}$ is an arbitrary pinch operator and σ is an operator such that $\sigma = \mathcal{P}[\sigma]$. Show that $\mathrm{Tr}[\rho\sigma] = \mathrm{Tr}[\mathcal{P}[\rho]\sigma]$ for all operators ρ.

Now we can give the definition of a quantum channel.

---

3 Karl Kraus, 1938–1988.

Definition 5.2 (Quantum channel). A *quantum channel* is a completely positive trace-preserving map. The set of quantum channels mapping $\mathrm{Lin}(\mathcal{H}_A)$ to $\mathrm{Lin}(\mathcal{H}_B)$ is denoted $\mathrm{Chan}(A, B)$.

We have already met two qubit channels, the depolarizing and dephasing channels. They are examples of *Pauli channels*, channels whose Kraus operators are (proportional to) Pauli operators. In general, a Pauli channel has the form $\mathcal{E} : \rho \mapsto \sum_{jk} P(j,k)\, \sigma_x^j \sigma_z^k \rho \sigma_z^k \sigma_x^j$ for some probability distribution $P(j,k)$. With probability $P(j,k)$, the operator $\sigma_x^j \sigma_z^k$ is applied to the qubit. Another interesting qubit channel is *amplitude damping*, a kind of quantum counterpart to the classical Z channel. Amplitude damping describes spontaneous emission, in which an excited state ($|1\rangle$) jumps to the ground state ($|0\rangle$) with some probability p. The channel is defined by the pair of Kraus operators

$$K(0) = \begin{pmatrix} 1 & 0 \\ 0 & \sqrt{1-p} \end{pmatrix} \quad \text{and} \quad K(1) = \begin{pmatrix} 0 & \sqrt{p} \\ 0 & 0 \end{pmatrix}. \tag{5.3}$$

Exercise 5.12. Show that the amplitude damping channel is trace-preserving.

The definitions of depolarizing and dephasing channels are already sensible for inputs of arbitrary dimension, with the slight adjustment that we should regard π as the maximally mixed state for input dimension d, i. e., $\pi = \frac{1}{d}\mathbb{1}$. We can also define the *quantum erasure channel* for arbitrary input dimension. It is the channel $\rho \mapsto (1-p)\rho + p\,\mathrm{Tr}[\rho]\,|?\rangle\langle ?|$, where $|?\rangle$ is orthogonal to all states of the input. Put differently, the output of the channel is a linear operator on the direct sum $\mathcal{H} \oplus |?\rangle$ for the input space $\mathcal{H}$.

Exercise 5.13. Show that the depolarizing, dephasing, and erasure channels are in fact channels. Which of these three channels are unital?

For arbitrary dimension, we can construct Kraus representations for depolarization and dephasing by making use of the Weyl–Heisenberg operators from Section 4.4.2.

Exercise 5.14. Suppose $\mathcal{P}$ is the pinch map with rank-one projections onto an orthonormal basis $\{|k\rangle\}_{k\in\mathbb{Z}_d}$. Show that, for V from (4.19),

$$\mathcal{P}[\rho] = \frac{1}{d}\sum_{k\in\mathbb{Z}_d} V^k \rho (V^k)^*. \tag{5.4}$$

Furthermore, for $\pi = \frac{1}{d}\mathbb{1}$, show that

$$\mathrm{Tr}[\rho]\,\pi = \frac{1}{d^2}\sum_{j,k\in\mathbb{Z}_d} U^j V^k \rho (U^j V^k)^* \tag{5.5}$$

and that for the pinch map $\tilde{\mathcal{P}}$ in the eigenbasis of U from (4.20),

$$\mathrm{Tr}[\rho]\,\pi = \tilde{\mathcal{P}} \circ \mathcal{P}[\rho]\,. \tag{5.6}$$

5.2 Everything is a channel

As in the classical case, quantum channels encompass states, effects, and measurements. Indeed, just as everything classical probability theory is described by compositions of stochastic maps, everything in quantum probability theory can be described by compositions of completely positive trace-preserving maps.

Formally, we may regard states on $\mathcal{H}$ as channels from $\mathrm{Lin}(\mathbb{C})$ to $\mathrm{Lin}(\mathcal{H})$, so that $\rho \in \mathrm{Stat}(\mathcal{H})$ is equivalent to the trivial-input channel $1 \mapsto \rho$. Note that the set $\mathrm{Stat}(\mathbb{C})$, the positive 1×1 matrices with trace 1, has just one element, the number 1. The simplest case occurs when $\rho = |\psi\rangle\langle\psi|$ is a pure state, for then the channel has just one Kraus operator, namely $|\psi\rangle$ itself. For arbitrary ρ, any pure state ensemble decomposition $\{(P(x), |\psi_x\rangle)\}$ gives rise to Kraus operators $K(x) = \sqrt{P(x)}|\psi_x\rangle$.

State preparation of some *other* system is also a channel, e. g., if A is already prepared in state ρ_A, then the preparation of B in some fixed state σ_B can be regarded as the channel $\rho_A \mapsto \rho_A \otimes \sigma_B$. Here the Kraus operators are just as before, constructed from an ensemble decomposition of σ_B, but each has an additional tensor factor $\mathbb{1}_A$.

A device that can prepare one of a set of quantum states can be described by a *classical-quantum channel*, or CQ channel, $\mathcal{E}_{A|X} : \mathrm{Lin}(\mathcal{H}_X) \to \mathrm{Lin}(\mathcal{H}_A)$, in which the input $|x\rangle\langle x|_X$ is mapped to the state $\rho_B(x)$, for an orthonormal basis $\{|x\rangle_X\}$. Properly speaking, this does not yet define a channel, since we have not specified the action on off-diagonal terms such as $|x\rangle\langle y|$ for $x \neq y$. But these can just be discarded by the channel. Supposing $\rho_B(x) = |\psi_x\rangle\langle\psi_x|_B$ is a pure state, Kraus operators $K_{B|A}(x) = |\psi_x\rangle_B\langle x|_A$ will deliver the desired result. As an example, the *pure state channel* $\mathrm{PSC}(f)$ takes classical binary inputs to pure states $|\theta_x\rangle$ for which $f = |\langle\theta_0|\theta_1\rangle|$.

For CQ channels with mixed output states $\rho_B(x) = \sum_y P_{Y|X}(y,x)|\psi_{y,x}\rangle\langle\psi_{y,x}|_B$, we simply use Kraus operators $K_{B|A}(x,y) = \sqrt{P_{Y|X}(y,x)}|\psi_{xy}\rangle_B\langle x|_A$. This includes the case that the $\rho_B(x)$ commute for all pairs of x values, meaning the channel output is essentially classical. Working in the standard basis and relabeling B as Y, a possible set of Kraus operators of a classical channel $W_{Y|X}$ is just given by $K_{Y|X}(y,x) = \sqrt{W_{Y|X}(y,x)}|y\rangle\langle x|$. The notation $\mathcal{E}_{A|X=x}$ is also sensible here since x is classical: $\mathcal{E}_{A|X=x}$ is a density operator.

A measurement, on the other hand, is a *QC channel*, as it produces a classical output from a quantum input. For a POVM with elements $\Lambda_A(x)$, the measurement can be regarded as a channel $\mathcal{M}_{X|A} : \mathrm{Lin}(\mathcal{H}_A) \to \mathrm{Lin}(\mathcal{H}_X)$ with the action

$$\mathcal{M}_{X|A} : \rho_A \mapsto \sum_{x\in\mathcal{X}} \mathrm{Tr}[\Lambda(x)\rho]\,|x\rangle\langle x|_X\,. \tag{5.7}$$

Since the trace is completely positive (see Exercise 5.9), so is each term in the sum. The map is trace-preserving due to the completeness relation of the POVM. The notation $\mathcal{M}_{X=x|A}$ is sensible here as well. In this case, $\mathcal{M}_{X=x|A}$ is the superoperator $\rho \mapsto \mathrm{Tr}[\Lambda_A(x)\rho]$. This map is completely positive but not trace-preserving. We will see in Section 5.4 that all CQ channels can be regarded as state preparation and all QC channels as measurement. The choice of Kraus operators above for a classical channel amount to measurement in the $\{|x\rangle\}$ basis followed by diagonal state preparation in the $\{|y\rangle\}$ basis. Note that a pinch map with rank-one projections corresponds to a measurement in the associated basis.

A channel like $\mathcal{M}_{X|A}$ only describes the classical output of a measurement but not the action on the quantum system. A channel that describes both together is called a *quantum instrument*. In the projective measurement of usual quantum mechanics, the projection postulate states that upon observing outcome x, a quantum system initially in the vector state $|\psi\rangle$ is transformed into the normalized version of $\Pi(x)|\psi\rangle$ for some projector $\Pi(x)$. We can express this transformation on density operators using the following quantum channel:

$$\mathcal{Q}_{XA|A} : \rho_A \mapsto \sum_x |x\rangle\langle x|_X \otimes \Pi_A(x)\rho_A\Pi_A(x)\,, \tag{5.8}$$

whose Kraus operators are simply $K_{XA|A}(x) = |x\rangle_X \otimes \Pi_A(x)$. The probability of outcome x is given by the trace of the second factor: $P_X(x) = \mathrm{Tr}[\Pi_A(x)\rho_A\Pi_A(x)]$. Thus the postmeasurement state for input ρ_A conditioned on measurement result x is $\rho'_A(x) = \Pi_A(x)\rho_A\Pi_A(x)/P_X(x)$; this is the quantum equivalent of (2.17).

Notice that by (5.7) $\mathrm{Tr}_A \circ \mathcal{Q}_{XA|A} = \mathcal{M}_{X|A}$ in the particular case that $\Lambda(x) = \Pi(x)$. In general, $\mathrm{Tr}_A \circ \mathcal{Q}_{XA|A} = \mathcal{M}_{X|A}$ for $\mathcal{Q}_{XA|A}$ defined using the Kraus operators $K_{XA|A}(x) = \Lambda_A(x)^{1/2} \otimes |x\rangle_X$, where $\Lambda^{1/2}$ is the square root of the operator Λ, which is well-defined for any $\Lambda \geq 0$ using the spectral decomposition. We also sometimes denote this as $\sqrt{\Lambda}$. Hence this choice of Kraus operators defines a possible instrument for the POVM with elements $\Lambda(x)$. For the observed measurement result, the conditional output state for this choice of Kraus operator is then $\Lambda_A(x)^{1/2}\rho_A\Lambda_A(x)^{1/2}/\,\mathrm{Tr}[\Lambda_A(x)\rho_A]$. We discuss more general forms of quantum instruments given a fixed POVM in Section 5.4.

Fixing a value of x in $\mathcal{Q}_{XB|A}$ gives a superoperator $\mathcal{Q}_{X=xB|A}$, which we may also write $\mathcal{Q}_{B|A}(x)$, so that $\mathcal{Q}_{XB|A} = \sum_x |x\rangle\langle x|_X \otimes \mathcal{Q}_{B|A}(x)$. In this way, any decomposition of a channel into completely positive superoperators can be associated with an instrument, just as any decomposition of a state leads to a CQ state. That is, if $\mathcal{E}_{B|A} = \sum_{x=1}^n \mathcal{E}_{B|A}(x)$ for some quantum channel $\mathcal{E}_{B|A}$ and completely positive superoperators $\mathcal{E}_{B|A}(x)$, then $\mathcal{Q}_{XB|A} = \sum_{x=1}^n |x\rangle\langle x|_X \otimes \mathcal{E}_{B|A}(x)$ is a valid channel. It takes a quantum input to a CQ output. In Section 5.4, we will see that every such channel is a quantum instrument in that it can be regarded as implementing a quantum measurement.

Exercise 5.15. Show that each of the superoperators $\mathcal{E}_{B|A}(x)$ in such a decomposition of a channel must be trace nonincreasing, i. e., its output must have a trace no larger than its input.

Unlike decompositions of states, however, decompositions of channels cannot be interpreted as ensembles. The original channel is not necessarily a convex combination of some renormalized channels associated with $\mathcal{E}_{B|A}(x)$ in the decomposition. The difficulty is that the $\mathcal{E}_{B|A}(x)$ cannot be renormalized independently of their inputs. For instance, take the channel $\mathcal{E}_A = \mathrm{Tr}_X \circ \mathcal{Q}_{XA|A}$ for $\mathcal{Q}_{XA|A}$ from (5.8). It is not possible to renormalize the individual superoperators $\mathcal{E}_{B|A}(x) : \rho_A \mapsto \Pi_A(x)\rho_A\Pi_A(x)$ to give a valid channel. By considering inputs in the kernel of $\Pi_A(x)$ versus those in its support it is clear that there is no constant c such that $c\mathcal{E}_{B|A}(x)$ is a trace-preserving map. Similarly, just renormalizing $\mathcal{E}_{B|A}(x)$ by dividing by the trace of the output also fails, as the result is not a linear map (or is not well-defined, as in the case of inputs in the kernel of the projector $\Pi_A(x)$).

5.3 The Choi isomorphism

In the classical case, we defined a channel by a set of conditional distributions, which we can regard as stemming from the joint distribution of channel inputs and outputs. In the quantum setting, that approach is not a priori possible, since we have no analog of conditional distributions, nor is it apparent what to use for the joint state of the input and output of a quantum channel. Instead, we have followed the approach required by convexity.

However, it turns out that we can define an analog of a joint input–output distribution of a quantum channel, and moreover we can describe the action of a quantum channel as marginalization over the input in almost the same manner as for a classical channel. To do so, we make use of the *Choi*[4] *isomorphism* between superoperators and bipartite operators. Under the *Choi map* C, every superoperator $\mathcal{E}_{B|A}$ is mapped to a *Choi operator* on $\mathrm{Lin}(\mathcal{H}_A \otimes \mathcal{H}_B)$; specifically, the result of the channel acting on half of an entangled state. In fact, we already encountered the Choi operator of the transpose operator back in (5.1). It turns out to be a little more convenient to use the unnormalized version of the entangled state, $|\Omega\rangle$, and we define the Choi map formally as follows.

Definition 5.3 (Choi map). For $\mathcal{H}_A \simeq \mathcal{H}_{A'}$, the *Choi map* C *for the basis* $\{|b_i\rangle\}_i$ is given by

$$\begin{aligned} \mathsf{C} : \mathrm{Map}(\mathcal{H}_A, \mathcal{H}_B) &\to \mathrm{Lin}(\mathcal{H}_A \otimes \mathcal{H}_B) \\ \mathcal{E}_{B|A} &\mapsto \mathcal{E}_{B|A'}[\Omega_{AA'}] \,. \end{aligned} \tag{5.9}$$

4 Man-Duen Choi.

Note that the Choi map depends on the choice of basis used to define the state $|\Omega\rangle_{A'A}$ of (4.24). The Choi operator associated with the identity channel is clearly just Ω_{AB}, whereas (5.1) shows that the Choi operator of the transpose channel is the swap operator Υ_{AB}.

The Choi map allows us to represent the action of any given superoperator in terms of composition of ordinary bipartite operators and application of the partial trace. This is formalized in the following:

Theorem 5.1 (Choi isomorphism of superoperators and bipartite operators). *The Choi map* C *is an isomorphism between* $\mathrm{Map}(\mathcal{H}_A, \mathcal{H}_B)$ *and* $\mathrm{Lin}(\mathcal{H}_A \otimes \mathcal{H}_B)$. *Its inverse* C^{-1} *takes any* $M_{AB} \in \mathrm{Lin}(\mathcal{H}_A \otimes \mathcal{H}_B)$ *to the superoperator* $\mathsf{C}^{-1}(M_{AB}) \in \mathrm{Map}(\mathcal{H}_A, \mathcal{H}_B)$ *defined by*

$$\mathsf{C}^{-1}(M_{AB}) : S_A \mapsto \mathrm{Tr}_A\big[\mathcal{T}_A[S_A]\, M_{AB}\big]. \tag{5.10}$$

It is perhaps a little surprising that $\Omega_{AA'}$, as opposed to $\mathbb{1}_{AA'}$, should be the operator associated with the identity channel, so let us verify that this is indeed the case. For $\mathsf{C}^{-1}(\Omega_{AA'})$ acting on an arbitrary S_A, we have

$$\mathrm{Tr}_A\big[\mathcal{T}_A[S_A]\,\Omega_{AA'}\big] = \mathrm{Tr}_A[S_A\,\Upsilon_{AA'}], \tag{5.11}$$

since we can apply the transpose to the system A of all operators inside the partial trace over A. Now using $\Upsilon^2_{AA'} = \mathbb{1}_{AA'}$, we obtain

$$\mathrm{Tr}_A[S_A\,\Upsilon_{AA'}] = \mathrm{Tr}_A[\Upsilon_{AA'}\Upsilon_{AA'}S_A\,\Upsilon_{AA'}] = \mathrm{Tr}_A[\Upsilon_{AA'}S_{A'}] = S_{A'}\,\mathrm{Tr}_A[\Upsilon_{AA'}]. \tag{5.12}$$

Since $\Upsilon_{AA'} = \mathcal{T}_{A'}[\Omega_{AA'}]$ and $\mathrm{Tr}_A[\Omega_{AA'}] = \mathbb{1}_{A'}$, the latter factor $\mathrm{Tr}_A[\Upsilon_{AA'}] = \mathbb{1}_{A'}$. So $\Omega_{AA'}$ does implement the identity operator, as intended.

Exercise 5.16. Show that $\mathrm{Tr}_B[\mathcal{T}_B[M_{AB}]\,\mathcal{T}_B[N_{BC}]] = \mathrm{Tr}_B[M_{AB}N_{BC}]$.

Proof of Theorem 5.1. The proof proceeds by showing that $\mathsf{C}^{-1} \circ \mathsf{C}(\mathcal{E}_{B|A}) = \mathcal{E}_{B|A}$ for all superoperators $\mathcal{E}_{B|A} \in \mathrm{Map}(\mathcal{H}_A, \mathcal{H}_B)$ and $\mathsf{C} \circ \mathsf{C}^{-1}(M_{AB}) = M_{AB}$ for all bipartite operators $M_{AB} \in \mathrm{Lin}(\mathcal{H}_A \otimes \mathcal{H}_B)$.

For the former, start by using $\mathrm{Tr}_{A'}[\Omega_{AA'}] = \mathbb{1}_A$ to obtain, for any $S_A \in \mathrm{Lin}(\mathcal{H}_A)$,

$$\mathcal{E}_{B|A'}[S_{A'}] = \mathcal{E}_{B|A'}[S_{A'}\,\mathrm{Tr}_A[\Omega_{A'A}]] = \mathrm{Tr}_A[\mathcal{E}_{B|A'}[(S_{A'} \otimes \mathbb{1}_A)\,\Omega_{A'A}]]. \tag{5.13}$$

Next, recall that $S_A^T \otimes \mathbb{1}_{A'}|\Omega\rangle_{AA'} = \mathbb{1}_A \otimes S_{A'}|\Omega\rangle_{AA'}$ from (4.31), and therefore

$$\begin{aligned}\mathcal{E}_{B|A'}(S_{A'}) &= \mathrm{Tr}_A\big[\mathcal{E}_{B|A'} \otimes \mathcal{I}_A\big[(\mathbb{1}_{A'} \otimes S_A^T)\,\Omega_{A'A}\big]\big] \\ &= \mathrm{Tr}_A\big[(\mathbb{1}_B \otimes \mathcal{T}_A[S_A])(\mathcal{E}_{B|A'} \otimes \mathcal{I}_A[\Omega_{A'A}])\big].\end{aligned} \tag{5.14}$$

We recognize $\mathsf{C}(\mathcal{E}_{B|A})$ as the second factor in the final expression, and we have established $\mathsf{C}^{-1} \circ \mathsf{C}(\mathcal{E}_{B|A}) = \mathcal{E}_{B|A}$.

For the latter, direct calculation gives

$$\mathsf{C} \circ \mathsf{C}^{-1}(M_{AB}) = \mathsf{C}^{-1}(M_{A'B})[\Omega_{AA'}] = \mathrm{Tr}_{A'}[\mathcal{T}_{A'}[\Omega_{AA'}]M_{A'B}] = M_{AB}, \tag{5.15}$$

since $\Omega_{AA'}$ is the Choi representative of the identity map. □

Exercise 5.17. Determine the Choi operator of an arbitrary Pauli channel.

Exercise 5.18. Consider a channel $\mathcal{E}_{B|A}$ whose output is a fixed density operator σ_B, i. e., $\mathcal{E}_{B|A}[\rho_A] = \sigma_B \,\mathrm{Tr}[\rho_A]$. Show that $\mathsf{C}(\mathcal{E}_{B|A}) = \sigma_B \otimes \mathbb{1}_A$.

Exercise 5.19. Show that $\mathsf{C}(\mathcal{E}^*) = \mathsf{C}(\mathcal{E})^T$ for every superoperator $\mathcal{E}$.

Exercise 5.20. Show that the Choi operator of a measurement channel $\mathcal{M}_{X|A}$ is simply $\mathsf{C}(\mathcal{M}_{X|A}) = \sum_x |x\rangle\langle x|_X \otimes \Lambda_A(x)^T$, where $\Lambda_A(x)$ are the POVM elements of $\mathcal{M}_{X|A}$.

Importantly, it turns out that the Choi operators of completely positive maps are positive semidefinite operators and vice versa. Thus the study of channels is reduced to the study of bipartite positive semidefinite operators, and we can regard these as the analogs of classical conditional distributions. The representation of channel action is also similar: In both cases the input state is combined with a bipartite "state" using the inner product to give the output.

Theorem 5.2 (Choi representation). *A superoperator $\mathcal{E}_{B|A} \in \mathrm{Map}(\mathcal{H}_A, \mathcal{H}_B)$ is completely positive iff $\mathsf{C}(\mathcal{E}_{B|A}) \geq 0$. It is trace-preserving iff $\mathrm{Tr}_B[\mathsf{C}(\mathcal{E}_{B|A})] = \mathbb{1}_A$.*

Proof. Take the complete positivity claim first. Clearly, $\mathsf{C}(\mathcal{E}_{B|A})$ is positive if $\mathcal{E}_{B|A}$ is completely positive. To establish sufficiency, suppose $\mathsf{C}(\mathcal{E}_{B|A})$ is positive and consider an arbitrary input state ρ_{AR} to $\mathcal{E}_{B|A} \otimes \mathcal{I}_R$. By superoperator linearity it suffices to consider pure $\rho_{AR} = |\varphi\rangle\langle\varphi|_{AR}$. Setting $K_{R|A} = \mathsf{V}^{-1}(|\varphi\rangle_{AR})$ using (4.26), we have $|\varphi\rangle_{A'R} = K_{R|A} \otimes \mathbb{1}_{A'}|\Omega\rangle_{AA'}$ (note the move from A to A' in $|\varphi\rangle$), and therefore

$$\mathcal{E}_{B|A'} \otimes \mathcal{I}_R[\rho_{A'R}] = K_{R|A}(\mathcal{I}_A \otimes \mathcal{E}_{B|A'}[\Omega_{AA'}])K_{R|A}^* = K_{R|A}\mathsf{C}(\mathcal{E}_{B|A})\,K_{R|A}^*. \tag{5.16}$$

Exercise 5.7 then implies that $\mathcal{E}_{B|A'} \otimes \mathcal{I}_R[\rho_{A'R}] \geq 0$.

Now consider the trace-preserving condition. By the Choi isomorphism with $O_{AB} = \mathsf{C}(\mathcal{E}_{B|A})$ we have $\mathrm{Tr}_B[\mathcal{E}_{B|A}[\rho_A]] = \mathrm{Tr}_A[\mathrm{Tr}_B[O_{AB}]\rho_A^T]$ for any $\rho_A \in \mathrm{Stat}(\mathcal{H}_A)$. Clearly, the trace is 1 if $\mathrm{Tr}_B[O_{AB}] = \mathbb{1}_A$. Conversely, since ρ_A is arbitrary, this condition must hold if the trace is 1. □

By the Choi representation theorem the set of channels from A to B is a convex set corresponding to positive operators on $\mathcal{H}_{AB}$ whose marginal on A is the identity operator. On the boundary of this set are the rank-deficient operators subject to the condition on the marginal. But the extreme points of the set are only a subset of the boundary, in contrast to, say, the Bloch sphere. For example, the dephasing channel has a Choi operator of rank two but is clearly a convex combination of two other channels.

Exercise 5.21. Consider the family of qubit mappings $\mathcal{E}_\lambda : \rho \mapsto \frac{1}{2}\operatorname{Tr}[\rho]\mathbb{1} + \lambda(\sigma_x\rho\sigma_z + \sigma_z\rho\sigma_x)$ for $\lambda \in [0,1]$. Determine the λ for which $\mathcal{E}_\lambda$ is positive. When is $\mathcal{E}_\lambda$ completely positive?

Exercise 5.22. Consider the qubit superoperator $\mathcal{E}_\lambda$ that reduces the y component by multiplying it with a factor λ. For what values of λ is $\mathcal{E}_\lambda$ completely positive? Suppose $\mathcal{F}_p$ is the depolarizing channel of probability p, as defined at the top of the chapter. For what values of λ and p is $\mathcal{E}_\lambda \circ \mathcal{F}_p$ completely positive?

Exercise 5.23. For what values of $a, b \in \mathbb{R}$ is the superoperator $\rho \mapsto a\operatorname{Tr}[\rho]\mathbb{1} - b\rho$ a valid quantum channel on $\operatorname{Lin}(\mathcal{H})$ with $|\mathcal{H}| = d$? What if we only require positivity instead of complete positivity? What about $\rho \mapsto a\operatorname{Tr}[\rho]\mathbb{1} - b\rho^T$?

Exercise 5.24. An *entanglement-breaking* channel is a channel $\mathcal{N}_{B|A}$ such that the output $\mathcal{N}_{B|A}[\rho_{AR}]$ is a separable state for every input density operator ρ_{AR}. Using the Choi isomorphism, show that every entanglement breaking channel is a composition of a measurement followed by state preparation.

Exercise 5.25. Consider the superoperator $\rho \mapsto A \odot \rho$, where $\odot$ represents entrywise multiplication, known as a Schur[5]–Hadamard[6] channel. Show that it is completely positive when $A \geq 0$. What is the condition for trace-preservation?

Exercise 5.26. Suppose a qubit channel $\mathcal{E}$ preserves the Pauli operators σ_x and σ_z in the sense that $\operatorname{Tr}[\sigma_x\,\mathcal{E}[\rho]] = \operatorname{Tr}[\sigma_x\rho]$ and similarly for σ_z, so that $\mathcal{E}^*[\sigma_x] = \sigma_x$ and $\mathcal{E}^*[\sigma_z] = \sigma_z$. Show that $\mathcal{E} = \mathcal{I}$.

Exercise 5.27. Show that for arbitrary $E_{AB}, F_{BC} \geq 0$, $\operatorname{Tr}_B[E_{AB}\mathcal{T}_B[F_{BC}]]$ is positive. Give an example of two positive operators such that $\operatorname{Tr}_B[E_{AB}F_{BC}]$ is not positive.

5.4 The Kraus representation

Using the Choi representation theorem, we can easily show that completely positive maps must have a Kraus representation. The Choi representation theorem ensures that the Choi state of a channel is positive semidefinite, and therefore we can consider its spectral decomposition or indeed any decomposition into pure states. Doing so enables the construction of Kraus operators.

---

5 Issai Schur, 1875–1941.

6 Jacques Salomon Hadamard, 1865–1963.

Theorem 5.3 (Kraus representation). *A superoperator $\mathcal{E}_{B|A} \in \mathrm{Map}(\mathcal{H}_A, \mathcal{H}_B)$ is completely positive iff there exists a set of operators $\{K(j) \in \mathrm{Lin}(\mathcal{H}_A, \mathcal{H}_B)\}_{j=1}^n$ such that*

$$\mathcal{E}_{B|A} : S_A \mapsto \sum_{j=1}^{n} K_{B|A}(j)\, S_A\, K_{B|A}(j)^* . \tag{5.17}$$

In addition, it is trace-preserving iff $\sum_{j=1}^n K_{B|A}(j)^* K_{B|A}(j) = \mathbb{1}_A$.

Proof. Suppose $\mathcal{E}_{B|A}$ has Kraus form. By the result of Exercise 5.7 it follows that $\mathcal{E}_{B|A}$ is completely positive. On the other hand, if $\mathcal{E}_{B|A}$ is completely positive, then we have a decomposition $\mathrm{C}(\mathcal{E}_{B|A}) = \sum_{j=1}^n |\psi_j\rangle\langle\psi_j|_{AB}$ for some nonnormalized vectors $|\psi_j\rangle_{AB}$. By the isomorphism of operators and bipartite states of Section 4.5, each vector can be expressed as $|\psi_j\rangle_{AB} = K_{B|A'}(j)|\Omega\rangle_{AA'}$ with $K_{B|A'}(j) = \mathrm{V}^{-1}(|\psi_j\rangle_{AB})$ from (4.26). Then, using the Choi representation, for arbitrary $S_A \in \mathrm{Lin}(\mathcal{H}_A)$, we have

$$\begin{aligned} \mathcal{E}_{B|A}[S_A] &= \mathrm{Tr}_A\Big[S_A^T \sum_j K_{B|A'}(j)\,\Omega_{AA'}\,K_{B|A'}(j)^*\Big] \\ &= \sum_j K_{B|A'}(j)\,\mathrm{Tr}_A[S_A^T \Omega_{AA'}]\,K_{B|A'}(j)^* = \sum_j K_{B|A}(j)\,S_A\,K_{B|A}(j)^* . \end{aligned} \tag{5.18}$$

This completes the characterization of completely positive maps.

By taking the trace of the output of (5.17) it is clear that $\sum_{j=1}^n K(j)^* K(j) = \mathbb{1}_A$ implies trace preservation. Conversely, if $\mathcal{E}_{B|A}$ is trace-preserving, then $\mathrm{Tr}[\mathcal{E}[\rho]] = \mathrm{Tr}[\sum_{j=1}^n K(j)^* K(j)\rho]$. Since this holds for arbitrary $\rho \in \mathrm{Lin}(\mathcal{H}_A)$, it implies that $\sum_{j=1}^n K(j)^* K(j) = \mathbb{1}$, completing the proof. □

Note that sets of Kraus operators are in one-to-one correspondence with pure state ensemble decompositions of the Choi operator. Hence Kraus representations are not unique. The minimal number of Kraus operators is set by the rank of the Choi operator, since there is no way to construct an ensemble decomposition of an operator with fewer pure states than its rank.

Exercise 5.28. Find a Kraus representation of the completely positive $\mathcal{E}_\lambda$ in Exercise 5.21.

Exercise 5.29. Find a Kraus representation of the Schur–Hadamard channel from Exercise 5.25.

Exercise 5.30. Consider a superoperator from qubit A to qubits A and B of the form

$$\mathcal{E}_{AB|A}[\rho_A] = \Pi_{AB}(\rho_A \otimes \theta_B)\Pi_{AB} \tag{5.19}$$

for some operator θ_B on B, where $\Pi_{AB} = \frac{1}{2}(\mathbb{1}_{AB} + \Upsilon_{AB})$ is the projector onto the symmetric subspace of two qubits. What form must θ_B have for $\mathcal{E}$ to be a valid quantum channel?

In Section 5.2, we saw that state preparation and measurement are described by CQ and QC channels. The Kraus representation theorem implies the converse: All CQ channels are state preparations, and all QC channels are measurements. To say that the channel has classical input or output is to say that only the diagonal part of the input or output matters (in some fixed basis). We can enforce this with a pinch map, that is, given an arbitrary quantum channel $\mathcal{E}_{B|A}$, the channel $\mathcal{E}'_{B|A} = \mathcal{E}_{B|A} \circ \mathcal{P}_A$ is a CQ channel. Let us regard the pinch map as a channel from X to A with $|X| = |A|$, to emphasize the classical nature of the input (though, strictly speaking, using the letter X only hints that the channel has a classical input; it is not a requirement) and write $\mathcal{E}'_{B|X} = \mathcal{E}_{B|A} \circ \mathcal{P}_{A|X}$. If $K_{B|A}(y)$ are the Kraus operators of $\mathcal{E}_{B|A}$, then $K'_{B|X}(x, y) = K_{B|A}(y)|x\rangle_A \langle x|_X$ are the Kraus operators of the CQ channel $\mathcal{E}_{B|X}$. Defining $|\psi_{x,y}\rangle_B = K_{B|A}(y)|x\rangle_A$, this is very nearly the general form of a state preparation channel we encountered in Section 5.2. The only difference is that here the probability for outputting $|\psi_{x,y}\rangle$ given x is encoded in the norm, as the normalization condition for Kraus operators implies that $\sum_y \langle\psi_{x,y}|\psi_{x,y}\rangle = 1$.

Similarly, applying the pinch map after the channel $\mathcal{E}_{B|A}$ results in a QC channel $\mathcal{M}_{X|A} = \mathcal{P}_{X|B} \circ \mathcal{E}_{B|A}$. The Kraus operators of $\mathcal{M}_{X|A}$ are given by $K_{X|A}(x, y) = |x\rangle_X \langle x|_B K_{B|A}(y)$. Therefore the action of $\mathcal{M}_{X|A}$ on a state ρ_A can be written as

$$\mathcal{M}_{X|A}[\rho_A] = \sum_{x,y} |x\rangle\langle x|_X \operatorname{Tr}[K_{B|A}(y)^* |x\rangle_B \langle x|_B K_{B|A}(y) \rho_A] , \tag{5.20}$$

which is just $\sum_x |x\rangle\langle x|_X \operatorname{Tr}[\Lambda_A(x)\rho_A]$ as in (5.7) for

$$\Lambda_A(x) = \sum_y K_{B|A}(y)^* |x\rangle_B \langle x|_B K_{B|A}(y) . \tag{5.21}$$

Note that the use of Kraus operators $\Lambda_A(x)^{1/2}$ for the instrument associated with a given POVM with elements $\Lambda_A(x)$, as discussed in Section 5.2, is not the most general option. As in (5.21), we merely require a set of $M_{B|A}(x, y)$ such that $\Lambda_A(x) = \sum_y M_{B|A}(x, y)^* M_{B|A}(x, y)$. Each summand defines a POVM element $\Gamma_A(x, y)$ that is a "fine-grained" version of $\Lambda_A(x)$ in that $\Lambda_A(x) = \sum_y \Gamma_A(x, y)$. Put the other way around, Λ_A is the "coarse-grained" version of Γ_A.

The Kraus representation theorem also implies that every quantum channel can be regarded as a measurement by some quantum instrument, followed by forgetting the measurement result. That is, for any $\mathcal{E}_{B|A}$, there exists an instrument $\mathcal{Q}_{XB|A}$ such that $\mathcal{E}_{B|A} = \operatorname{Tr}_X \circ \mathcal{Q}_{XB|A}$. To see this, suppose $\{K(x) \in \operatorname{Lin}(\mathcal{H}_A, \mathcal{H}_B)\}_{x=1}^n$ is a set of Kraus operators for $\mathcal{E}_{B|A}$. Observe that the operators $\hat{K}(x) \in \operatorname{Lin}(\mathcal{H}_A, \mathcal{H}_B \otimes \mathcal{H}_X)$ with $|\mathcal{H}_X| = n$ defined by $\hat{K}(x) = K(x) \otimes |x\rangle$ for an orthonormal basis $\{|x\rangle\}_{x=1}^n$ of $\mathcal{H}_X$ are also a valid set of Kraus operators. Denote the associated channel $\mathcal{Q}_{XB|A}$. For a given input ρ_A, its output is

$$\mathcal{Q}_{XB|A}[\rho_A] = \sum_{x=1}^{n} |x\rangle\langle x|_X \otimes K_{B|A}(x)\, \rho_A\, K_{B|A}(x)^* . \tag{5.22}$$

Clearly, $\mathcal{E}_{B|A} = \mathrm{Tr}_X \circ \mathcal{Q}_{XB|A}$. On the other hand, discarding B leaves a classical state in X:

$$\mathrm{Tr}_B[\mathcal{Q}_{XB|A}[\rho_A]] = \sum_{x=1}^{n} \mathrm{Tr}[K(x)^* K(x)\rho]\, |x\rangle\langle x|_X \,. \tag{5.23}$$

This is precisely the form of a measurement channel with POVM elements $\Lambda_A(x) = K_{B|A}(x)^* K_{B|A}(x)$. Such a set of operators indeed forms a POVM. Each $\Lambda_A(x)$ is necessarily positive, and by the trace-preserving condition their sum is $\mathbb{1}_A$.

Although Kraus representations are only available to completely positive maps, a similar-looking form can be constructed for arbitrary superoperators. Consider the Choi operator $\mathsf{C}(\mathcal{E}_{B|A})$ of an arbitrary superoperator $\mathcal{E}_{B|A}$. We can express it as $\mathsf{C}(\mathcal{E}_{B|A}) = \sum_{j=1}^{n} |\varphi_j\rangle\langle\vartheta_j|_{AB}$, for some unnormalized vectors $|\varphi_j\rangle_{AB}$ and $|\vartheta_j\rangle_{AB}$ (see (B.4)). But then we can define $L_{B|A}(j) = \mathsf{V}^{-1}(|\varphi_j\rangle_{AB})$ and $R_{B|A} = \mathsf{V}^{-1}(|\vartheta_j\rangle)$ and proceed as in (5.18) to obtain

$$\mathcal{E}_{B|A}[M_A] = \sum_{j=1}^{n} L_{B|A}(j)\, M_A\, R_{B|A}(j)^* \,. \tag{5.24}$$

As with the Kraus representation of completely positive maps, this representation is not unique.

Exercise 5.31. Given a representation of $\mathcal{E}_{B|A}$ as in (5.24), show that its Choi operator satisfies

$$\mathsf{C}(\mathcal{E}_{B|A}) = \sum_{j=1}^{n} \mathsf{V}(L_{B|A}(j))\, (\mathsf{V}(R_{B|A}(j)))^* \,. \tag{5.25}$$

Exercise 5.32. Find a generalized Kraus representation of the transposition map.

5.5 Two further isomorphisms

5.5.1 The Jamiołkowski isomorphism

In the contemporary jargon of quantum information theory, it is common for the Choi isomorphism to be called the "Choi–Jamiołkowski[7]" isomorphism. However, this conflates two closely related but distinct isomorphisms. The *Jamiołkowski map* J is defined similarly to the Choi map but using Υ_{AB} instead of Ω_{AB}:

$$\begin{aligned} \mathsf{J} : \mathrm{Map}(\mathcal{H}_A, \mathcal{H}_B) &\to \mathrm{Lin}(\mathcal{H}_A \otimes \mathcal{H}_B) \\ \mathcal{E}_{B|A} &\mapsto \mathcal{E}_{B|A'}[\Upsilon_{AA'}] \,. \end{aligned} \tag{5.26}$$

7 Andrzej Jamiołkowski, born 1946.

Its inverse J^{-1} takes any $M_{AB} \in \mathrm{Lin}(\mathcal{H}_A \otimes \mathcal{H}_B)$ to the superoperator $\mathsf{J}^{-1}(M_{AB}) \in \mathrm{Map}(\mathcal{H}_A, \mathcal{H}_B)$, whose action is specified by

$$\mathsf{J}^{-1}(M_{AB}) : S_A \mapsto \mathrm{Tr}_A\big[(S_A \otimes \mathbb{1}_B)M_{AB}\big]\,. \tag{5.27}$$

Alternatively, J can be defined as the map from $\mathrm{Map}(\mathcal{H}_A, \mathcal{H}_B)$ to $\mathrm{Lin}(\mathcal{H}_A \otimes \mathcal{H}_B)$ such that, for all $S_A \in \mathrm{Lin}(\mathcal{H}_A)$ and $T_B \in \mathrm{Lin}(\mathcal{H}_B)$,

$$\langle S_A^* \otimes T_B, \mathsf{J}(\mathcal{E}_{B|A})\rangle = \langle T_B, \mathcal{E}_{B|A}[S_A]\rangle\,. \tag{5.28}$$

Exercise 5.33. Show that J satisfies (5.28).

The characterization by inner products implies that the Jamiołkowski isomorphism is *natural* in that it does not depend on any basis choice. This fact can also be appreciated in the original definition, since the swap operator $\Upsilon_{AA'}$ is independent of any basis choice for $\mathcal{H}_A$ and $\mathcal{H}_{A'}$.

Since $\Upsilon_{AA'}$ is the partial transpose of $\Omega_{AA'}$, the Jamiołkowski map is closely related to the Choi map. Indeed, using the transposition map $\mathcal{T}_A$ in the basis defining $|\Omega\rangle_{AA'}$, we have

$$\mathsf{J}(\mathcal{E}_{B|A}) = \mathcal{T}_A[\mathsf{C}(\mathcal{E}_{B|A})]. \tag{5.29}$$

The lack of transposition as compared to the Choi representation is appealing, but we lose the simple characterization of completely positive superoperators.

Exercise 5.34. Show that the Choi map C satisfies $\langle \overline{S}_A \otimes T_B, \mathsf{C}(\mathcal{E}_{B|A})\rangle = \langle T_B, \mathcal{E}_{B|A}[S_A]\rangle$ for all $S_A \in \mathrm{Lin}(\mathcal{H}_A)$ and $T_B \in \mathrm{Lin}(\mathcal{H}_B)$. Here the dependence on the basis choice for $\mathcal{H}_A$ is immediately apparent.

5.5.2 The Liouville isomorphism

Since superoperators are linear maps, it must be possible to represent their composition by ordinary operator composition, essentially matrix multiplication. This is often called the Liouville isomorphism. It is easily defined using the operator-vector isomorphism V from Section 4.5. Suppose we have an operator representative $\mathsf{L}(\mathcal{E}_{B|A})$ of a superoperator $\mathcal{E}_{B|A}$ that satisfies the condition

$$\mathsf{L}(\mathcal{E}_{B|A})\mathsf{V}(M_A) = \mathsf{V}(\mathcal{E}_{B|A}[M_A]) \tag{5.30}$$

for all M_A. Evidently, $\mathsf{L}(\mathcal{E}_{B|A})$ must be an element of $\mathrm{Lin}(\mathcal{H}_A \otimes \mathcal{H}_{A'}, \mathcal{H}_B \otimes \mathcal{H}_{B'})$, where $\mathcal{H}_{A'} \simeq \mathcal{H}_A$ and $\mathcal{H}_{B'} \simeq \mathcal{H}_B$. It then follows that $\mathsf{L}(\mathcal{F}_{C|B} \circ \mathcal{E}_{B|A}) = \mathsf{L}(\mathcal{F}_{C|B})\mathsf{L}(\mathcal{E}_{B|A})$. The Liouville map is the unique L that satisfies (5.30), and it is an isomorphism. It must be unique because varying M_A on the left-hand side of (5.30) results in arbitrary vectors in $\mathcal{H}_A \otimes \mathcal{H}_{A'}$.

However, it is not so straightforward to express $\mathsf{L}(\mathcal{E}_{E|A})$ in terms of the action of $\mathcal{E}_{B|A}$ itself, as done in the Choi and Jamiołkowski isomorphisms. But we can do so by making use of the generalized Kraus representation in (5.24). Consider the result of applying the operator $\mathcal{E}_{B|A}[M_A] \in \mathrm{Lin}(\mathcal{H}_B)$ to half of the state $|\Omega\rangle_{BB'}$ for some operator M_A. The result of Exercise 4.19 gives

$$\mathsf{V}(\mathcal{E}_{B|A}[M_A]) = \sum_{j=1}^{n} \mathsf{V}(L_{B|A}(j)\, M_A\, R_{B|A}(j)^*) = \left(\sum_{j=1}^{n} L_{B|A} \otimes \overline{R}_{B'|A'}\right) \mathsf{V}(M_A)\,. \tag{5.31}$$

The statement in Exercise 4.19 is ambiguous in this case since the operator M_A maps $\mathcal{H}_A$ to itself. Here we make the replacement $(A, B, C, D) \to (B', A', A, B)$, i. e., we treat M_A as $M_{A|A'}$, Thus, for a superoperator with generalized Kraus operators $L_{B|A}(j)$ and $R_{B|A}(j)$, we have

$$\mathsf{L}(\mathcal{E}_{B|A}) = \sum_{j=1}^{n} L_{B|A}(j) \otimes \overline{R}_{B'|A'}(j)\,. \tag{5.32}$$

The isomorphism can also be understood in terms of the usual method of creating a matrix representation of a linear transformation. Abstractly, for transformation $T : U \to V$, we choose bases $\{u_j\}$ and $\{v_k\}$ of U and V, respectively, and then use a suitable inner product in V to define $[T]_{jk} = \langle v_j, T u_k\rangle$. Here we use the Hilbert–Schmidt inner product and choose bases $\{|b_j\rangle\langle b_k|\}_{jk}$ for $\mathrm{Lin}(\mathcal{H}_A)$ and $\{|b'_j\rangle\langle b'_k|\}_{jk}$ for $\mathrm{Lin}(\mathcal{H}_B)$, where $\{|b_k\rangle\}_k$ is a basis of $\mathcal{H}_A$, and $\{|b'_k\rangle\}_k$ is a basis of $\mathcal{H}_B$. Then the Liouville map L is defined by

$$\mathsf{L}(\mathcal{E}_{B|A}) = \sum_{jk,\ell m} [\mathsf{L}(\mathcal{E}_{B|A})]_{jk,\ell m} |b'_j\rangle_B |b'_k\rangle_{B'} \langle b_\ell|_A \langle b_m|_{A'}, \tag{5.33}$$

$$[\mathsf{L}(\mathcal{E}_{B|A})]_{jk,\ell m} = \mathrm{Tr}\big[|b'_k\rangle\langle b'_j|\, \mathcal{E}_{B|A}[|b_\ell\rangle\langle b_m|]\big]\,. \tag{5.34}$$

Note that the order of j and k is inverted in the expression for the matrix elements due to the adjoint in the Hilbert–Schmidt inner product.

Exercise 5.35. Show that the matrix elements of (5.32) are those from (5.34).

The Liouville representation is also an isomorphism of superoperators and bipartite operators, which can be appreciated by showing that $\mathsf{L}(\mathcal{E}_{B|A})$ is just a reshuffled version of the Choi operator. In particular,

$$[\mathsf{L}(\mathcal{E}_{B|A})]_{jk,\ell m} = (\langle b_\ell|_A \otimes \langle b'_j|_B)\mathsf{C}(\mathcal{E}_{B|A})(|b_m\rangle_A \otimes |b'_k\rangle_B)\,. \tag{5.35}$$

Exercise 5.36. Show (5.35).

Exercise 5.37. What is the condition for trace preservation of $\mathcal{E}$ in terms of $\mathsf{L}(\mathcal{E})$?

Exercise 5.38. What is the Liouville representation of the transposition map?

We are free to make other choices of bases in (5.33) and (5.34). Choosing the Weyl–Heisenberg operators leads to a different representation, often called the *process matrix* (especially, in the context of qubits), but we will not pursue this further.

5.6 Notes and further reading

The quote is from Feynman's 1964 Messenger lecture at Cornell University, transcribed in [99]. The mathematical structure relevant for open quantum systems that we have traced here were developed in the works of, among many others, Naimark [209], Stinespring [275], Haag and Kastler [114], Hellwig and Kraus [135, 136], de Pillis [75], Jamiołkowski [152], and Choi [59]. For more detailed treatments, see the earlier works by Davies [70] and especially Kraus [171], as well as the more recent treatments by Peres [220], Werner [299], Holevo [145], and Wolf [307]. The term "pinching" was coined by Davis [73], while "Liouville representation" stems from Fano [98] (see also Sudarshan et al. [278]). Leifer and Spekkens [186] explored the analogy of Choi operators as the analog of conditional probability.

6 Purification

The best that most of us can hope to achieve in physics is simply to misunderstand at a deeper level.

Wolfgang Pauli

In Chapter 4, we established two “interpretations” of density operators, either as mixed states or as marginal states. This raises the question of whether every density operator ρ_A can be regarded as the marginal state of a pure state $|\Psi\rangle_{AB}$, that is, whether every mixed state is mixed because it is only part of a composite pure state. This issue turns out to be related to the famous “measurement problem” of quantum mechanics, whether or not measurement can be understood as a dynamical process, subject to unitary evolution according to the Schrödinger equation. Despite the apparent incompatibility of measurement and unitary dynamics, we will find that not just measurement, but any channel can be thought of as unitary dynamics involving additional degrees of freedom.

6.1 Purification of density operators

First, let us consider the issue of marginal versus mixed states. For a given density operator ρ_A on system A, any state θ_{AB} on systems A and B such that $\mathrm{Tr}_B[\theta_{AB}] = \rho_A$ is an *extension* of ρ_A. A purification is an extension to a pure state.

Definition 6.1 (Purification). A *purification* of $\rho_A \in \mathrm{Stat}(\mathcal{H}_A)$ is a normalized $|\Psi\rangle_{AB} \in \mathcal{H}_A \otimes \mathcal{H}_B$ for some $\mathcal{H}_B$ such that $\rho_A = \mathrm{Tr}_B[|\Psi\rangle\langle\Psi|_{AB}]$. System B is often called the purifying system.

Purifying a mixed state, regarding it as the marginal of a bipartite pure state, is common in the study of quantum information theory and is jokingly referred to as “going to the church of the larger Hilbert space”.

Indeed, all mixed states are marginals of pure states, since every density operator has a purification as we now show. Suppose ρ_A has the eigendecomposition $\rho_A = \sum_{k=1}^r P(k)|\xi_k\rangle\langle\xi_k|_A$, where r is the number of nonzero eigenvalues (the rank). Then, for any choice of orthonormal vectors $|b_k\rangle_B$, the bipartite state

$$|\Psi\rangle_{AB} = \sum_{k=1}^{r} \sqrt{P(k)}\,|\xi_k\rangle_A \otimes |b_k\rangle_B \tag{6.1}$$

is a purification of ρ_A. Hence the purifying system B need only have dimension equal to the rank of ρ_A. This works nicely when ρ_A is already pure, as then B is one-dimensional, i. e., trivial. Moreover, the notion of purifications completely subsumes extensions, since any extension of ρ_A can itself be purified.

https://doi.org/10.1515/9783110570250-006

We can construct a purification from any pure state ensemble decomposition. Given an ensemble decomposition of a density operator $\rho_A = \sum_{z=1}^n P_Z(z)|\varphi_z\rangle\langle\varphi_z|_A$, simply invent an additional system B of dimension at least n and define

$$|\Psi\rangle_{AB} = \sum_{z=1}^{n} \sqrt{P_Z(z)}\,|\varphi_z\rangle_A \otimes |b_z\rangle_B\,, \tag{6.2}$$

where $|b_z\rangle_B$ is a basis for $\mathcal{H}_B$. Furthermore, the CQ state $\rho_{AZ} = \sum_z P_Z(z)|b_z\rangle\langle b_z|_Z \otimes |\varphi_z\rangle\langle\varphi_z|_A$ corresponding to the ensemble is immediately recovered by measuring system B in the basis $|b_z\rangle$. That is, $\rho_{AZ} = \mathcal{P}_{Z|B}[\Psi_{AB}]$, where $\mathcal{P}_{Z|B}$ is a pinch map.

The *canonical purification*

$$|\Psi\rangle_{AA'} = \sqrt{\rho_A} \otimes \mathbb{1}_{A'}|\Omega\rangle_{AA'} \tag{6.3}$$

is especially convenient in calculations.

Exercise 6.1. Show that the canonical purification is indeed a purification.

The above shows that purifications and pure-state ensembles are two different descriptions of essentially the same thing. In the framework of quantum theory, we have two different options for interpreting an ensemble of states on system A. One is the CQ state ρ_{AZ}, where the particular state of A is correlated with the classical random variable Z. The other is to regard the randomness of the ensemble as arising from measurement of one half of the entangled bipartite state in (6.2). Here we are not attempting to make any claim about which version is the "actual" state of affairs, merely that they are equivalent for our statistical purposes. This shift in perspective is commonly used to understand the properties of various information processing tasks.

6.2 Ensembles and purifications

6.2.1 Schmidt decomposition

In Chapter 4, we saw that general density operators have many possible pure state ensemble decompositions. The results of the previous section imply that they have several possible purifications. It turns out that all possible purifications and, equivalently, pure state ensembles are related in a simple way. To see how, we make use of the *Schmidt decomposition* of a bipartite vector.

Proposition 6.1 (Schmidt decomposition). *For any $|\Psi\rangle_{AB} \in \mathrm{Lin}(\mathcal{H}_A \otimes \mathcal{H}_B)$, there exist orthonormal bases $\{|\xi_j\rangle_A\}_{j=1}^{d_A}$ and $\{|\eta_k\rangle_B\}_{k=1}^{d_B}$ and $n \le \min(d_A, d_B)$* Schmidt coefficients *$s_k > 0$ such that*

$$|\Psi\rangle_{AB} = \sum_{k=1}^{n} s_k|\xi_k\rangle_A \otimes |\eta_k\rangle_B\,. \tag{6.4}$$

Proof. The proof is an application of the singular value decomposition (see Section B.6). By (4.26) we can write $|\Psi\rangle_{AB} = (K_{A|B'} \otimes \mathbb{1}_B)|\Omega\rangle_{B'B}$ for $K_{A|B'} = {}_{BB'}\langle\Omega|\Psi\rangle_{AB}$. Now consider the singular value decomposition $K_{A|B'}$. Specifically, denote the n singular values s_k and define $\mathcal{H}_F$ to have dimension n. Then, in accordance with Lemma B.4, $K_{A|B'}$ can be expressed as $K_{A|B'} = U_{A|F}D_F(V_{B'|F})^*$ for some isometries $U_{A|F}$ and $V_{B'|F}$, where D_F is the diagonal matrix with entries s_k. Using (4.31), we have

$$\begin{aligned} |\Psi\rangle_{AB} &= U_{A|F}D_F(V_{B'|F})^*|\Omega\rangle_{B'B} = U_{A|F}D_F \otimes \overline{V}_{B|F'}|\Omega\rangle_{FF'} \\ &= \sum_{k=1}^{n} s_k U_{A|F}|b_k\rangle_F \otimes \overline{V}_{B|F'}|b_k\rangle_{F'} . \end{aligned} \tag{6.5}$$

Now define $|\xi_k\rangle_A = U_{A|F}|b_k\rangle_F$ and $|\eta_k\rangle_B = \overline{V}_{B|F'}|b_k\rangle_{F'}$ for $k = 1, \ldots, n$. Since U and V (and therefore $\overline{V}$) are isometries, these sets of vectors are each orthonormal. When $n \leq d_A, d_B$, the sets can each be arbitrarily extended to full orthonormal bases of $\mathcal{H}_A$ and $\mathcal{H}_B$. □

Observe that the marginal density operators $\Psi_A = \mathrm{Tr}_B[\Psi_{AB}]$ and $\Psi_B = \mathrm{Tr}_A[\Psi_{AB}]$ are given by $\sum_{k=1}^{r} s_k^2|\xi_k\rangle\langle\xi_k|_A$ and $\sum_{k=1}^{r} s_k^2|\eta_k\rangle\langle\eta_k|_B$, respectively. These are eigendecompositions of the marginals, and we immediately see that both marginals have the same eigenvalues. The number of Schmidt coefficients is equal to the rank of either of the marginals. Hence a purification $|\Psi\rangle_{AB}$ of a given ρ_A can only exist if $\dim(\mathcal{H}_B) \geq r = \mathrm{rank}(\rho_A)$. By the relation of purifications and ensembles in (6.2) it is evident that every ensemble decomposition of a given ρ_A has at least r elements. Purifications with $\dim(\mathcal{H}_B) = r$ or ensembles with r elements are called *minimal.*

Now suppose that $|\Psi\rangle_{AB}$ and $|\Psi'\rangle_{AC}$ are two purifications of ρ_A that has rank r. Because the two states have the same marginal on $\mathcal{H}_A$, using the Schmidt form yields $|\Psi\rangle_{AB} = \sum_{k=1}^{r} s_k|\xi_k\rangle_A \otimes |\eta_k\rangle_B$ and $|\Psi'\rangle_{AC} = \sum_{k=1}^{r} s_k|\xi_k\rangle_A \otimes |\eta'_k\rangle_C$. If $r < \dim(\mathcal{H}_B)$, then arbitrarily extend the set of $|\eta_k\rangle_B$ to a complete orthonormal basis. Then define the map $V_{C|B} : \mathcal{H}_B \to \mathcal{H}_C$ by the action

$$V_{C|B}|\eta_k\rangle_B = \begin{cases} |\eta'_k\rangle_C, & k = 1, \ldots, r, \\ 0, & k = r+1, \ldots, \dim(\mathcal{H}_B). \end{cases} \tag{6.6}$$

By construction $|\Psi'\rangle_{AC} = \mathbb{1}_A \otimes V_{C|B}|\Psi\rangle_{AB}$. Since $\{|\eta_k\rangle_B\}$ and $\{|\eta'_k\rangle_C\}$ are sets of orthonormal vectors, $V_{C|B}$ is a partial isometry, an isometry on its support.

Alternately, suppose that $\dim(\mathcal{H}_C) \geq \dim(\mathcal{H}_B)$, which holds without loss of generality. In this case, we may extend both sets of r vectors $|\eta_k\rangle_B$ and $|\eta'_k\rangle_C$ by additional $\dim(\mathcal{H}_B) - r$ orthonormal elements and define the isometry

$$\hat{V}_{C|B}|\eta_k\rangle_B = |\eta'_k\rangle_C , \tag{6.7}$$

for which we also have $|\Psi'\rangle_{AC} = \mathbb{1}_A \otimes \hat{V}_{C|B}|\Psi\rangle_{AB}$. Altogether, we have shown the following:

Proposition 6.2 (Existence and nonuniqueness of purifications). *For any $\rho_A \in \mathrm{Stat}(\mathcal{H}_A)$ of rank r, there exists a purification in $\mathcal{H}_A \otimes \mathcal{H}_B$ if and only if $\dim(\mathcal{H}_B) \geq r$. For any two purifications $|\Psi\rangle_{AB}$ and $|\Psi'\rangle_{AC}$, there exists a partial isometry $V_{C|B}$ such that $|\Psi'\rangle_{AC} = (\mathbb{1}_A \otimes V_{C|B})|\Psi\rangle_{AB}$. If $\dim(\mathcal{H}_C) > \dim(\mathcal{H}_B)$, then $V_{C|B}$ can be taken to be an isometry, or unitary in the case of equality.*

Using the canonical purification of (6.3), every purification of ρ has the form

$$|\Psi\rangle_{AR} = \sqrt{\rho_A} \otimes V_{R|A'}|\Omega\rangle_{AA'} \tag{6.8}$$

for some partial isometry $V_{R|A}$ whose kernel is the kernel of ρ_A.

Exercise 6.2. For any given state ρ_{AB}, suppose that $|\psi\rangle_{AC}$ is a purification of ρ_A. Show that there exists an isometry $V_{BC|C}$ such that $V_{BC|C}|\psi\rangle_{AC}$ purifies ρ_{AB}.

The Schmidt decomposition and unitary relation of purifications justifies our choice of defining $|\Phi\rangle$ to be the state of maximal entanglement. It is a state whose Schmidt values are completely uniform, i. e., whose reduced density operator is the maximally mixed state. By Proposition 6.2 all other states with this reduced state are equivalent by local actions on one subsystem, which does not change the entanglement.

6.2.2 Steering

Since pure state ensembles can be recovered from purifications by projective measurement, the relation of purifications translates into a relation of ensembles.

Proposition 6.3 (Unitary relation of ensemble decompositions). *For any density operator ρ, let $\{(p_k, |\varphi_k\rangle)\}_{k=1}^n$ and $\{(q_j, |\psi_j\rangle)\}_{j=1}^m$ be pure state ensemble decompositions, and set $\ell = \max(n, m)$. Then there exists an $\ell \times \ell$ unitary matrix U with components U_{jk} such that*

$$\sqrt{q_j}\,|\psi_j\rangle = \sum_{k=1}^{\ell} U_{jk}\sqrt{p_k}\,|\varphi_k\rangle \quad \forall j \in \{1, \ldots, m\}. \tag{6.9}$$

The smallest ensemble has a number of elements equal to the rank of ρ.

Proof. By assumption, $\rho = \sum_{k=1}^n p_k|\varphi_k\rangle\langle\varphi_k| = \sum_{j=1}^m q_j|\psi_j\rangle\langle\psi_j|$. Let ℓ be the larger of n and m, and define $p_k = 0$ for $k = n+1, \ldots, m$ when $n \leq m$ or $q_j = 0$ for $j = m+1, \ldots, n$ when $m \leq n$. Now construct the purifications

$$|\Psi_1\rangle_{AB} = \sum_{k=1}^{\ell} |\varphi_k\rangle_A \otimes |b_k\rangle_B \quad \text{and} \quad |\Psi_2\rangle_{AB} = \sum_{j=1}^{\ell} |\psi_j\rangle_A \otimes |b_j\rangle_B\,. \tag{6.10}$$

As these pure states are purifications of the same density operator, by Proposition 6.2 there must be a unitary $U_B \in \mathrm{Lin}(\mathcal{H}_B)$ such that $\mathbb{1}_A \otimes U_B|\Psi_1\rangle_{AB} = |\Psi_2\rangle_{AB}$. Applying $\langle b_k|_B$ to both sides of this equation yields the claimed result. □

In fact, not only can we recover an ensemble from a suitably constructed purification; as in the beginning of this section, *every* purification can generate *every* possible pure state ensemble by suitable measurement of the purifying system. For instance, measuring system B of (6.10) with projectors $\Pi(k) = |b_k\rangle\langle b_k|$ produces the ensemble $\{(p_k, |\varphi_k\rangle)\}$, whereas measuring with projectors $\Pi'(j) = U^*|b_j\rangle\langle b_j|U$ produces $\{(q_j, |\psi_j\rangle)\}$. Since all pure state ensemble decompositions are related by unitaries (after suitable embedding), any of them can be obtained in this way. Put differently, for any pure state decomposition of a density operator ρ_A, the associated CQ state can be generated by suitable measurement of the purifying system of a fixed purification.

Exercise 6.3. Find purifications of minimal dimension for the qubit ensembles (here specified by their Bloch vectors) $\{\frac{1}{2}, \vec{r}_j\}_{j=1}^2$ with any $\vec{r}_1 = -\vec{r}_0$ and $\{\frac{1}{3}, \vec{s}_j\}_{j=1}^3$ with $\vec{s}_j$ forming an equilateral triangle. Construct a partial isometry taking the second purification to the first. Find a measurement on (the purifying system of) the second purification that yields the first ensemble.

Schrödinger called this phenomenon *steering*, as it appears that a quantum system can be "steered or piloted into one or the other type of state at the experimenter's mercy in spite of his having no access to it". However, this is not quite correct, since the particular pure state that results from the measurement procedure is random. The important distinction here is between the CQ state ρ_{XA} representing the ensemble, in which the measurement result X is recorded, and the marginal state ρ_A. This distinction makes clear that steering does not allow measurement on the purifying system B to itself transmit information to A, as we already saw in Section 4.4.3, because the marginal state of A is unaffected by action on B.

The above is concerned with pure state ensemble decompositions, but in fact *all* ensemble decompositions of ρ_A can be obtained from any given purification. We can give an explicit measurement to steer the canonical purification $|\Psi\rangle_{AA'} = \sqrt{\rho_A} \otimes \mathbb{1}_{A'}|\Omega\rangle_{AA'}$ of ρ_A to the CQ state $\rho_{XA} = \sum_{XA} = P_X(x)|x\rangle\langle x|_X \otimes \rho_A(x)$ associated with any ensemble. In particular, the required measurement is a version of the *pretty good measurement* (also called the square-root measurement) defined as follows.

Definition 6.2 (Pretty good measurement). For any CQ state $\rho_{XA} = \sum_x P_X(x)|x\rangle\langle x|_X \otimes \rho_A(x)$, the *pretty good measurement* is the POVM with elements $\Lambda_A(x) \in \mathrm{Lin}(\mathcal{H}_A)$ given by

$$\Lambda_A(x) = \rho_A^{-1/2} P_X(x)\, \rho_A(x)\, \rho_A^{-1/2}, \tag{6.11}$$

where the inverse of the square root is defined on the support of ρ_A.

Observe that all the $\rho_A(x)$ are supported in the support of the average state ρ_A, i. e., $\ker(\rho_A) \subseteq \ker(\rho_A(x))$ for all x. If ρ_A is not full rank, then strictly speaking, $\Lambda_A(x)$ does not form a POVM, as their sum is equal to the projection Π_A onto the support of ρ_A. However, we can include one additional POVM element $\mathbb{1}_A - \Pi_A$.

Exercise 6.4. Show that the pretty good measurement for an ensemble consisting of linearly independent pure states is a projective measurement, that is, the pretty good measurement constructs an orthonormal basis for the span of the pure states.
Hint: Write $M = \sum_{k=1}^{n} |\varphi_k\rangle\langle k|$ for the pure states $|\varphi_k\rangle$ and an orthonormal set $|k\rangle$ and use the singular value decomposition to compute $(MM^*)^{-1/2}M$.

To steer the canonical purification $|\Psi\rangle_{AA'}$ to the CQ state ρ_{XA} by measurement on A', it suffices to make the measurement with POVM elements $\Lambda_{A'}(x)^T$. The state of A given outcome x is indeed $\rho_A(x)$ with probability $P_X(x)$:

$$\begin{aligned}\mathrm{Tr}_{A'}\left[\sqrt{\rho_A}\,\Omega_{AA'}\sqrt{\rho_A}\,\Lambda_{A'}(x)^T\right] &= \mathrm{Tr}_{A'}\left[\sqrt{\rho_A}\Omega_{AA'}\,\Lambda_A(x)\,\sqrt{\rho_A}\right] \\ &= \sqrt{\rho_A}\,\Lambda_A(x)\,\sqrt{\rho_A} = P_X(x)\rho_A(x)\,.\end{aligned} \tag{6.12}$$

From (6.8), any other purification has the form $|\Psi\rangle_{AB} = \sqrt{\rho_A} \otimes V_{B|A'}|\Omega\rangle_{AA'}$, meaning that in the general case, we can simply use the measurement with elements $\Gamma_B(x) = V_{B|A}\,\Lambda_A(x)^T\,V_{B|A}^*$.

Proposition 6.4 (Steering). *Suppose $|\Psi\rangle_{AB}$ is a purification and $\rho_{XA} = \sum_x P_X(x)|x\rangle\langle x|_X \otimes \rho_A(x)$ is a CQ extension of a quantum state ρ_A. Then there exists a measurement $\mathcal{M}_{X|B}$ such that $\mathcal{M}_{X|B}[\Psi_{AB}] = \rho_{XA}$. In particular, let $V_{B|A}$ be the partial isometry such that $|\Psi\rangle_{AB} = \sqrt{\rho_A} \otimes V_{B|A'}|\Omega\rangle_{AA'}$, and let $\{\Lambda_A(x)\}$ be the pretty good measurement associated with ρ_{XA}. Then the POVM elements of $\mathcal{M}_{X|B}$ are $\Gamma_B(x) = V_{B|A}\,\Lambda_A(x)^T\,V_{B|A}^*$.*

Again, we include an additional POVM element in case of rank-deficient ρ_A.

Exercise 6.5. Fix a density operator ρ and let $|\psi\rangle$ be any vector in the support of ρ, i. e., $\rho|\psi\rangle \neq 0$. Show that the probability p associated with $|\psi\rangle$ in any *minimal* pure state ensemble decomposition of ρ satisfies $p\,\langle\psi|\rho^{-1}|\psi\rangle = 1$.

6.3 Dilation of channels

The results of the previous section also apply to channels via the Choi isomorphism. Continuing with the analogy to density operators, we can consider for the purification of the Choi operator. For an arbitrary channel $\mathcal{E}_{B|A}$, let E_{BA} be the associated Choi operator, and let $|\Psi\rangle_{ABR}$ be a purification of E_{BA}. Using (4.26), we can write $|\Psi\rangle_{ABR} = \mathbb{1}_A \otimes V_{BR|A'}|\Omega\rangle_{AA'}$ for some map $V_{BR|A} : \mathcal{H}_A \to \mathcal{H}_B \otimes \mathcal{H}_R$. The fact that the channel is trace-preserving implies that V is an isometry, since

$$\begin{aligned}\mathbb{1}_A &= \mathrm{Tr}_B[E_{BA}] = \mathrm{Tr}_{BR}[|\Psi\rangle\langle\Psi|_{ABR}] = \mathrm{Tr}_{BR}\left[V_{BR|A'}\Omega_{AA'}V_{BR|A'}^*\right] \\ &= \mathrm{Tr}_{A'}[(V^*V)_{A'}\Omega_{AA'}] = (V^*V)_A^T\,.\end{aligned} \tag{6.13}$$

Using the purification in the Choi isomorphism gives

$$\begin{aligned}\mathcal{E}_{B|A}[\rho_A] &= \mathrm{Tr}_A[E_{BA}\,\rho_A^T] = \mathrm{Tr}_{AR}[V_{BR|A'}\,\Omega_{A'A}\,(V_{BR|A'})^*\rho_A^T] \\ &= \mathrm{Tr}_{AR}[V_{BR|A'}\,\Omega_{A'A}\,\rho_{A'}\,(V_{BR|A'})^*] = \mathrm{Tr}_R[V_{BR|A}\,\rho_A\,(V_{BR|A})^*]\,.\end{aligned} \tag{6.14}$$

Now the channel action is represented by the action of an isometry $V_{BR|A}$ and the partial trace over the purifying system R. This is called the Stinespring[1] representation, and the operator $V_{BR|A}$ is called a *dilation* of the channel $\mathcal{E}_{B|A}$.

Theorem 6.1 (Stinespring representation). *A map $\mathcal{E}_{B|A}$ is completely positive if and only if there exist $\mathcal{H}_R$ and $V_{BR|A} \in \mathrm{Lin}(\mathcal{H}_A, \mathcal{H}_B \otimes \mathcal{H}_R)$ with*

$$\mathcal{E}_{B|A}[S_A] = \mathrm{Tr}_R\big[V_{BR|A}\,S_A\,V_{BR|A}^*\big] \quad \forall\, S_A \in \mathrm{Lin}(\mathcal{H}_A). \tag{6.15}$$

The smallest possible d_R is no larger than $d_A d_B$. A completely positive $\mathcal{E}_{B|A}$ is trace-preserving if and only if V is an isometry, i. e., $V_{BR|A}^* V_{BR|A} = \mathbb{1}_A$.

In the trivial case of the identity map $\mathcal{I}_A$, a Stinespring dilation is clearly just $\mathbb{1}_A$ with a trivial (one-dimensional) system R. This accords with the construction above since the Choi state is already pure. Another simple case is the pinch map $\mathcal{P}_A$, for which a Stinespring dilation is just

$$V_{RA|A} = \sum_x |x\rangle_R \otimes |x\rangle\langle x|_A\,. \tag{6.16}$$

The Stinespring representation reveals that quantum channels can be regarded as unitary operations involving additional systems. Any channel $\mathcal{E}_{A|A}$ acting on a system A can be dilated to an isometry $V_{AR|A}$, which can then be extended to a unitary U_{AR} on AR. Concretely, the isometry $V_{AR|A}$ determines the action of U_{AR} on a fixed vector in R, call it $|0\rangle_R$, by $V_{AR|A}|\varphi\rangle_A = U_{AR}|\varphi\rangle_A|0\rangle_R$. We are free to define the action of U_{AR} on inputs $|\varphi\rangle_A|k\rangle_R$ for $k \neq 0$ as we like, subject to the unitary constraint. For channels $\mathcal{E}_{B|A}$ that map one system to another, we can also include a fixed vector of B at the input. In this case, we use the isometry $V_{BR|A}$ to define the action of a unitary map U_{ABR} on A, B, and R by $V_{BR|A}|\varphi\rangle_A = U_{ABR}|\varphi\rangle_A|0\rangle_B|0\rangle_R$.

Theorem 6.1 is the quantum analog of the equivalence between the two alternate definitions of classical channels given in Section 3.2. Our definition of quantum channels in Definition 5.2 is based on convexity, in precisely the same spirit as the first option in Section 3.2. The second approach there is motivated by physics rather than statistics. Since classical dynamics is deterministic, any possible classical channel must be a mixture of deterministic dynamics, where the mixture results from ignoring some degrees of freedom. Deterministic dynamics corresponds to unitary operations in the quantum case, and ignoring degrees of freedom to partial trace, the analog

1 William Forrest "Woody" Stinespring, 1929–2012.

of the second option is to define quantum channels by (6.15). Then, after extending isometries to unitaries as just explained above, Theorem 6.1 can be understood as establishing that the two possible definitions of quantum channels are equivalent.

6.4 Relationship of Choi, Kraus, and Stinespring

The relation between Choi, Kraus, and Stinespring representations is precisely that of density operator, ensemble, and purification. As with the Kraus representation, the Stinespring representation is not unique either. But, like purifications, all Stinespring dilations are related by partial isometries. We merely need to use the partial isometry of (6.6) on different purifications of the Choi operator to infer the following analog of Proposition 6.2.

Proposition 6.5 (Isometric relation of Stinespring dilations). *For any two dilations $V_{BR|A}$ and $V'_{BR'|A}$ of $\mathcal{E}_{B|A}$, there exists a partial isometry $W_{R'|R}$ such that $V'_{BR'|A} = W_{R'|R}V_{BR|A}$. If $\dim(\mathcal{H}_{R'}) > \dim(\mathcal{H}_R)$, then $W_{R'|R}$ can be taken to be an isometry, or unitary in the case of equality.*

Exercise 6.6. Show that all possible Stinespring representations of the identity channel $\mathcal{I}_A$ have the form $V_{RA|A} = |\varphi\rangle_R \otimes \mathbb{1}_A$ for some normalized vector $|\varphi\rangle_R \in \mathcal{H}_R$.

The Stinespring representation also leads to the notion of the *complement* of a given quantum channel $\mathcal{E}_{B|A}$. Instead of tracing out R to get back to $\mathcal{E}_{B|A}$ in (6.15), we could trace out B. This defines the complement $\widehat{\mathcal{E}}_{R|A}$. More properly, this construction leads to a whole set of complementary channels, since the Stinespring dilation is not unique. However, since all dilations are related by isometries involving the purifying system, all the possible complementary channels are isometrically related, making them essentially equivalent.

Exercise 6.7. Consider a channel $\mathcal{E}_{B|A}$ that has a fixed output independent of the input, i. e., $\mathcal{E}_{B|A} : \rho_A \mapsto \sigma_B$ for some σ_B and all ρ_A. Show that its complement is, up to equivalence, the identity channel $\mathcal{I}_A$.

Exercise 6.8. Show that the complement of the pinch map $\mathcal{P}_A$ with rank-one projectors is again $\mathcal{P}_A$.

Exercise 6.9. Show that the complement of the quantum erasure channel is again an erasure channel. What is the relationship between the erasure probabilities of the two channels?

Exercise 6.10. Show that for some appropriate normalized states $|\theta_x\rangle_R$, the complement of the qubit dephasing channel has Kraus operators $K_{R|A}(x) = |\theta_x\rangle_R \langle b_x|_A$, where $x \in \{0,1\}$ and $\{|b_x\rangle_A\}_{x=0}^1$ is the standard basis of A. Hence the complement is a CQ channel.

Exercise 6.11. Show that the complement of an entanglement-breaking channel is a Schur–Hadamard channel and vice versa. (See Exercises 5.25 and 5.24.)

Kraus representations, meanwhile, are based on pure state ensemble decompositions of the Choi operator. This has several implications. First, just as in (6.2), we can directly construct a Stinespring dilation from set of Kraus operators as follows. Suppose $\{K_{B|A}(x)\}_{x=1}^n$ is a set of Kraus operators of a channel $\mathcal{E}_{B|A}$. Then, for $\mathcal{H}_R$ of dimension n and an orthonormal basis $\{|b_x\rangle_R\}_{x=1}^n$,

$$V_{BR|A} = \sum_{x=1}^{n} K_{B|A}(x) \otimes |b_x\rangle_R \tag{6.17}$$

is a Stinespring dilation of $\mathcal{E}_{B|A}$. It is easy to verify that $V_{BR|A}$ leads to the correct channel action and that the trace-preserving condition from Theorem 5.3 implies that $V_{BR|A}$ is an isometry. The second immediate implication is that all Kraus representations must be unitarily related. The smallest number of Kraus operators is the rank of the Choi state, and such a Kraus representation is called *minimal*. Combining Proposition 6.3 with the Kraus operator construction in the proof of Theorem 5.3, we have the following:

Proposition 6.6 (Unitary relation of Kraus representations). *Let $\{K(i)\}_{i=1}^n$ and $\{K'(j)\}_{j=1}^m$ be two Kraus representations of the same superoperator $\mathcal{E}$, and set $\ell = \max(n,m)$. Then there exists an $\ell \times \ell$ unitary U with components U_{ji} such that $K'(j) = \sum_i U_{ji} K(i)$ for all $j \in \{1,\dots,m\}$. Furthermore, the minimal number of Kraus operators is no larger than $d_A d_B$.*

Exercise 6.12. Show that a Kraus representation is minimal iff the $K(j)$ are linearly independent.

However, we do not have to confine ourselves to pure state decompositions of the Choi operator. A general decomposition of $\mathrm{C}(\mathcal{E}_{B|A})$ gives, via the Choi isomorphism, a decomposition of $\mathcal{E}_{B|A}$ into a set $\{\mathcal{E}_{B|A}(x)\}_x$ of completely positive maps such that $\mathcal{E}_{B|A} = \sum_x \mathcal{E}_{B|A}(x)$. As at the end of Section 5.2, any such decomposition of $\mathcal{E}_{B|A}$ immediately gives an instrument $\mathcal{Q}_{XB|A}$ with classical X such that $\mathcal{E}_{B|A} = \mathrm{Tr}_X \circ \mathcal{Q}_{XB|A}$.

In analogy with the general steering result in Proposition 6.4, we can show that all possible instrument extensions $\mathcal{Q}_{XB|A}$ of $\mathcal{E}_{B|A}$ can be obtained by measurement on the purifying system R of any Stinespring dilation $V_{BR|A}$.

Proposition 6.7 (Steering channel decompositions). *Given a channel $\mathcal{E}_{B|A}$, let $V_{BR|A}$ be a Stinespring dilation, and let $\mathcal{Q}_{XB|A} = \sum_x |x\rangle\langle x|_X \otimes \mathcal{E}_{B|A}(x)$ be an instrument extension. Then there exists a measurement $\mathcal{M}_{X|R}$ such that for all $S_A \in \mathrm{Lin}(\mathcal{H}_A)$,*

$$\mathcal{Q}_{XB|A}[S_A] = \mathcal{M}_{X|R}\left[V_{BR|A}\, S_A\, V_{BR|A}^*\right]. \tag{6.18}$$

Proof. Let $|\Psi\rangle_{ABR} = V_{BR|A'}|\Omega\rangle_{AA'}$ be a purification of the Choi operator, and let $T_{AB}(x) = \mathsf{C}(\mathcal{E}_{B|A}(x))$. Since $\sum_x \mathcal{E}_{B|A}(x) = \mathcal{E}_{B|A}$, $\mathsf{C}(\mathcal{E}_{B|A}) = \sum_x T_{AB}(x)$. By Proposition 6.4 there exists a measurement $\Gamma_R(x)$ on R that realizes $T_{AB}(x)$ from Ψ_{ABR}, i. e., $T_{AB}(x) = \mathrm{Tr}_R[\Gamma_R(x)\,\Psi_{ABR}]$. Using this expression for $T_{AB}(x)$ in the Choi isomorphism, the output of $\mathcal{E}_{B|A}(x)$ applied to any S_A is

$$\begin{aligned}\mathrm{Tr}_{AR}[S_A^T \otimes \Gamma_R(x)\,\Psi_{ABR}] &= \mathrm{Tr}_{AR}[S_A^T \otimes \Gamma_R(x)\,V_{BR|A'}\Omega_{AA'}(V_{BR|A'})^*] \\ &= \mathrm{Tr}_R[\Gamma_R(x)\,V_{BR|A}\,S_A\,(V_{BR|A})^*]\,.\end{aligned} \tag{6.19}$$

Hence the sought after $\mathcal{M}_{X|R}$ is the measurement with POVM elements $\Gamma_R(x)$. □

By a similar argument steering provides us with a means of implementing any POVM by projective measurement in a larger space, a result known as the *Naimark*[2] *extension.* Let $\Lambda_A(x)$ be the POVM elements of $\mathcal{M}_{X|A}$ and define the isometry $V_{RA|A} = \sum_{x=1}^n |b_x\rangle_R \otimes \Lambda_A(x)^{1/2}$ as in (6.17). Then $\mathcal{M}_{X|A}[\rho_A] = \mathrm{Tr}_A \circ \mathcal{P}_{X|R}[V_{RA|A}\,\rho_A\,V_{RA|A}^*]$. The pinch map $\mathcal{P}_{X|A}$ and the partial trace Tr_A correspond to projective measurement by the projectors $\Pi_{AR}(x) = \mathbb{1}_A \otimes |b_x\rangle\langle b_x|_R$.

The original formulation of the Naimark extension is the statement that any POVM can be extended to a projection measurement in a larger space, where the projectors may be of arbitrary rank, but the larger space need not come from the tensor product of the original space with an *ancilla* (extra) system. In our presentation the projectors in the larger space all have rank equal to the dimension of A, since they are of the form $\mathbb{1}_A \otimes |b_x\rangle\langle b_x|$. Moreover, in the finite-dimensional case we are studying, it is also possible to find a Naimark extension of any POVM to a projective measurement consisting of *rank-one* elements, but we will not go into this here.

Exercise 6.13. Consider the mapping $V_{AR|A}$ defined by

$$\begin{aligned}|0\rangle_A &\to \tfrac{1}{\sqrt{2}}(|0\rangle_A|1\rangle_R + |0\rangle_A|2\rangle_R)\,, \\ |1\rangle_A &\to \tfrac{1}{\sqrt{6}}(2|1\rangle_A|0\rangle_R + |0\rangle_A|1\rangle_R - |0\rangle_A|2\rangle_R)\,.\end{aligned} \tag{6.20}$$

Confirm that $V_{AR|A}$ is an isometry and show that applying $V_{AR|A}$ to system A and then projectively measuring R in the standard basis implements the three-outcome measurement from Exercise 4.6.

6.5 Information disturbance

According to the uncertainty principle, quantum measurements invariably "disturb" or alter the system being measured in some way. Using the formalism we have now

---

2 Mark Aronovich Naimark, 1909–1978. Also transliterated as Neumark.

developed, we can illustrate two extreme versions of this phenomena. The first states that measurements that do not disturb the system at all do not reveal any information about it in the sense that the measurement result is completely independent of the input state.

Proposition 6.8 (No disturbance implies no information gain). *For every quantum instrument $\mathcal{Q}_{XA|A}$ satisfying $\mathrm{Tr}_X \circ \mathcal{Q}_{XA|A} = \mathcal{I}_A$, there exists a probability distribution P_X such that $\mathcal{Q}_{XA|A} = P_X \otimes \mathcal{I}_A$.*

Proof. Recall from Section 5.2 that we can regard the instrument $\mathcal{Q}_{XA|A}$ as the set $\{\mathcal{Q}_A(x)\}_x$ so that $\mathcal{Q}_{XA|A} = \sum_x |x\rangle\langle x|_X \otimes \mathcal{Q}_A(x)$. Since $\mathrm{Tr}_X \circ \mathcal{Q}_{XA|A} = \mathcal{I}_A$, the $\mathcal{Q}_A(x)$ form a decomposition of the identity channel: $\mathcal{I}_A = \sum_x \mathcal{Q}_A(x)$. The Choi operators $Q_{AA'}(x)$ associated with $\mathcal{Q}_A(x)$ must therefore satisfy $\Omega_{AA'} = \sum_x Q_{AA'}(x)$. Rescaling by the dimension of A transforms this to the statement that $\{\frac{1}{d}Q_{AA'}(x)\}_x$ form an ensemble decomposition of $\Phi_{AA'}$, where the probability $P_X(x)$ for the xth element is just $P_X(x) = \frac{1}{d}\mathrm{Tr}[Q_{AA'}(x)]$. Since pure states are extreme points, this can only hold if $Q_{AA'}(x) = P_X(x)\Omega_{AA'}$. Thus the instrument $\mathcal{Q}_{XA|A}$ has the claimed form. □

There cannot be any meaningful converse statement that no information gain implies no disturbance. For instance, both the classical and quantum outputs of an instrument could be fixed and independent, e. g., $\mathcal{Q}_{XB|A} : \rho_A \mapsto P_X \otimes \theta_B$ for some distribution P_X and state θ_B and all states ρ_A.

Exercise 6.14. Prove Proposition 6.8 by appealing to steering to recover the instrument from the Stinespring dilation, which for the identity channel has a simple form according to Exercise 6.6.

The second information-disturbance statement is the fact that rank-one projective measurements are completely disturbing in the sense that the output quantum state can always be generated from the measurement result itself. Thus the projection postulate is in some sense redundant.

Proposition 6.9 (Rank-one projective measurements are maximally disturbing). *Let $\mathcal{Q}_{XB|A}$ be a quantum instrument such that $\mathrm{Tr}_B \circ \mathcal{Q}_{XB|A} = \mathcal{P}_{X|A}$ for a pinch map $\mathcal{P}_{X|A}$ having rank-one projectors. Then there exists a set of density operators $\varphi_B(x)$ such that $\mathcal{Q}_{XB|A} = \mathcal{E}_{XB|X} \circ \mathcal{P}_{X|A}$ for the CQ channel*

$$\mathcal{E}_{XB|X} : |x\rangle\langle x|_X \mapsto |x\rangle\langle x|_X \otimes \varphi_B(x)\,. \tag{6.21}$$

Proof. The proof hinges on the fact that any Stinespring dilation $V'_{XBR|A}$ of $\mathcal{Q}_{XB|A}$ is also a dilation of $\mathcal{P}_{X|A}$, and therefore the two are isometrically related. Let $\mathcal{P}_{X|A}$ have Kraus operators $|x\rangle_X\langle x|_A$; a straightforward dilation is $V_{XA|A} = \sum_x |x,x\rangle_{XA}\langle x|_A$ as in (6.16). As $\mathcal{Q}_{XB|A}$ is a quantum instrument, it can be expressed as $\mathcal{Q}_{XB|A} = \sum_x |x\rangle\langle x|_X \otimes \mathcal{Q}_{B|A}(x)$ for some completely positive maps $\mathcal{Q}_{B|A}(x)$. Now suppose $U_{BR|A}(x)$ is a dilation of $\mathcal{Q}_{B|A}(x)$. A possible dilation of $\mathcal{Q}_{XB|A}$ itself is then $V'_{XX'BR|A} = \sum_x |x,x\rangle_{XX'} \otimes U_{BR|A}(x)$. This must also be a dilation of $\mathcal{P}_{X|A}$, and hence by Proposition 6.5 there exists a partial isometry $W_{X'BR|A}$ for which $V'_{XX'BR|A} = W_{X'BR|A} V_{XA|A}$. Applying $\langle x, x|_{XX'}$ to both sides yields $U_{BR|A}(x) = |\varphi(x)\rangle_{BR}\langle x|_A$, where we define $|\varphi(x)\rangle_{BR} = {}_{X'}\langle x|W_{X'BR|A}|x\rangle_A$. The trace-preserving condition for V' implies that the vectors $|\varphi(x)\rangle_{BR}$ are each normalized. Thus $\mathcal{Q}_{B|A}(x) : \rho_A \mapsto \varphi_B(x)\,\langle x|\rho|x\rangle_A$, and the desired result follows. □

Exercise 6.15. Consider the projective measurement instrument $\mathcal{Q}_{XA|A}$ from (5.8) with $\Pi_A(j) = |j\rangle\langle j|_A$. A Stinespring isometry is simply $W_{RA|A} = \sum_{j=1}^{d} |j\rangle_R \otimes \Pi_A(j)$, where $d = \dim(\mathcal{H}_A)$. Show that $W_{RA|A} = \sum_{x=1}^{d} |\tilde{x}\rangle_R \otimes V_A^x$ for V from (4.19) and $|\tilde{x}\rangle = \frac{1}{\sqrt{d}} \sum_{z=1}^{d} \omega^{xz}|z\rangle$ from (4.18).

6.6 Coherent classical information (?)

Now we turn to a potentially confusing issue in our formalism that also touches on one of the big outstanding issues in quantum mechanics, the measurement problem. The issue is the status of classical information in a quantum description. On the one hand, we treat classical information by diagonal (incoherent) quantum systems and measurements or instruments as producing such states. On the other hand, CQ states can always be purified, and by the Stinespring representation all QC channels can be dilated to fully coherent unitary transformations. Thus the status of the encoded information as "classical" is perhaps not so clear.

6.6.1 Classical information via copying

Let us examine the purification of a CQ state. To purify the state ρ_{XA} in (4.23), simply invent two additional systems B and X' and define

$$|\Psi\rangle_{XX'AB} = \sum_{x=1}^{n} \sqrt{P_X(x)}\,|b_x\rangle_X \otimes |b_x\rangle_{X'} \otimes |\varphi_x\rangle_{AB}\,, \tag{6.22}$$

where $|\varphi_x\rangle_{AB}$ is a purification of $\varphi_A(x)$. This is a coherent description, which is to say that there are no mixtures, only superpositions. System X' is identical in size to X and ensures that the marginal state ρ_{XA} is a CQ state with classical X. Equally well, then the state of $X'A$ is a CQ state with the classical information stored in X'. Thus classical information is information that can be, and in some sense has already been, copied to another system. Since there is nothing stopping us from appending another system X'' to generate an even larger purification, the number of additional systems storing the classical information is in principle unlimited. This accords nicely with the fact that classical information can be freely copied.

For quantum instruments, the situation is similar. Given an instrument $\mathcal{Q}_{XB|A}$ with Kraus operators $K_{B|A}(x) \otimes |x\rangle_X$ as in (5.22), following the recipe in (6.17) leads to the

Stinespring isometry

$$U_{XX'B|A} = \sum_x |x\rangle_X \otimes |x\rangle_{X'} \otimes K_{B|A}(x)\,, \tag{6.23}$$

so that $\mathcal{Q}_{XB|A}[\rho_A] = \mathrm{Tr}_{X'}[U_{XX'B|A}\,\rho_A\,U^*_{XX'B|A}]$. Tracing out X' again ensures that the X part of the output is diagonal. Indeed, we just made use of this in the proof of Proposition 6.9.

In this way the incoherent and coherent treatments of classical information are entirely consistent. Either classical information is directly treated as a system in an incoherent state, or it can be treated coherently by having (at least) two coherent copies of the information. Tracing out either copy restores the incoherent description, and purifying the incoherent description gives the coherent description. Modeling classical information in a coherent manner by having at least two copies is common in the literature on quantum information theory. In this view, classical information is epitomized by having an arbitrary number (possibly macroscopic) number of copies.

Nevertheless, we may feel that this picture cannot or ought not be fundamental. One important objection, though it is beyond the scope of this work, is that the picture does not work in the case of continuous variable systems. Apart from various technical issues, a more conceptual problem in that setting is that exactly copying a continuous value is not particularly realistic.

6.6.2 Classical information via observable restriction

A different, more fundamental perspective on the above treatment of classical information is that what we are really doing is restricting our attention to certain observables on all the "coherently classical" systems. To illustrate, consider the purification $|\Psi\rangle_{XX'AB}$ from (6.22) again. Tracing out X' in $|\Psi\rangle_{XX'AB}$ leaves the X part of the state classical, which is to say that only diagonal (and therefore commuting) observables on X are relevant. But there is really no need to go through the whole procedure of coherently appending copies just to trace them all away; we can just directly restrict attention to a commuting subset of observables without changing the state. This accords nicely with the discussion in Section 4.3, particularly (4.13).

For instance, observe that $\mathcal{Q}_{XB|A}$ could equally well be obtained by using the dilation $V_{BR|A}$ from (6.17) directly and then measuring the R system. The R measurement can be performed with the pinching operation, so that for all ρ_A, we have

$$\mathcal{Q}_{XB|A}[\rho_A] = \mathcal{P}_{X|R}[V_{BR|A}\,\rho_A\,V^*_{B|RA}]\,. \tag{6.24}$$

The important point is that the effect of measuring R is achieved by just restricting attention to diagonal observables on R. It is not necessary to actually have multiple copies around to treat classical information as we have done above, nor it is necessary to measure the R system itself with the pinching operation. Restricting to commuting observables on the system storing the classical information is sufficient to have a meaningful coherent (fully quantum) treatment of classical information.

That said, in the present finite-dimensional setting, there is really no difference between these two approaches. Note that everything ties together nicely because by treating the projection measurement on R coherently we recover $U_{XX'B|A}$ of (6.23) from the dilation of $\mathcal{Q}_{XB|A}$. Specifically,

$$U_{XX'B|A} = W_{XX'|R}V_{BR|A} \tag{6.25}$$

for the dilation $W_{XX'|R}$ of the pinch map from (6.16).

6.6.3 Consistency

In his treatise of quantum mechanics, von Neumann[3] gave two descriptions of the measurement process, of which (6.24) is the second, called Process 2. Process 1 corresponds to our original description using a quantum instrument and is the statistical description. Regarding the isometry $V_{BR|A}$ as arising from a unitary interaction of the system A to be measured and the measurement device R, Process 2 is the more dynamical, physical description of measurement. The fact that there are two descriptions is a version of the measurement problem. Of course, Process 2 also contains the projection onto R, which is a Process 1 measurement description. But (6.25) shows that the Process 1 and Process 2 are consistent, as the projective measurement on R can just as well be described by (another) Process 2.

In the description of measurement using $V_{BR|A}$, i. e., Process 2, it is crucial to include the restriction to commuting observables. This can be done with the pinching operation by appending an additional copy of the R system, or simply by only considering observables and operations on R that commute with these two operations. Not doing so yields a reversible description of the measurement process, which does not accord with our physical notion of measurement.

The difficulty is that the action of $V_{BR|A}$ can in principle be inverted using $V^*_{BR|A}$, though note that $V^*_{BR|A}$ does not give a Stinespring representation as it is only a partial isometry. To properly treat this case, we can first regard $V_{BR|A}$ as a unitary U_{ABR} acting on a fixed input $|0\rangle_B|0\rangle_R$. Then the action of the adjoint U^*_{ABR} defines a valid channel. For instance, in the simple case of a projective measurement with $V_{RA|A} = \sum_x |x\rangle_R \otimes |x\rangle\langle x|_A$, a possible unitary extension is $U_{AR} = \sum_{x=0}^{d-1} |x\rangle\langle x|_A \otimes |y+x\rangle\langle y|_R$, where addition inside the ket is modulo d, the dimension of system A. In the qubit case, this is just the CNOT gate. Notice that now the U^*_{AR} does not commute with the projectors $|y\rangle\langle y|_R$

3 John von Neumann, born Neumann János Lajos, 1903–1957.

needed for the full Process 2 measurement description, and therefore U_{AR} by itself cannot be regarded as a description of measurement.

6.6.4 The quantum eraser

It is a beautiful irony of quantum mechanics that if we do not include the restriction to commuting observables on R in the description of measurement, then the original measurement can be undone not just by inverting the measurement unitary as just discussed, but by a subsequent measurement of system R! This is called the *quantum eraser*, and it will actually be useful to us later in the construction of various quantum information processing protocols.

Consider an arbitrary pure qubit state $|\psi\rangle_A = a|0\rangle_A + b|1\rangle_A$ for $a, b \in \mathbb{C}$, which is measured in the standard basis by coupling to an ancilla qubit R and applying the CNOT gate. This results in the state $|\Psi\rangle = a|00\rangle_{AR} + b|11\rangle_{AR}$. Now consider a measurement of σ_x on system R. Projecting onto eigenstate $|\pm\rangle_R$ gives

$$_R\langle\pm|\Psi\rangle_{AR} = \tfrac{1}{\sqrt{2}}\big(a|0\rangle \pm b|1\rangle\big)_A \,. \tag{6.26}$$

The normalization of the state gives the probability, so the outcome of the σ_x measurement is completely random. Moreover, if the outcome is +, then system A has returned to the initial state $|\psi\rangle_A$. On the other hand, if the outcome is −, then an operation of σ_z on A will restore the initial state $|\psi\rangle_A$.

Thus U^*_{AR} is not necessary to reverse the action of U_{AR}, and instead a local measurement and a local unitary suffice. Again in this case the operation of the eraser does not commute with the projection operators $|x\rangle_R$, so we do not regard $|\Psi\rangle_{AR}$ as describing the output of a measurement process. The restriction to commuting observables on R allows us to avoid the apparent paradox that quantum measurements can be erased by further measurement.

Exercise 6.16. Give a fully quantum description of the quantum eraser.

Exercise 6.17. Consider a measurement device whose Process 2 description again utilizes the CNOT gate but starts with an ancilla prepared in the state $|\theta\rangle = \cos\theta|0\rangle + \sin\theta|1\rangle$. Compute the Kraus operators of the measurement and show that the channel $\mathcal{M}_{X|A}$ describing just the measurement result is a mixture of an ideal projective measurement and a measurement with constant output (i. e., independent of the input). Is the associated instrument a mixture of an ideal projective measurement and some other measurement?

6.7 Notes and further reading

The quotation from Pauli is from in the epigraph of [204]. The notion of purification goes back to Powers and Størmer [229], while the Schmidt decomposition is from [248]. Schrödinger's discussion of steering is from [249], and his subsequent work [250] showed the unitary freedom of ensemble decompositions. This was rediscovered by Jaynes [154], Gisin [111], and Hughston et al. [150]. The pretty good measurement was introduced by Belavkin [17] and rediscovered by Hausladen and Wootters [123], who coined the name. The use of the pretty good measurement for steering can be found in Wolf [307]. The Stinespring representation theorem is from [275]. Interested readers can find a proof along the lines of the original in Theorem 6.9 of [146] by Holevo. For more details on the Naimark's extension, see Peres [220, § 9-6] or Preskill [230, § 3.1.4]. Proposition 6.8 is adapted from Werner [299]. Von Neumann's two descriptions of measurement are from his treatise on quantum mechanics [293]. See also Bub [48] for a discussion. The quantum eraser was proposed by Scully and Drühl [256].

7 Quantum mysteries

> No reasonable definition of reality could be expected to permit this.
>
> Albert Einstein, Boris Podolsky, and Nathan Rosen

Now that we have completed the presentation of the formalism of quantum theory, we can look back and examine the differences from usual classical theory in more detail. The uncertainty principle, the structure of entangled states, and the lack of a unified Boolean structure all point to irreconcilable differences between quantum and classical. However, might there be some way, despite these apparent differences, to reconcile the two? To see quantum mechanics in the framework of classical mechanics and probability theory? The purpose of this chapter is to carefully explain that there is no simple, more classical formalism that encompasses all of quantum mechanics.

7.1 Complementarity

Let us begin by examining the earliest conceptual difference between classical and quantum mechanics, namely complementarity, and show how it can be concretely modeled in our quantum formalism. The notion of complementarity stems from the rather ancient question about the nature of light, whether it is a particle or a wave. But first consider the difference between bits and qubits. Both bits and qubits are defined as systems with two well-defined "levels" or configurations. What distinguishes a bit from a qubit is the ability to realize superposition states and measurements of noncommuting observables. This distinction is generally referred to as "coherence".

For instance, even though the qubit state $|+\rangle$ looks like a probabilistic mixture of $|0\rangle$ and $|1\rangle$, and does give uniform probability of outcome when measuring with projectors in this basis, it is not a mixture at all, since the probability of the projection measurement of $|+\rangle$ versus $|-\rangle$ (i. e., measurement of σ_x) is certain. This distinction is what is meant by saying that $|+\rangle$ is a *coherent* combination of $|0\rangle$ and $|1\rangle$. An incoherent combination of these alternatives is simply the probabilistic mixture, which would have $\Pr[+] = 1/2$. In classical probability a coherent mixture cannot occur: If a classical bit has probability $1/2$ to take either value, then there is no other elementary event, like $\Pi(+)$, for which the probability is one. Quantum mechanically, however, we can observe the relative phase $+1 = e^{i0}$ or $-1 = e^{i\pi}$ between the two elementary states $|0\rangle$ and $|1\rangle$ by making the σ_x measurement.

Another way to describe this phenomena is to say that σ_x and σ_z are *complementary* observables or *conjugate* observables. We have just seen that an eigenstate of one of these observables has no definite value of the other, as a measurement in the other basis gives a completely random result. Indeed, there is no vector that is an eigenstate of both σ_x and σ_z, since they do not commute. One property or the other can be realized, but not both, making them complementary.

https://doi.org/10.1515/9783110570250-007

However, quantum mechanics is even stranger than this. Alternately measuring the two observables can lead to a situation in which a once-certain outcome is made random. After a σ_z measurement of $|+\rangle$, for instance, the state is either $|0\rangle$ or $|1\rangle$, both of which have only a probability of $1/2$ of returning + in a subsequent measurement of σ_x. Without the intervening σ_z measurement, the result would of course be + with certainty. This is quite unlike the classical case, where we may imagine that each measurement just reveals more and more properties about the system (or to a higher and higher precision) and need not itself cause any change to the system. Moreover, in the classical framework, making a measurement and forgetting the result can be thought of as not having performed the measurement at all, by the law of total probability.

Putting things the other way around, we cannot so easily use the probabilistic framework we have developed to model the complementarity of σ_x and σ_z. Suppose X and Z are random variables that describe the values of σ_x and σ_z, respectively. Then the intervening measurement of σ_z fixes the value of Z, and we will have some conditional distributions $P_{X|Z=z}$ for the result of the final σ_x measurement. Apparently, these are both just uniform distributions. To model the absence of the intervening measurement, we would just average the conditional distributions of X over the value of $Z = z$. This clearly does not produce the deterministic value of X.

The complementarity of σ_x and σ_z is the same complementarity that manifests itself in the famous double-slit experiment, devised by Young[1] to demonstrate the wave nature of light. Complementarity of the particle and wave nature of light is one of the most well-known examples of the difference between classical and quantum mechanics. Indeed, Feynman[2] starts off his treatment of quantum mechanics in his famous lectures with a treatment of the double-slit experiment, stating

> ...we shall tackle immediately the basic element of the mysterious behavior in its most strange form. We choose to examine a phenomenon which is impossible, *absolutely* impossible, to explain in any classical way, and which has in it the heart of quantum mechanics. [100]

Let us see how the double slit illustrates what Feynman calls the "basic peculiarities of all quantum mechanics" in our general quantum formalism. For simplicity, consider a single photon in a balanced Mach[3]–Zehnder[4] interferometer, depicted in Figure 7.1. The two paths the photon could take through the interferometer constitute a basis for a qubit. Let us label the paths 0 and 1 so that a basis is just $|0\rangle$ and $|1\rangle$. The paths are defined as if the beamsplitters were mirrors. After the first beamsplitter, the photon is in a superposition of the two paths described by the state $|+\rangle$. This is the analog of the light immediately after the slits in the usual double-slit experiment. We can model

1 Thomas Young, 1773–1829.

2 Richard Phillips Feynman, 1918–1988.

3 Ludwig Mach, 1868–1951.

4 Ludwig Louis Albert Zehnder, 1854–1949.

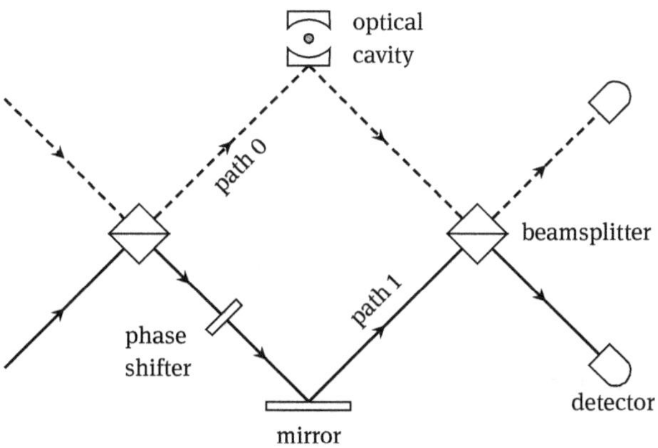

Figure 7.1: Schematic of a Mach–Zehnder interferometer, nominally consisting of two mirrors and two beamsplitters. This defines two optical paths, indicated by the dashed and solid lines. The phase convention of the beamsplitters is such that photons exit the interferometer on the same path they entered, to be detected by the photodetectors. Inserting a phase shifter into one path alters this behavior. By replacing one mirror with an optical cavity containing a trapped atom, a passing photon can in principle be detected without destroying it, determining the path taken by the photon.

the beamsplitter by the Hadamard unitary operator $H = \frac{1}{\sqrt{2}}\left(\begin{smallmatrix} 1 & 1 \\ 1 & -1 \end{smallmatrix}\right)$ and the preparation of the state as inputting a single photon to one port, $|0\rangle$ of the beamsplitter, since $|+\rangle = H|0\rangle$.

We arrange the second beamsplitter so that when the light recombines there, it exits the interferometer in the same path that it entered. Then we can also model the second beamsplitter by a Hadamard operator, since $H^2 = \mathbb{1}$. If we insert a π phase shifter into one arm prior to the second beamsplitter, then the light will exit the interferometer along the other path. This is modeled by describing the action of the phase shifter (a dynamical evolution) by the unitary operator σ_z, which creates the state $|-\rangle$. Then the Hadamard operation describing the second beamsplitter produces the state $|1\rangle$ at the output.

Suppose we determine which path the photon exits the interferometer by placing photodetectors in both outgoing paths. This is the analog of the photographic film in the double-slit experiment. Altogether, the second beamsplitter and photodetectors act as a σ_x measurement, determining the relative phase between the two paths. The fact that the relative phase takes a well-defined value is the coherence of the quantum state.

On the other hand, if we measure which path the photon takes in the interferometer before allowing the two paths to recombine at the second beamsplitter, then the state is changed to either $|0\rangle$ or $|1\rangle$, corresponding to the measurement outcome. Such a measurement can in principle be made without destroying the photon, for instance, by replacing one of the mirrors with a small optical cavity containing a trapped atom and then appropriately measuring the atom. The second beamsplitter will now produce a superposition at the output of the interferometer, which will end up as a random outcome of the photodetectors since there is only one photon.

Thus the measurement of which path the photon took washes out the interference pattern. One says that the coherence has been destroyed, because there is no longer any way to determine the relative phase of the input state. As anticipated, path and phase are complementary observables, and this reflects the wave–particle duality of quantum mechanics since a definite path is a particle property, whereas a definite phase is a wave property. Of course, as we know from Section 6.6, the coherence is not actually destroyed; it has merely been converted to entanglement with the measurement apparatus, e. g., as in the CNOT example in Section 6.6.4.

7.2 Hidden variables

The double slit experiment vividly highlights the essential question that stems from the phenomenon of complementarity. How should we think of the values of σ_x and σ_z or of path and phase information at different stages of the interferometer? Does the photon possess both properties, and complementarity means that it will never tell us both, one property remaining forever hidden? Or is it the case that somehow the photon does not have both properties at the same time? Is our attention to properties of individual photons perhaps itself incorrect? As we saw above, a straightforward probabilistic model cannot reproduce the appearance of interference as a mixture of the two possible interferometer outputs given a particular path.

7.2.1 Hidden variables for the interferometer

A simple model for the first possibility above is that any measurement device just unavoidably "kicks" the system being measured in some way, so that both complementary properties exist but are never simultaneously revealed. For the interferometer, we can imagine that the path measurement somehow alters the photon. It is actually not too difficult to come up with a *nondeterministic* classical probabilistic model for the above phenomenon. Consider again two random variables X and Z but now suppose that $|0\rangle$ corresponds to $Z = 0$ and random X, i. e., $P_{XZ}(x,z) = \frac{1}{2}\delta_{z,0}$, $|1\rangle$ to $Z = 1$ and random X, $|+\rangle$ to $X = 0$ and random Z, and $|-\rangle$ to $X = 1$ and random Z. The random variables are meant to correspond to the values of σ_z and σ_x, and the model is nondeterministic because the various quantum states are not represented by deterministic random variables.

If we stipulate that a σ_x measurement reveals the value of X but randomizes the value of Z, and vice versa for σ_z, then we can recover the interference effect. Measuring the state $|+\rangle$ in the basis of σ_x corresponds to asking for the value of X in the distribution $P_{XZ}(x,z) = \frac{1}{2}\delta_{x,0}$, so the outcome is deterministic. A σ_z measurement on the same state results in a random value of Z and randomizes the value of X in the process. Therefore the final σ_x measurement will also be random, just as in the quantum case.

But notice one consequence of this model: The "back-action" of the measurement is not local, it must occur in both arms. In this model the state of the system after the first beamsplitter is a fixed value $X = 0$ of the phase and a random value of Z, the path. If the photon travels through the cavity making the path measurement, then the notion that the cavity affects the value of X is not problematic. However, the interference is also destroyed when the cavity detects no photon, meaning it traveled along the other arm of the interferometer. Now we are in an awkward situation that there is no apparent physical mechanism to perform the randomization of X in the other arm. Thus we say that the effect is nonlocal. Although complementarity can be modeled by classical random variables, doing so appears to involve some degree of nonlocality. Note that this is not saying that quantum mechanics itself is somehow nonlocal, only that this particular hidden variable model is. This state of affairs is summed up nicely by Einstein:[5]

> Dirac... rightly points out that it appears, for example, to be by no means easy to give a theoretical description of a photon that shall contain within it the reasons that determine whether or not the photon will pass a polarizator set obliquely in its path. [89]

7.2.2 Local hidden variables for the interferometer

However, the nonlocality of the model arises because we too quickly focused on the single-photon aspect of the experiment. There we regard the hidden variables as path and phase properties of the photon that traverses the interferometer, even though these suffer from nonlocal influences. But a more straightforward starting point is to shift the focus to the properties of the modes and only afterward restrict to the case of having a single photon. Instead of labeling the paths 0 and 1, let us now call them A and B, and instead of $|0\rangle_P$ and $|1\rangle_P$ representing the photon traveling either in arm A or B, we can use states $|1\rangle_A|0\rangle_B$ and $|0\rangle_A|1\rangle_B$ for the same thing. For systems A and B, $|0\rangle$ represents the vacuum state in the mode, and $|1\rangle$ represents a single photon. In principle, there are also states corresponding to two, three, and more photons, but we will not need those here.

To formally translate the previous single-photon description of the interferometer to the two-mode description, we can use a CNOT gate plus an ancilla qubit Q in the $|1\rangle$ state. Controlling on system P and associating the P output with B and the Q output with A, the CNOT gate $U_{AB|PQ}$ exactly produces $|1\rangle_A|0\rangle_B$ from $|0\rangle_P$ and $|0\rangle_A|1\rangle_B$ from $|1\rangle_P$, as intended. We can use this to translate all the other elements of the single-photon

5 Albert Einstein, 1879–1955.

description to the two-mode description. A phase shift by π in one of the arms is represented by σ_z (on P) in the single-photon model and now becomes $U_{AB|PQ}(\sigma_z)_P U^*_{AB|PQ} = (\sigma_z)_A$. Confined to the single-photon sector, this has the same action as $(\sigma_z)_B$, which reflects the fact that only the relative phase between the arms matters. Similarly, the action of either beamsplitter is now $U_{AB|PQ} H_P U^*_{AB|PQ}$. This produces $|\Psi_+\rangle_{AB}$ for input $|1\rangle_A|0\rangle_B$ and maps $|\Psi_+\rangle_{AB}$ back to $|1\rangle_A|0\rangle_B$ at the second beamsplitter. A π phase shift in either arm produces $|\Psi_-\rangle$, which is then transformed by the second beamsplitter into $|0\rangle_A|1\rangle_B$. Measuring the photon number in either arm, again with the optical cavity, will collapse the state to either $|0\rangle_A|1\rangle_B$ or $|1\rangle_A|0\rangle_B$, and then the interference at the output disappears.

Exercise 7.1. Confirm the details of the translation.

Now consider a pair of binary random variables X and Z for each mode, with the same association of individual quantum states and distributions as before. The value of Z encodes the presence of a photon in the mode, which we call the amplitude of the field (even though "intensity" might be more precise). The value of X is less clear at this point, because it has to do with the relative phase of the vacuum and single photon states, but let us just think of it as the phase of the field. This is borne out by the action of a π phase shifter in a mode. Above we saw that this is described by σ_z acting on either A or B. This interchanges $|+\rangle$ and $|-\rangle$ and so corresponds to flipping the value of X.

Next, we need to model the action of the beamsplitter in terms of the random variables. It is a reversible operation, so the random variables associated with the output will be deterministic functions of the input random variables. As our random variables are meant to be the values of σ_x and σ_z, we can read off the proper transformation from the action of $U_{AB|PQ} H_P U^*_{AB|PQ}$. Then, for instance, $(\sigma_x)_B$ at the output of the beamsplitter corresponds to $(\sigma_x)_B$ at the input, and so the value of X_B is unchanged. On the other hand, $(\sigma_x)_A$ at the output corresponds to $(\sigma_z)_A \otimes (\sigma_x)_B$, so X_A is transformed to $Z_A + X_B$ (modulo 2). Actually, let us modify this slightly and say that X_A is transformed to $Z_A + X_B + 1$. The two Z variables are transformed as follows: $Z_A \to X_A + X_B + 1$ and $Z_B \to X_A + X_B + Z_A + Z_B + 1$.

Exercise 7.2. Confirm that this mapping is its own inverse.

Our interpretation of X as the phase of the mode is further bolstered by the fact that the amplitude in the output of A is determined by the relative phase of the two inputs. Also note that the total photon number is preserved, as it should be. For an input of $Z_A = 1$ and $Z_B = 0$, with the phases random, the output of the first beamsplitter will have Z_A and Z_B anticorrelated but completely random, and similarly X_A and X_B are correlated but completely random.

Exercise 7.3. Show that if we drop the +1s from the transformation rules, then X_A and X_B end up anticorrelated after the beamsplitter.

The second beamsplitter will simply reverse this operation, restoring the original input. Furthermore, a phase shift in one arm will shift the value of $X_A + X_B$, and therefore the photon will end up in the other output mode.

Now consider the effects of a measurement of the photon number in one arm, say A. The value of Z_A will be determined, and the value of X_A will be randomized. If the photon is found in arm A, so that $Z_A = 1$, then we can infer that $Z_B = 0$. The back-action of the measurement decorrelates the phases, so that each is now random. Therefore the photon will exit the interferometer in a randomly chosen mode.

That is all well and good when the photon is found by the measurement, but what happens when the measurement reveals the photon is in the other arm? How does the back-action of the measurement *locally* ensure that the photon interference is destroyed? In the standard quantum-mechanical treatment, the wavefunction collapses to the photon traveling in the other arm. Then the photon meets the vacuum at the beamsplitter, a combination that results in exit of the interferometer through a random output port. Here the story is precisely the same: The vacuum carries the random phase information to the second beamsplitter, producing a random path at the output of the interferometer.

Now it is easy to see that, by focusing to narrowly on the photon properties, the earlier model simply did not have enough degrees of freedom to make the measurement back-action work locally. Locality is restored by broadening the possibilities of the model to include the vacuum doing something meaningful. We have found a local hidden variable model for the double slit experiment, at least a simplified version of it. Feynman's claim that the phenomenon of the double slit experiment is absolutely impossible to explain in any classical way is starting to look shaky. True, we have made use of measurement back-action, which is also not classical, but it seems that we have come a long way to "understanding" quantum mechanics.

7.3 Bell's theorem and the CHSH inequality

Alas, this kind of model cannot be extended to cover arbitrary quantum phenomena. This is a consequence of Bell's theorem, which states that quantum mechanics is capable of statistical predictions that are incompatible with classical probabilistic models that are local. A simplified version of the argument put forth by Clauser,[6] Horne,[7] Shimony,[8] and Holt[9] already demonstrates this conclusion for a modification of the

6 John Francis Clauser, born 1942.

7 Michael A. Horne, 1943–2019.

8 Abner Shimony, 1928–2015.

9 Richard Arnold Holt.

two-mode interferometer setup. Their argument makes use of the so-called *CHSH inequality*, which can be illustrated using the *CHSH game*.

The CHSH game involves two players, Alice and Bob. They each receive a binary input and are to each produce a binary output without communicating with each other. Call Alice's input x and output a, Bob's input y and output b, and for simplicity, take them all to be elements of $\mathbb{Z}_2$. The goal of the game is to produce outputs a and b that satisfy $a + b = xy$.

Suppose that each of their outputs is a deterministic function of the input, and denote Alice's output for input x as a_x and so forth. It is easy to see that in this case, there is no way to win the game for all possible inputs x and y. The constraint $a_x = b_y$ for $x \neq y$ requires that all values are equal, which then violates the constraint $a_x \neq b_y$ for $x = y$. That is, from the first we have $a_0 = b_0$, $a_0 = b_1$, and $a_1 = b_0$, which implies $a_0 = a_1 = b_0 = b_1$. At best, Alice and Bob can win with probability 3/4 by satisfying three of the constraints. For instance, $a_0 = 1$, $a_1 = 0$, $b_0 = 0$, and $b_1 = 1$ obeys $a_x + b_y = xy$ in only three cases, with $x = y = 0$ giving $a_0 + b_0 = 1$. Similarly, $a_0 = 0$, $a_1 = 1$, $b_0 = 0$, and $b_1 = 1$ fails in three cases, only $x = y = 0$ being correct.

More generally, we might consider that Alice's and Bob's outputs are not completely determined by their respective inputs, but are instead given by a conditional probability distribution $P_{AB|X=x,Y=y}$. To model local but correlated strategies, we suppose that the conditional distribution takes the form

$$P_{AB|X=x,Y=y}(a,b) = \sum_{\lambda} P_\Lambda(\lambda) P_{A|X=x,\Lambda=\lambda}(a) P_{B|Y=y,\Lambda=\lambda}(b) \tag{7.1}$$

for some random variable Λ. The additional Λ models the choices they might make before starting the game, and the fact that a (b) is generated only using x and λ (y and λ) reflects the fact that they cannot communicate during the game. When the inputs x and y are chosen uniformly at random, independently of Λ, the overall winning probability takes the form

$$\Pr[A + B = X \cdot Y] = \tfrac{1}{4} \sum_{a,b,x,y} \delta_{a+b=x\cdot y} \sum_{\lambda} P_\Lambda(\lambda)\, P_{A|X=x,\Lambda=\lambda}(a)\, P_{B|Y=y,\Lambda=\lambda}(b)\,. \tag{7.2}$$

By Proposition 3.3 Alice's and Bob's conditional distributions (channels) are convex combinations of deterministic functions, and therefore we can just as well attribute the randomness in the choice of deterministic function to Λ. That is, there exists a random variable Z, independent of X and Λ, such that $P_{A|X=x,\Lambda=\lambda}(a) = \sum_z P_Z(z) 1[f_z(x,\lambda) = a]$. The same is true at Bob's end using an independent random variable Z'. Then we can just define Λ' to be the collection of Λ, Z, and Z', at which point (7.2) becomes a mixture of local, deterministic strategies. Therefore the guessing probability in this probabilistic setting inherits the previous bound on the winning probability. This is the CHSH inequality:

$$\Pr[A + B = XY] \leq \frac{3}{4}. \tag{7.3}$$

The CHSH inequality can be violated in quantum mechanics by making use of entangled states of just two qubits. In this setting, it is more convenient for the outcomes to be labeled by ±1, so we define $a'_x = (-1)^{a_x}$ and $b'_y = (-1)^{b_y}$. The observables corresponding to a'_x and b'_y are associated with Bloch vectors $\hat{a}_x$ and $\hat{b}_y$ so that $a'_x = \hat{a}_x \cdot \vec{\sigma}$ and similarly for b'_y. For a given joint state $|\Psi\rangle_{AB}$ and choice of x and y, let us write $\langle a'_x b'_y\rangle$ for $\langle\Psi|(\hat{a}_x \cdot \vec{\sigma}_A)(\hat{b}_y \cdot \vec{\sigma}_B)|\Psi\rangle_{AB}$. Then $\langle a'_x b'_y\rangle = \langle(-1)^{a_x+b_y}\rangle$. When $x = y = 1$, the value of $(-1)^{a_x+b_y}$ should be -1 to win, but in every other case, it should be $+1$. If we denote the probability that $a_x + b_y = x \cdot y$ for given x and y by p_{xy}, in terms of a'_x and b'_y, we have $p_{xy} = \frac{1}{2}(1 + \langle a'_x b'_y\rangle)$ for the case that $x \cdot y = 0$, but $p_{11} = \frac{1}{2}(1 - \langle a'_1 b'_1\rangle)$. Therefore the overall probability of winning is just

$$\Pr[A + B = XY] = \tfrac{1}{8}(4 + \langle a'_0 b'_0\rangle + \langle a'_0 b'_1\rangle + \langle a'_1 b'_0\rangle - \langle a'_1 b'_1\rangle). \tag{7.4}$$

Now suppose $|\Psi\rangle_{AB} = \frac{1}{\sqrt{2}}(|01\rangle - |10\rangle)_{AB}$, which we also defined as $|\Phi_{11}\rangle$. It is not difficult to verify that $|\Psi\rangle$ has the property that for any unitary U with determinant equal to one,

$$U_A \otimes U_B|\Psi\rangle_{AB} = |\Psi\rangle_{AB}. \tag{7.5}$$

(This is the statement that $|\Psi\rangle$, which is the spin singlet, is rotationally invariant.) A simple calculation gives

$$\langle\Psi|(\hat{a} \cdot \vec{\sigma}_A)(\hat{b} \cdot \vec{\sigma}_B)|\Psi\rangle_{AB} = -\hat{a} \cdot \hat{b}. \tag{7.6}$$

Exercise 7.4. Show (7.5) and (7.6).

Therefore from (7.4) we obtain

$$\Pr[A + B = XY] = \tfrac{1}{8}(4 - \hat{a}_0 \cdot \hat{b}_0 - \hat{a}_0 \cdot \hat{b}_1 - \hat{a}_1 \cdot \hat{b}_0 + \hat{a}_1 \cdot \hat{b}_1). \tag{7.7}$$

Choosing $\hat{a}_0 = \hat{x}$, $\hat{a}_1 = \hat{y}$, $\hat{b}_0 = \frac{1}{\sqrt{2}}(\hat{x} + \hat{y})$, and $\hat{b}_1 = \frac{1}{\sqrt{2}}(\hat{x} - \hat{y})$ gives $\Pr[A + B = X \cdot Y] = \frac{1}{2}(1 + \frac{1}{\sqrt{2}})$. Since this is larger than 3/4, we have Bell's theorem: *No local probabilistic model like that of* (7.2) *can reproduce all of the predictions of quantum mechanics.*

Note that if we apply Proposition 3.3 to a general $P_{AB|X=x,Y=y}$, then we decompose it into functions $f_z : (x, y) \mapsto (a, b)$, which require communication between Alice and Bob. For instance, clearly, the conditional distribution $P_{AB|X=x,Y=y}$ in which $a = 0$ and $b = xy$ will lead to a winning probability of one, as would $b = 0$ and $a = xy$. Both of these are nonlocal in that they allow for instantaneous signaling from one party to the other. For instance, in the former case, changing the value of x will change the value of b. This enables Alice to send a one bit message to Bob, and since nothing in the measurement setup specifies that Bob cannot be spacelike separated from Alice, in principle, this would allow for superluminal communication.

Interestingly, the game can be always be won by mixing the above two deterministic conditional distributions, but the result does *not* allow signaling. Their mixture is

$$Q_{AB|X=x,Y=y}(a,b) = \tfrac{1}{2}\delta_{a+b,xy}\,, \tag{7.8}$$

which satisfies $Q_{A|X=x,Y=y}(a) = Q_{A|X=x}(a)$, and similarly for B.

Exercise 7.5. Show this.

Therefore Alice's output A only depends on her input X, and no signaling is possible from Bob to Alice. As much holds for the Alice to Bob direction as well. Nevertheless, there is something nonlocal about this distribution, as it cannot be written in the form of (7.2).

Exercise 7.6. Show that the CHSH inequality also holds if we require the conditional distribution to satisfy

$$P_{A|B=b,X=x,Y=y}(a) = P_{A|X=x}(a) \quad \text{and} \quad P_{B|A=a,X=x,Y=y}(b) = P_{B|Y=y}(b)\,, \tag{7.9}$$

or if we include hidden variables and require

$$\begin{aligned} &P_{A|B=b,X=x,Y=y,\Lambda=\lambda}(a) = P_{A|X=x,\Lambda=\lambda}(a) \quad \text{and} \\ &P_{B|A=a,X=x,Y=y,\Lambda=\lambda}(b) = P_{B|Y=y,\Lambda=\lambda}(b)\,. \end{aligned} \tag{7.10}$$

7.3.1 Further implications

In Section 4.3, we saw that the quantum formalism is not described by a single Boolean algebra, but by many, one for every possible measurement. Indeed, the quantum formalism itself is the means of relating all the algebras together. As we saw with the double-slit experiment, demanding some simpler, more classical relation immediately runs into trouble. We cannot average the behavior of the interferometer when the photon takes a fixed path to reproduce the interference fringes.

Bell's theorem definitively rules out the simplest thing we could hope for, a single overarching Boolean algebra that can describe all measurements (as in the classical case), since this would satisfy the CHSH inequality. True, we can avoid this conclusion by imagining that the action of measurement necessarily has a back-action on the probability distribution, but this also takes us away from the simplest thing we could hope for, just in a different direction. Even then we will have to be content with the underlying properties specified by the events of the algebra being nonlocal (except in some simple cases, like the Mach–Zehnder interferometer, where the properties are again local).

In either case, we see that Bell's theorem is saying something quite profound: *It is ultimately incorrect to draw any implications from the results of measurements that are not actually performed.* That is to say, there is no guarantee that doing so will make sense, as in general, there is simply no overarching Boolean algebra in which to carry out the logical implications of a particular measurement having a certain result. If there always were, we would again arrive at the CHSH inequality. In employing the quantum formalism (or a nonlocal hidden variable theory), we are forced to decide whether to include a measurement under consideration—and its attendant back-action—or not, but we cannot do both. Peres[10] put it succinctly in the title of a paper on the subject of Bell inequalities: *Unperformed experiments have no results* [219].

7.4 Notes and further reading

The quote at the start of the chapter is from Einstein, Podolsky, and Rosen's seminal 1935 paper questioning the completeness of quantum mechanics [90]. The importance of complementarity in quantum mechanics was famously stressed by Bohr [41], but it was Einstein [88] who first realized that both wave and particle aspects of light are necessary for a correct thermodynamical treatment (specifically, of fluctuations in radiation pressure). For more on Einstein's role, see Stone's delightful history [276].

The Mach–Zehnder interferometer was originally introduced by Zehnder [314] and improved by Mach [197]. Our "qubit" treatment of complementarity in terms of the Mach–Zehnder interferometer is indebted to Englert [94]. See also Busch and Shilladay [50]. For the possibilities of measuring the photon nondestructively, see Reiserer, Ritter, and Rempe [236]. The nonlocal hidden variable model for the interferometer is an instance of Spekkens' "toy theory" from [270]. He presented the local hidden variable model for the interferometer in [271].

The literature on Bell inequalities is vast. It begins with Bell's original inequality [19] and discussion of earlier purported hidden variable no-go theorems in [20]. The CHSH appeared shortly thereafter [61]. For more, the two conference proceedings [36, 37] are a good place to start; the contribution by Wiseman and Cavalcanti in the latter was particularly illuminating to the author. The example of (7.8) is from Popescu and Rohrlich [226].

---

10 Asher Peres, 1934–2005.

Part II: **Resource measures**

8 Basic resources

> The algebraic sum of all the transformations occurring in a cyclical process can only be positive, or, as an extreme case, equal to nothing.[1]
>
> Rudolf Clausius

In analyzing any given information processing task from the resource simulation approach, there are two main questions. First, what resources are absolutely necessary for performing the task? Second, can we actually construct protocols that achieve these limits? In the world of information theory the latter is usually referred to as an achievability statement or achievability bound, while the former is termed a "converse" bound in the sense of being a converse to the achievability statement. The terminology goes back to Shannon. A converse bound quantifies the properties needed by any collection of actual resources that can simulate a chosen ideal resource. The reader is undoubtedly already familiar with the two fundamental converse and achievability statements in thermodynamics: The second law, which limits the efficiency of heat engines, and the corresponding achievability statement of the Carnot[2] cycle.

The purpose of this chapter is to examine the simplest and most immediate converses on classical and quantum communication. For instance, we certainly expect that a single bit classical channel, i. e., one transforming one bit to one bit, cannot reliably transmit more than one bit of information. Similarly, we expect a single qubit channel cannot reliably transmit more than one qubit worth of information. We expect that a classical bit channel cannot reliably transmit any number of qubits, and perhaps we are unsure what to think about whether or not a qubit channel can transmit more than one classical bit. The first two statements would appear to be true on their face, but the latter two are not as immediate. Fortunately, all have simple answers, which we can already formulate.

Moreover, we can also consider the case of either kind of communication assisted by additional resources such as shared randomness or shared entanglement. Nominally, there are eight possible simple scenarios corresponding to the choice of which kind of information (classical or quantum), which kind of channel (classical or quantum), and which kind of assistance (shared randomness or shared entanglement). However, it happens that shared randomness never helps, leaving just four cases. Even then, shared entanglement only helps in the two "mixed" scenarios of classical communication over quantum channels and quantum communication over classical channels. The former is realized by the protocol of superdense coding and the latter by teleportation.

---

1 Die algebraische Summe aller in einem Kreisprocesse vorkommenden Verwandlungen kann nur positiv oder als Gränzfall Null seyn. [62, p. 109]

2 Nicolas Léonard Sadi Carnot, 1796–1832.

https://doi.org/10.1515/9783110570250-008

Finally, we also consider a task whose goal is not communication but rather "distillation", namely the production of maximally entangled states from available bipartite resource states. Here we imagine that the two constituents of the resource states are some distance from each other, so that performing joint quantum operations on both at once is not possible. Indeed, being able to do so would trivialize the problem, as we can simply create the desired entangled states directly. Instead, the protocol should distill the entanglement already present in the resource states by performing local operations on each of the constituent systems and making use of classical communication between the two locations. This set of operations is usually abbreviated LOCC. Here we will show that there is a sensible notion of "amount of entanglement" of some given resources in the sense that LOCC protocols cannot output a maximally entangled state of larger dimension than the resources used to create it.

8.1 Converses for classical communication

8.1.1 Over classical channels

Let us first consider the case of transmitting classical information over classical noisy channels, as depicted in Figure 1.1. According to the formalism developed in Part I, the messages and channel inputs and outputs are modeled by random variables, while the encoder, noisy channel, and decoder are modeled by classical channels. Denoting the input (output) message random variables by M (M') and by X (Y) the channel input (output), for a given noisy channel $N : X \to Y$, the sender and receiver would like to find encoding and decoding operations $E : M \to X$ and $D : Y \to M'$ such that the combination $D \circ N \circ E$ is essentially the identity operation from M to M'. This is depicted below; the equality symbol = denotes the identity channel.

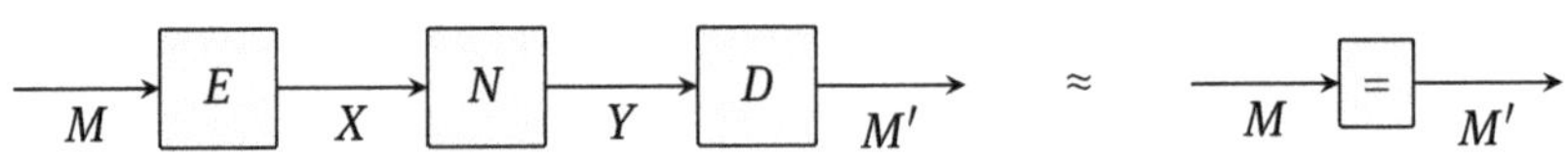

The approximate equality ≈ in the diagram indicates that we do not expect to simulate the ideal channel exactly. We will develop the tools needed to properly treat the approximation and be able to state how good the approximation can be for a fixed noisy channel in the following chapters. However, we can already establish the simple and intuitive statement that an identity channel on an alphabet $\mathcal{X}$ cannot reliably transmit more than $|\mathcal{X}|$ messages.

To make a concrete statement, let us consider the average probability that the input to a channel agrees with its output. In some sense, this measures how close the channel is the identity channel, for which the agreement probability is obviously equal to unity. For an arbitrary classical channel $W_{Y|X}$ with $\mathcal{Y} = \mathcal{X}$, the agreement probability $P_{\text{agree}}(W_{Y|X})$ is naturally defined as

$$P_{\text{agree}}(W_{Y|X}) := \frac{1}{|X|} \sum_{x\in\mathcal{X}} W_{Y|X}(x,x)\,. \tag{8.1}$$

Now we can consider the communication scenario with $W_{M'|M} = D_{M|Y} \circ N_{Y|X} \circ E_{X|M}$. Then

$$P_{\text{agree}}(W_{M'|M}) = \frac{1}{|M|} \sum_{m\in\mathcal{M}} \sum_{x\in\mathcal{X}} \sum_{y\in\mathcal{Y}} D_{M'|Y}(m,y)\, N_{Y|X}(y,x)\, E_{X|M}(x,m)\,. \tag{8.2}$$

The encoder $E_{X|M}$ outputs an element of Prob(X) for every input $M = m$, so it follows that $E_{X|M}(x,m) \le 1$ for all x and m. Since the other contributions to the summation are all positive, using this inequality in (8.2) gives

$$P_{\text{agree}}(W_{M'|M}) \le \frac{1}{|M|} \sum_{m\in\mathcal{M}} \sum_{x\in\mathcal{X}} \sum_{y\in\mathcal{Y}} D_{M'|Y}(m,y)\, N_{Y|X}(y,x) = \frac{|X|}{|M|}\,. \tag{8.3}$$

For the last equality, we use the fact that $\sum_{m\in\mathcal{M}} D_{M'|Y}(m,y) = 1$ for all y, the normalization condition of stochastic matrices, and the corresponding statement for $N_{Y|X}$.

Exercise 8.1. By bounding $N_{Y|X}$ instead of $E_{X|M}$ show that $P_{\text{agree}}(D \circ N \circ E) \le \frac{|Y|}{|M|}$.

Thus, for all choices of encoder and decoder, the average agreement probability satisfies

$$P_{\text{agree}}(D_{M|Y} \circ N_{Y|X} \circ E_{X|M}) \le \frac{\min(|X|, |Y|)}{|M|}\,. \tag{8.4}$$

This bound formalizes the intuition that an k-bit channel cannot meaningfully transmit more than k bits of information. For the n-bit identity channel $N_{Y|X}$ and $|M| = 2^k$, the bound gives

$$P_{\text{agree}}(D_{M|Y} \circ N_{Y|X} \circ E_{X|M}) \le 2^{n-k}\,. \tag{8.5}$$

In the exact case of zero error the channel must therefore have at least a k-bit input and a k-bit output. Moreover, the bound can be achieved for $k > n$ by just transmitting the first n bits of the input message and having the decoder simply guess the remainder. The bound in (8.5) is known as the *strong converse* for the identity channel, as the error rate is not only nonzero when communicating above its "capacity" (see more in Chapter 15), but goes exponentially to 1 (the agreement rate goes exponentially to 0) in the quantity $k - n$.

Exercise 8.2. Show that (8.4) also holds when the encoder and decoder make use of arbitrary common randomness, that is, when the encoder has access to U and the decoder has access to V for any bipartite probability distribution P_{UV}. The setup is depicted below.

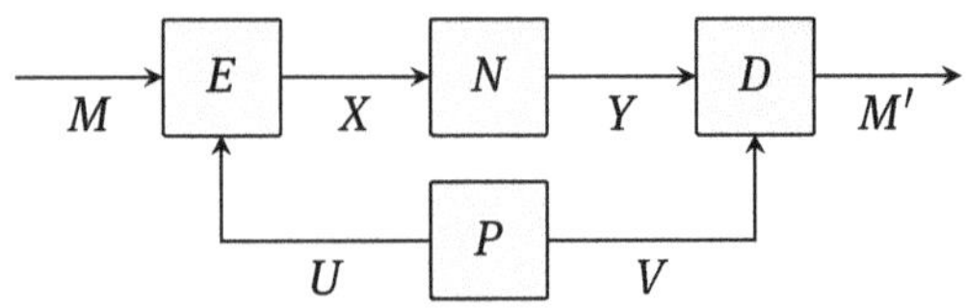

8.1.2 Over quantum channels

The case of transmitting classical information over quantum channels is actually not much different. The goal is still to simulate the classical identity channel, but using a quantum channel. Hence the encoder and decoder are CQ and QC channels, respectively. That is, as depicted below, the encoder will prepare a quantum state to be input to the quantum channel, the specific state depending on the message, and the decoder will make a measurement of the output of the quantum channel. Here we use dashed lines to indicate quantum systems and solid lines for classical systems.

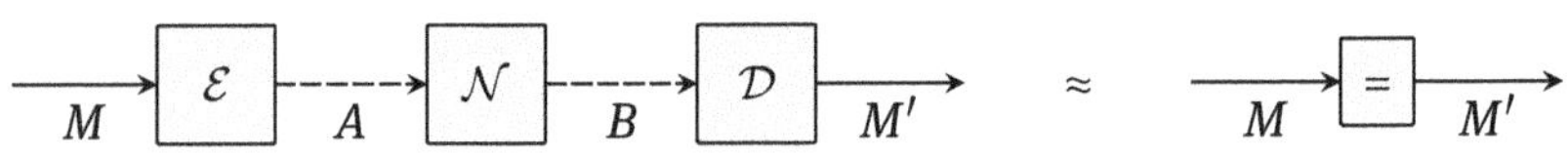

For quantum channel $\mathcal{N}_{B|A}$, encoder $\mathcal{E}_{A|M}$, and decoder $\mathcal{D}_{M'|B}$, we are interested in the agreement probability $P_{\text{agree}}(\mathcal{D}_{M'|B} \circ \mathcal{N}_{B|A} \circ \mathcal{E}_{A|M})$. Again, we can denote $\mathcal{D}_{M'|B} \circ \mathcal{N}_{B|A} \circ \mathcal{E}_{A|M}$ by $W_{M'|M}$. Supposing the encoder produces states $\rho_A(m)$, i. e., $\mathcal{E}_{A|M=m} = \rho_A(m)$, and the decoder is described by POVM elements $\Lambda_B(m)$, i. e., $\mathcal{D}_{M'|B} : \theta_B \mapsto \mathrm{Tr}[\Lambda_B(m)\,\theta_B]$, we can write the agreement probability as

$$\begin{aligned} P_{\text{agree}}(W_{M'|M}) &= \frac{1}{|M|}\sum_m \mathcal{D}_{M'=m|B} \circ \mathcal{N}_{B|A} \circ \mathcal{E}_{A|M=m}] \\ &= \frac{1}{|M|}\sum_m \mathrm{Tr}[\Lambda_B(m)\,\mathcal{N}_{B|A}[\rho_A(m)]] \,. \end{aligned} \tag{8.6}$$

In the previous case of a classical channel, we used the bound $E_{X|M}(x,m) \le 1$, which holds because $E_{X|M=m}$ is a normalized probability distribution. Similarly, in the quantum case, we have $\rho \le \mathbb{1}$ for any density operator ρ, implying $\rho_A(m) \le \mathbb{1}_A$ for all m. Since $\mathcal{N}_{B|A}$, $\mathcal{D}_{M'|B}$, and Tr are completely positive maps, they will preserve this inequality. Therefore

$$\begin{aligned} P_{\text{agree}}(W_{M'|M}) &\le \frac{1}{|M|}\sum_m \mathrm{Tr}[\Lambda_B(m)\,\mathcal{N}_{B|A}[\mathbb{1}_A]] \\ &= \frac{1}{|M|}\mathrm{Tr}[\mathbb{1}_B \mathcal{N}_{B|A}[\mathbb{1}_A]] = \frac{1}{|M|}\mathrm{Tr}[\mathbb{1}_A] = \frac{|A|}{|M|} \,. \end{aligned} \tag{8.7}$$

Comparing with (8.4), we see that the dimension of A plays the same role as the size of the input alphabet in the classical bound. Thus we have shown that a qubit channel (for which $|A| = |B| = 2$) cannot reliably transmit more than one bit of classical information. Perhaps unfortunately, in this regard, quantum channels are not more powerful than classical channels.

Exercise 8.3. Extend the above argument to establish that for any channel $\mathcal{N}_{B|A}$, encoder $\mathcal{E}_{A|M}$, and decoder $\mathcal{D}_{M'|B}$,

$$P_{\text{agree}}(\mathcal{D}_{M'|B} \circ \mathcal{N}_{B|A} \circ \mathcal{E}_{A|M}) \leq \frac{\min(|A|, |B|)}{|M|} . \tag{8.8}$$

8.2 Converses for quantum communication

To handle the case of quantum communication, we need a quantity analogous to the agreement probability to quantify how close the resulting channel is to the identity quantum channel. We will investigate this question in more detail in the coming chapters, but for now let us take an intuitive approach.

One thing an ideal quantum channel can certainly do is transmit entanglement; that is, if the sender Alice prepares an entangled state $|\Phi\rangle_{AB}$, then she can send the subsystem B through an ideal channel to the receiver Bob, so that they now share the entangled state. An imperfect version of this process produces some other state ρ_{AB}, and we can compare how close ρ_{AB} is to the pure state via the quantity $\langle\Phi|\rho_{AB}|\Phi\rangle_{AB} = \mathrm{Tr}[\Phi_{AB}\rho_{AB}]$. Since Φ_{AB} is a rank-one density operator, we can consider the POVM that tests if the state is Φ_{AB} or not, and $\mathrm{Tr}[\Phi_{AB}\rho_{AB}]$ is the probability of passing the test. Let us therefore define the "agreement probability" for a quantum channel $\mathcal{N}_Q$ as

$$P_{\text{agree}}(\mathcal{N}_Q) := \mathrm{Tr}[\Phi_{QQ'}\, \mathcal{N}_Q[\Phi_{QQ'}]] . \tag{8.9}$$

Here we abruptly switch from A and B to Q and Q' as Q and Q' play the role of M and M' in the classical case, whereas A and B are analogous to the input X and output Y of the classical channel, respectively.

8.2.1 Over classical channels

Let us first consider the case of sending quantum information over an arbitrary classical channel $N_{Y|X}$, which we expect will not be very successful. The encoder needs to transform quantum information into classical information, so it must be a measurement $\mathcal{M}_{X|Q}$. The decoder needs to perform the opposite transformation, meaning it is a state preparation $\mathcal{E}_{Q|Y}$. Then we have $\mathcal{N}_Q = \mathcal{E}_{Q|Y} \circ N_{Y|X} \circ \mathcal{M}_{X|Q}$, as shown below.

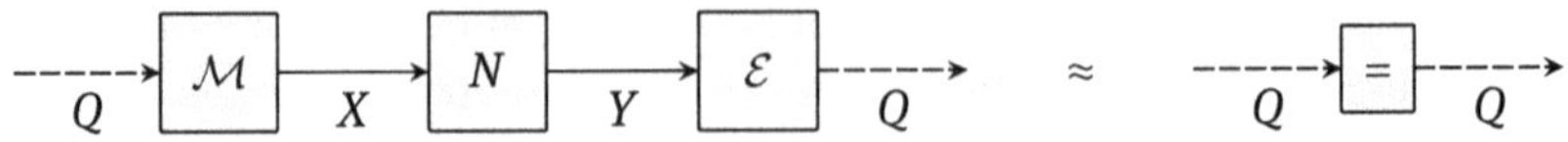

Suppose that $\mathcal{E}_{Q|Y=y} = \rho_Q(y)$ and $\mathcal{M}_{X=x|Q}$ is the map $\rho_Q \mapsto \mathrm{Tr}[\Lambda_Q(x)\rho_Q]$. Then

$$\mathcal{N}_Q[\Phi_{QQ'}] = \sum_{xy} N_{Y|X}(y,x)\,\rho_Q(y) \otimes \mathrm{Tr}_Q[\Lambda_Q(x)\,\Phi_{QQ'}]\,. \tag{8.10}$$

This is a separable state, since $\rho_Q(y)$ and $\mathrm{Tr}_Q[\Lambda_Q(x)\,\Phi_{QQ'}]$ are positive operators and $N_{Y|X}(y,x)$ is a positive function. Therefore we can appeal to the following result, which shows that separable states cannot have a large overlap with the maximally entangled state. In fact, this is even true for states with positive partial transpose, called *PPT states*, i. e., states ρ_{AB} such that $\mathcal{T}_A[\rho_{AB}] \geq 0$. Clearly, separable states are PPT, but it turns out that not all PPT states are separable.

Proposition 8.1 (PPT bound). *For any PPT state σ_{AB} and the maximally entangled state Φ_{AB},*

$$\mathrm{Tr}[\Phi_{AB}\,\sigma_{AB}] \leq \frac{1}{|A|}\,. \tag{8.11}$$

Proof. By Exercise 5.4 we have $\mathcal{T}_A[\Phi_{AB}] = \frac{1}{|A|}\Upsilon_{AB}$. Therefore, since $\sigma'_{AB} = \mathcal{T}_A[\sigma_{AB}] \geq 0$ by assumption and $\Upsilon_{AB} \leq \mathbb{1}_{AB}$,

$$\begin{aligned} \mathrm{Tr}[\Phi_{AB}\sigma_{AB}] &= \mathrm{Tr}[\mathcal{T}_A[\Phi_{AB}]\,\mathcal{T}_A[\sigma_{AB}]] \\ &= \frac{1}{|A|}\,\mathrm{Tr}[\Upsilon_{AB}\,\sigma'_{AB}] \;\leq\; \frac{1}{|A|}\,\mathrm{Tr}[\sigma'_{AB}] \;=\; \frac{1}{|A|}\,. \end{aligned} \tag{8.12}$$

The inequality is $\mathrm{Tr}[\sigma'_{AB}(\mathbb{1}_{AB} - \Upsilon_{AB})] \geq 0$, which follows from the positivity of σ'_{AB} since the trace of the product of positive operators is positive. □

It follows immediately that for every classical channel $N_{Y|X}$ and all choices of encoder $\mathcal{M}_{X|Q}$ and decoder $\mathcal{E}_{Q|Y}$,

$$P_{\text{agree}}(\mathcal{E}_{Q|Y} \circ N_{Y|X} \circ \mathcal{M}_{X|Q}) \leq \frac{1}{|Q|}\,. \tag{8.13}$$

Thus quantum information cannot be usefully transmitted by classical channels, as entanglement is not transmitted faithfully even by the ideal classical channel. Indeed, quantum information cannot be reliably transmitted by any quantum channel $\mathcal{N}_{B|A}$ whose output $\omega_{BR} = \mathcal{N}_{B|A}[\rho_{AR}]$ is certain to be a PPT state for any ρ_{AR}. This class includes entanglement-breaking channels discussed in Exercise 5.24.

Exercise 8.4. Use the PPT condition to determine for which parameters the qubit depolarizing, dephasing, amplitude damping, and erasure channels are useless for transmitting quantum information.

In the case of the identity classical channel, the bound can be saturated by preparing the correlated classical state $\rho_{QQ'} = \frac{1}{|Q|}\sum_x |xx\rangle\langle xx|_{QQ'}$ and using the classical channel to transmit Q from sender to receiver. Since the state being sent is $|x\rangle\langle x|_Q$ for some x, it is essentially classical. It is easy to confirm that $\mathrm{Tr}[\Phi_{QQ'}\rho_{QQ'}] = \frac{1}{|Q|}$: $\rho_{QQ'}$ is $1/|Q|$ times a projection operator that acts trivially on $\Phi_{QQ'}$.

8.2.2 Over quantum channels

Finally, let us confirm that no qubit channel can reliably transmit more than one qubit. For a generic channel $\mathcal{N}_{B|A}$, encoder $\mathcal{E}_{A|Q}$, and decoder $\mathcal{D}_{Q|B}$, we are interested in $P_{\text{agree}}(\mathcal{D}_{Q|B} \circ \mathcal{N}_{B|A} \circ \mathcal{E}_{A|Q})$:

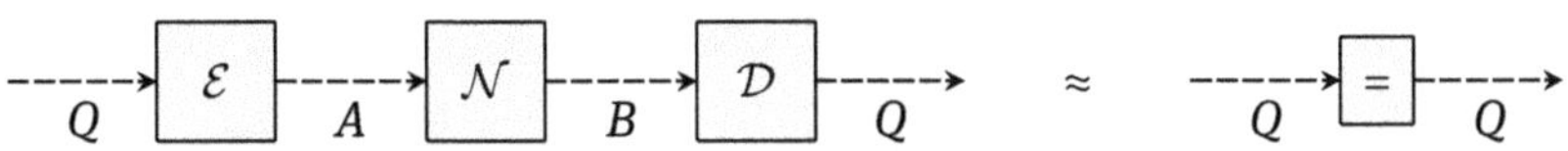

Unlike the case of transmitting classical information, we cannot immediately upper bound $\mathcal{E}_{A|Q}[\Phi_{QQ'}]$. However, the bound $\rho_A \le \mathbb{1}_A$ can be extended to one subsystem of a bipartite operator as follows. First, we establish two important inequalities.

Proposition 8.2. *Fix a system A with dimension $|A|$ and let $\mathcal{P}_A$ be a pinching map in an arbitrary orthonormal basis. Then, for any $S_{AB} \ge 0$,*

$$S_{AB} \le |A|\,\mathcal{P}_A[S_{AB}]\,. \tag{8.14}$$

Moreover, for any CQ operator S_{AB} with classical A (equivalently, $S_{AB} = \mathcal{P}_A[S_{AB}]$),

$$S_{AB} \le \mathbb{1}_A \otimes S_B\,. \tag{8.15}$$

Proof. Using the Kraus representation of $\mathcal{P}_A$ given in (5.4), we immediately have $|A|\,\mathcal{P}_A[S_{AB}] - S_{AB} = \sum_{k=1}^{|A|-1} V_A^k S_{AB}(V_A^k)^* \ge 0$, which is the first inequality.

The pinched S_{AB} can also be written $\mathcal{P}_A[S_{AB}] = \sum_{k=0}^{d-1} |k\rangle\langle k|_A \otimes \mathrm{Tr}_A[|k\rangle\langle k|_A\, S_{AB}]$, where $|k\rangle$ is the orthonormal basis of the pinch map. Since $S_B = \sum_{k=0}^{d-1} \mathrm{Tr}_A[|k\rangle\langle k|_A\, S_{AB}]$ and each term in the previous sum is positive, we have $\mathrm{Tr}_A[|k\rangle\langle k|_A\, S_{AB}] \le S_B$, the second inequality. □

The first of these is often called the "pinching inequality". Combining the two yields the following very useful inequality, valid for all positive S_{AB}:

$$S_{AB} \le |A|\mathbb{1}_A \otimes S_B\,. \tag{8.16}$$

As an aside to the current discussion, but one that will be useful later, a generalized pinching inequality holds for pinching maps defined using projection operators of arbitrary rank. Suppose $\{\Pi_A(x)\}_{x=1}^n$ is a complete set of mutually disjoint projection operators on a system A of dimension $|A| \geq n$. Then $\Pi_A(x)$ are Kraus operators of a quantum channel $\mathcal{P}_A$, and for all $S_{AB} \geq 0$, we have

$$S_{AB} \leq n\mathcal{P}_A[S_{AB}]. \tag{8.17}$$

Exercise 8.5. Construct a Kraus representation of the generalized pinch map $\mathcal{P}_A$ using operators $V_A(k)$ of the form $V_A(k) = \sum_{x=1}^n c(k)\Pi_A(k)$ for appropriate coefficients $c(k) \in \mathbb{C}$ and prove (8.17).

With (8.16) in hand, we can proceed as in Section 8.1.1. Start with $\mathcal{E}_{A|Q}[\Phi_{QQ'}] \leq |A|\mathbb{1}_A \otimes \Phi_{Q'} = \frac{|A|}{|Q|}\mathbb{1}_{AQ'}$. Abbreviating $\mathcal{D}_{Q|B} \circ \mathcal{N}_{B|A} \circ \mathcal{E}_{A|Q}$ as $\mathcal{F}_Q$, we have

$$\begin{aligned} P_{\text{agree}}(\mathcal{F}_Q) &\leq \frac{|A|}{|Q|}\operatorname{Tr}_{QQ'}[\Phi_{QQ'}\,\mathcal{D}_{Q|B} \circ \mathcal{N}_{B|A}[\mathbb{1}_{AQ'}]] \\ &= \frac{|A|}{|Q|^2}\operatorname{Tr}_Q[\mathcal{D}_{Q|B} \circ \mathcal{N}_{B|A}[\mathbb{1}_A]] \\ &= \frac{|A|}{|Q|^2}\operatorname{Tr}_B[\mathcal{N}_{B|A}[\mathbb{1}_A]] = \frac{|A|}{|Q|^2}\operatorname{Tr}_A[\mathbb{1}_A] = \frac{|A|^2}{|Q|^2}. \end{aligned} \tag{8.18}$$

This is precisely analogous to (8.4), albeit with a square.

Exercise 8.6. Extend the above argument to show that

$$P_{\text{agree}}(\mathcal{D}_{Q|B} \circ \mathcal{N}_{B|A} \circ \mathcal{E}_{A|Q}) \leq \frac{\min(|A|^2, |B|^2)}{|Q|^2} \tag{8.19}$$

for all noisy channels $\mathcal{N}_{B|A}$, encoders $\mathcal{E}_{A|Q}$, and decoders $\mathcal{D}_{Q|B}$.

When $\mathcal{N}$ is the identity channel on n qubits, i. e., $|A| = |B| = 2^n$, and Q is m qubits, i. e., $|Q| = 2^m$, the bound can be saturated. The encoder simply transmits the first n qubits of m via the identity channel, and they are recovered by the decoder. For each of the remaining qubits, the encoder discards the input qubit, transmitting nothing, and the decoder simply generates a fixed state θ at the output. Thus, for each of these $m-n$ inputs, the shared state is $\theta \otimes \pi$ with $\pi = \frac{1}{2}\mathbb{1}$. It is easy to work out that $\operatorname{Tr}[\Phi\,\theta \otimes \pi] = \frac{1}{4}$, and so the overall value of P_{agree} is simply $2^{-2(m-n)}$, precisely in agreement with (8.18).

8.3 Assisted communication: dense coding and teleportation

Now we turn to the case that the communication task is assisted by additional resources, specifically shared entanglement. Instead of trying to simulate the ideal channel just using the noisy channel, the encoder and decoder now also have access to some distributed quantum state $\rho_{TT'}$. System T is available to the encoder, system T' to the decoder, and the state $\rho_{TT'}$ is arbitrary. Then the encoder has two inputs, one for the message or quantum information and one for the assistance T. Similarly, the decoder has two inputs, one for the channel output and one for the assistance T'.

The case of shared classical randomness is included in this description simply by taking the shared state $\rho_{TT'}$ to be diagonal and characterized by a probability distribution P_{XY}: $\rho_{TT'} = \sum_{xy} P_{XY}(x,y)|x\rangle\langle x|_T \otimes |y\rangle\langle y|_{T'}$. However, this assistance is never of any help. We can illustrate the reason in the case of classical communication over quantum channels. The encoder $\mathcal{E}_{A|MX}$ will now be a map from two classical inputs to one quantum output, while the decoder $\mathcal{D}_{M'|BY}$ takes a CQ input, with classical Y and quantum B, to a classical output. The combined channel is just $W_{M'|M} = \mathcal{D}_{M'|BY} \circ \mathcal{N}_{B|A} \circ \mathcal{E}_{A|MX}[P_{XY}]$. The agreement probability of $W_{M'|M}$ is just the average over the agreement probability for fixed inputs $X = x$ and $Y = y$ of the assistance:

$$P_{\text{agree}}(W_{M'|M}) = \sum_{xy} P_{XY}(x,y) P_{\text{agree}}(\mathcal{D}_{M'|B,Y=y} \circ \mathcal{N}_{B|A} \circ \mathcal{E}_{A|M,X=x}) \,. \tag{8.20}$$

Since the maps $\mathcal{E}_{A|M,X=x}$ and $\mathcal{D}_{M'|B,Y=y}$ are themselves legitimate channels, (8.8) applies for each value of x and y. Hence we recover the same bound in the case of classical assistance. The same argument holds for quantum communication, as again the assistance will show up as an average of the agreement probability.

Shared entanglement, by contrast, does help in the two "mixed" scenarios of classical communication over quantum channels or quantum communication over classical channels. In the former, we can use the *superdense coding* protocol and *teleportation* in the latter. The usefulness of shared entanglement in teleportation is especially dramatic, since without it quantum communication is impossible over classical channels. Both protocols demonstrate the stark difference between classical and quantum information, that the physical properties of quantum systems have no fixed "values" (which are simply revealed by measurement). If they did, then the averaging argument just above would go through, and shared entanglement would not be useful in these two communication tasks.

The protocol of superdense coding transmits two classical bits over a noiseless one-qubit quantum channel, making use of the two-qubit shared entangled state $|\Phi\rangle_{TT'}$. Take the message to be two bits (j,k). The encoder applies the operator $(\sigma_x^j \sigma_z^k)_T$ to the share T of the entangled state and then transmits T over the noiseless channel. Upon receipt of T, the decoder jointly measures the TT' system in the Bell basis from (4.15). The Bell states are labeled by two bits, and the output of the decoder is the measurement result. The protocol works as intended because the Bell states can be generated from $|\Phi\rangle_{TT'}$ by the Pauli operators, as shown in (4.16).

Teleportation puts the Bell measurement at the sender and the Pauli operation at the receiver. The teleportation protocol uses a two-bit noiseless classical channel and the same two-qubit shared entangled state $|\Phi\rangle_{TT'}$ to transmit one qubit of information. Here the encoder takes the quantum input A to be transmitted and measures the combined system AT in the Bell basis. Again, this results in two classical bits (j,k). These are transmitted via the classical channel. The decoder subsequently applies the Pauli operator $(\sigma_x^j\sigma_z^k)_{T'}$ to the share T' of the entangled state, and the resulting state is the output of the protocol.

Confirming that the teleportation protocol works as intended requires a little more work than superdense coding. The protocol amounts to the quantum channel

$$\begin{aligned}\mathcal{E}_{T'|A}[\rho_A] &= \sum_{jk}(\sigma_x^j\sigma_z^k)_{T'}\operatorname{Tr}_{AT}[|\Phi_{jk}\rangle\langle\Phi_{jk}|_{AT}\,\rho_A\otimes\Phi_{TT'}]\,(\sigma_x^j\sigma_z^k)_{T'}^*\\ &= \sum_{jk}(\sigma_x^j\sigma_z^k)_{T'}\langle\Phi_{jk}|_{AT}\,|\Phi\rangle_{TT'}\,\rho_A\,\langle\Phi|_{TT'}\,|\Phi_{jk}\rangle_{AT}(\sigma_x^j\sigma_z^k)_{T'}^*,\end{aligned} \tag{8.21}$$

where we have used Dirac notation to move the bras and kets around from their more usual appearance. Doing so reveals that the channel has Kraus operators $K_{T'|A}(j,k) = (\sigma_x^j\sigma_z^k)_{T'}\langle\Phi_{jk}|_{AT}\,|\Phi\rangle_{TT'}$, but by (4.16) and (4.28) we have

$$\begin{aligned}K_{T'|A}(j,k) &= (\sigma_x^j\sigma_z^k)_{T'}\langle\Phi|_{AT}\,|\Phi\rangle_{TT'}\,(\sigma_x^j\sigma_z^k)_A^* = \tfrac{1}{2}(\sigma_x^j\sigma_z^k)_{T'}\mathbb{1}_{T'|A}(\sigma_x^j\sigma_z^k)_A^*\\ &= \tfrac{1}{2}(\sigma_x^j\sigma_z^k)_{T'}(\sigma_x^j\sigma_z^k)_{T'}^*\mathbb{1}_{T'|A} = \tfrac{1}{2}\mathbb{1}_{T'|A}\,.\end{aligned} \tag{8.22}$$

Hence the teleportation channel is simply $\mathcal{I}_{T'|A}$ as intended.

Exercise 8.7. In the teleportation protocol, what is the state of Bob's system T' averaged over the measurement result?

Exercise 8.8. Using Proposition 6.8, show that the probability distribution of Alice's measurement results must be independent of the input state $|\psi\rangle$. What is the particular output distribution in the protocol?

8.4 Converses for assisted classical communication

To address the optimality of superdense coding and teleportation, we require converses for assisted communication. Fortunately, these can be established using the tools developed for the unassisted case. In this section, we address converses for classical communication tasks, and we take up quantum communication tasks in the following section.

8.4.1 Over classical channels

For classical information transmission over a classical channel $N_{Y|X}$, the encoder $\mathcal{E}_{X|MT}$ is a channel with a CQ input, classical M and quantum T, and classical output X. Similarly, the decoder $\mathcal{D}_{M'|YT'}$ also has a CQ input, classical Y and quantum T', and classical output M':

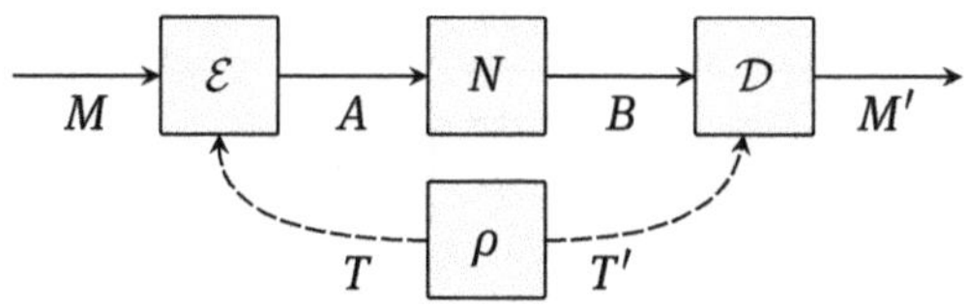

Given an arbitrary assistance state $\rho_{TT'}$, the protocol induces a classical channel $W_{M'|M} = \mathcal{D}_{M'|YT'} \circ N_{Y|X} \circ \mathcal{E}_{X|MT}[\rho_{TT'}]$, whose agreement probability is

$$P_{\text{agree}}(W_{M'|M}) = \frac{1}{|M|} \sum_m \mathcal{D}_{M'=m|YT'} \circ N_{Y|X} \circ \mathcal{E}_{X|M=m,T}[\rho_{TT'}] . \tag{8.23}$$

For every input m, the state $\mathcal{E}_{X|M=m,T}[\rho_{TT'}]$ is a CQ state with classical X and quantum T'. Therefore we can appeal to (8.15) to infer $\mathcal{E}_{X|M=m,T}[\rho_{TT'}] \le \mathbb{1}_X \otimes \rho_{T'}$ for all m. Then we have

$$\begin{aligned} P_{\text{agree}}(W_{M'|M}) &\le \frac{1}{|M|} \sum_m \mathcal{D}_{M'=m|YT'} \circ N_{Y|X}[\mathbb{1}_X \otimes \rho_{T'}] \\ &= \frac{1}{|M|} \operatorname{Tr}_{YT'}[N_{Y|X}[\mathbb{1}_X \otimes \rho_{T'}]] = \frac{1}{|M|} \operatorname{Tr}_X[\mathbb{1}_X] = \frac{|X|}{|M|} . \end{aligned} \tag{8.24}$$

This is precisely (8.3), so even quantum assistance makes absolutely no difference. The distinction between assistance by $\rho_{TT'}$ and a channel from T to T' is that here $\rho_{T'}$ is completely independent of m.

Exercise 8.9. Extend the argument to show

$$P_{\text{agree}}(\mathcal{D}_{M'|YT'} \circ N_{Y|X} \circ \mathcal{E}_{X|MT}[\rho_{TT'}]) \le \frac{\min(|X|, |Y|)}{|M|} . \tag{8.25}$$

Thus (8.4) holds even with arbitrary assistance $\rho_{TT'}$.

8.4.2 Over quantum channels

Although *classical* assistance does not help with classical communication over quantum channels, *quantum* assistance does, as we saw in the superdense coding protocol above. To specify the general setup more concretely, we again have a shared entangled state $\rho_{TT'}$, while the channel $\mathcal{N}_{B|A}$ has quantum inputs and outputs. Therefore the encoder $\mathcal{E}_{A|MT}$ outputs a quantum system from a classical and quantum input, whereas the decoder $\mathcal{D}_{M'|BT'}$ outputs a classical value from two quantum inputs:

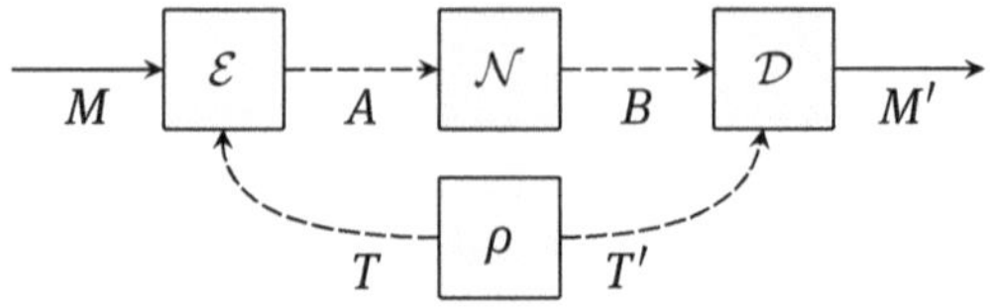

We are interested in the agreement probability $P_{\text{agree}}(\mathcal{D}_{M'|BT'} \circ \mathcal{N}_{B|A} \circ \mathcal{E}_{A|MT}[\rho_{TT'}])$, which we abbreviate by P_{agree}. In this case, we will show

$$P_{\text{agree}} \leq \frac{1}{|M|} \min(|A|^2, |B|^2, |A||T|, |B||T|, |A||T'|, |B||T'|). \tag{8.26}$$

The converse bound is necessarily complicated because we could just use the quantum channel directly and not bother with the assistance, or we could perform a superdense coding scheme, or some combination of the two. The first two and last two bounds follow from methods we have already used.

Exercise 8.10. Use inequality (8.16) to obtain the first two bounds. Use the even cruder inequality $\rho \leq \mathbb{1}$ for a suitable choice of systems to obtain the last two.

To show the middle two upper bounds, we use both (8.14) and (8.15). First, using (8.14), pinch T to obtain $\rho_{TT'} \leq |T| \sum_x |x\rangle\langle x|_T \otimes \varphi_{T'}(x)$ for some subnormalized $\varphi_{T'}(x)$. Observe that $\sum_x \varphi_{T'}(x) = \rho_{T'}$. Applying the encoder, we have $\mathcal{E}_{A|TM=m}[\rho_{TT'}] \leq |T| \sum_k \mathcal{E}_{A|TM=m}[|k\rangle\langle k|_T] \otimes \varphi_{T'}(k)$. Since the right-hand side is a CQ state, (8.15) implies $\mathcal{E}_{A|TM=m}[\rho_{TT'}] \leq |T| \sum_k \mathbb{1}_A \otimes \varphi_{T'}(k) = |T| \mathbb{1}_A \otimes \rho_{T'}$. This gives the third bound. To obtain the fourth, apply the same argument for the channel $\mathcal{N}_{B|A} \circ \mathcal{E}_{A|TM=m}$ instead of $\mathcal{E}_{A|TM=m}$ itself.

Superdense coding saturates (8.26). Setting $|A| = |B|$ and $|T| = |T'|$, the bound reduces to $P_{\text{agree}} \leq \frac{|A|}{|M|} \min(|A|, |T|)$. A single round of the protocol corresponds to the choice $|M| = 4$, $|A| = 2$, $|T| = 2$. Of course, we can choose $|T| = 1$ and just transmit classical information directly over the quantum channel with no assistance. In this case the bound reduces to (8.8). These bounds are perhaps more insightful in the scenario of multiple rounds, with $|A| = 2^n$, i. e., n qubits, $|T| = 2^k$, and $|M| = 2^{2m}$ (note the 2 in the exponent). Then the converse implies that $2n \geq 2m$ and $n + k \geq 2m$, so that to send $2m$ bits, we will need at least m classical channels and a combination of assistance and channels numbering $2m$.

Notice that a separable assistance state $\rho_{TT'}$ will simply result in a state of the form $\sum_k P(k) \mathcal{N}_{B|A} \circ \mathcal{E}_{A|TM=m}[\sigma_T(k)] \otimes \theta_{T'}(k)$ at the decoder. Hence whatever agreement probability can be obtained will just be the average over k of using product assistance states. Then we could just pick the states $\sigma_T(k)$ and $\theta_{T'}(k)$ with the best agreement probability and build them into the encoder and decoder, resulting in an unassisted scheme.

Therefore entanglement in $\rho_{TT'}$ is required to increase the agreement probability over the unassisted case.

8.5 Converses for assisted quantum communication

8.5.1 Over quantum channels

Just as with classical communication over classical channels, quantum communication over quantum channels is not enhanced by shared entanglement. Here the encoder is a quantum channel $\mathcal{E}_{A|QT}$ with two quantum inputs, Q and T, and output A. The decoder is a quantum channel $\mathcal{D}_{Q|BT'}$, also with two quantum inputs, B and T', and output Q.

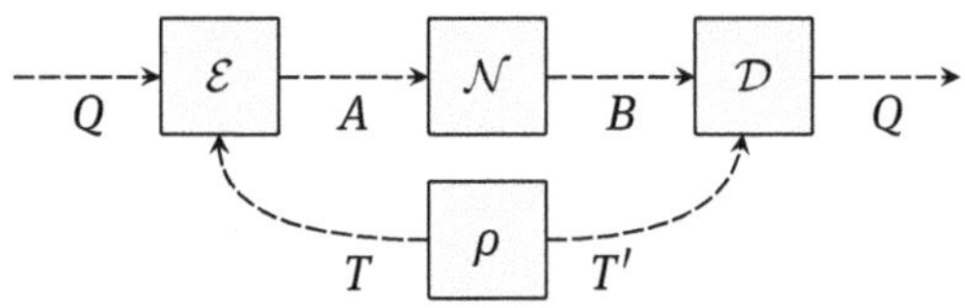

Again, we abbreviate the overall channel on Q as $\mathcal{F}_Q = \mathcal{D}_{Q|BT'} \circ \mathcal{N}_{B|A} \circ \mathcal{E}_{A|QT}[\rho_{TT'}]$. Then the "agreement" probability is

$$P_{\text{agree}}(\mathcal{F}_Q) = \operatorname{Tr}[\Phi_{QQ'}(\mathcal{D}_{Q|BT'} \circ \mathcal{N}_{B|A} \circ \mathcal{E}_{A|QT}[\Phi_{QQ'} \otimes \rho_{TT'}])] . \tag{8.27}$$

We can proceed just as in the unassisted case, applying (8.16) to the state $\mathcal{E}_{A|QT}[\Phi_{QQ'} \otimes \rho_{TT'}]$. This gives

$$\begin{aligned} \mathcal{E}_{A|QT}[\Phi_{QQ'} \otimes \rho_{TT'}] &\leq |A| \mathbb{1}_A \otimes \operatorname{Tr}_A[\mathcal{E}_{A|QT}[\Phi_{QQ'} \otimes \rho_{TT'}]] \\ &= |A| \mathbb{1}_A \otimes \operatorname{Tr}_{QT}[\Phi_{QQ'} \otimes \rho_{TT'}] = \frac{|A|}{|Q|} \mathbb{1}_A \otimes \mathbb{1}_{Q'} \otimes \rho_{T'} . \end{aligned} \tag{8.28}$$

Then for any encoder, decoder, and shared entanglement, we have

$$\begin{aligned} P_{\text{agree}}(\mathcal{F}_Q) &\leq \frac{|A|}{|Q|} \operatorname{Tr}_{QQ'}[\Phi_{QQ'} (\mathcal{D}_{Q|BT'} \circ \mathcal{N}_{B|A}[\mathbb{1}_A \otimes \rho_{T'}])] \\ &= \frac{|A|}{|Q|^2} \operatorname{Tr}_Q[\mathcal{D}_{Q|BT'} \circ \mathcal{N}_{B|A}[\mathbb{1}_A \otimes \rho_{T'}]] \\ &= \frac{|A|}{|Q|^2} \operatorname{Tr}_{BT'}[\mathcal{N}_{B|A}[\mathbb{1}_A \otimes \rho_{T'}]] = \frac{|A|^2}{|Q|^2} , \end{aligned} \tag{8.29}$$

which is precisely the same bound as the unassisted case in (8.18). Applying the same argument to the state $\mathcal{N}_{B|A} \circ \mathcal{E}_{A|QT}[\Phi_{QQ'} \otimes \rho_{TT'}]$ gives the bound $|B|^2/|Q|^2$, so we recover (8.19). Neither shared randomness nor shared entanglement helps with quantum communication over quantum channels.

8.5.2 Over classical channels

Finally, we turn to quantum communication over classical channels. Shared entanglement is most certainly helpful, as demonstrated by teleportation. Now the encoder $\mathcal{E}_{X|QT}$ is essentially a measurement of the two quantum inputs Q and T, whereas the decoder $\mathcal{D}_{Q|YT'}$ outputs a quantum system from one classical and one quantum input.

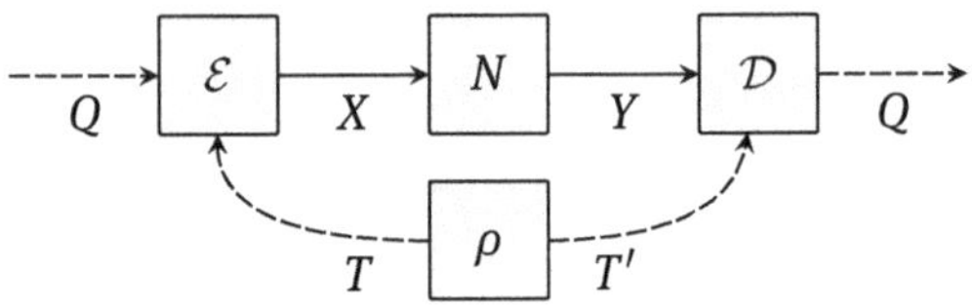

The converse bound in this case is not quite as elaborate as (8.26), but does have four cases:

$$P_{\text{agree}}(\mathcal{D}_{Q|YT'} \circ N_{Y|X} \circ \mathcal{E}_{X|QT}[\rho_{TT'}]) \leq \min\left(\frac{|X|}{|Q|^2}, \frac{|Y|}{|Q|^2}, \frac{|T|}{|Q|}, \frac{|T'|}{|Q|} \right). \tag{8.30}$$

The argument is not quite as complicated as that of superdense coding, either. There we have the possibility of ignoring the shared entanglement and just using the quantum channel. For the present task, the analogous bound is (8.13), $1/|Q|$. But this is implied by the last of the two bounds in (8.30) by taking $|T| = |T'| = 1$, so there is no need for a separate statement as there is in (8.26). The first two bounds are straightforward.

Exercise 8.11. Use (8.15) to obtain the first two bounds in (8.30).

For the latter two, we follow the approach in superdense coding and pinch the shared entangled state. Pinching T gives $\rho_{TT'} \leq |T|\mathcal{P}_T[\rho_{TT'}] = |T| \sum_z |z\rangle\langle z|_T \otimes \varphi_{T'}(z)$, which is a separable state. Then observe that since the encoder is a measurement, the result of applying the encoder is also separable,

$$\begin{aligned} \mathcal{E}_{X|TQ}[\mathcal{P}_T[\rho_{TT'}]\Phi_{QQ'}] &= \sum_{xz} |x\rangle\langle x|_X \operatorname{Tr}_{TQ}[\Lambda_{TQ}(x)(|z\rangle\langle z|_T \otimes \Phi_{QQ'})]\varphi_{T'}(z) \\ &= \sum_{xz} |x\rangle\langle x|_X \theta_{Q'}(x,z)\varphi_{T'}(z), \end{aligned} \tag{8.31}$$

where we have defined the subnormalized $\theta_{Q'}(x,z) = \operatorname{Tr}_{TQ}[\Lambda_{TQ}(x)(|z\rangle\langle z|_T \otimes \Phi_{QQ'})]$ and omitted some tensor product symbols. The classical channel and decoder will act on X and T', but not Q', so the output of the protocol will be separable between Q and

Q'. Then employing (8.11) gives the third bound. The fourth follows by the same argument, starting from a pinch applied to T'.

Observe that the argument goes through immediately without pinching if the initial assistance state $\rho_{TT'}$ is separable. Hence, as expected, only entangled states can be of use here.

In teleportation, we have $|T'| = |T|$ and $|Y| = |X|$, and the converse bound reduces to $\min(\frac{|X|}{|Q|^2}, \frac{|T|}{|Q|})$. A single round of teleportation is the case $|X| = 4$, $|T| = 2$, and $|Q| = 2$, which saturates the bound. The converse implies that transmitting m qubits ($|Q| = 2^m$) using k qubits of shared entanglement ($|T| = |T'| = 2^k$) and n single-bit classical identity channels ($|X| = |Y| = 2^n$) requires $n \geq 2m$ and $k \geq m$.

8.6 Entanglement distillation

Finally, we establish that LOCC operations cannot increase the amount of entanglement shared by two separated parties. In general, an LOCC operation may consist of many rounds of local operations and classical communication from one party to the other. Let us first consider the case of just one round, a local operation by Alice, forward communication from her to Bob, and finally a local operation by Bob.

Suppose Alice and Bob share a state ρ_{AB} and they would like to create the state $\Phi_{QQ'}$ for some $|Q|$. Alice applies the instrument $\mathcal{E}_{QY|A}$ with quantum output Q and classical output Y, and then communicates the classical result to Bob. (We keep with the symbol $\mathcal{E}$ for Alice's operation out of inertia.) Bob then performs a "decoding" operation $\mathcal{D}_{Q'|BY}$, and hopefully the state of the QQ' system is close to being maximally entangled:

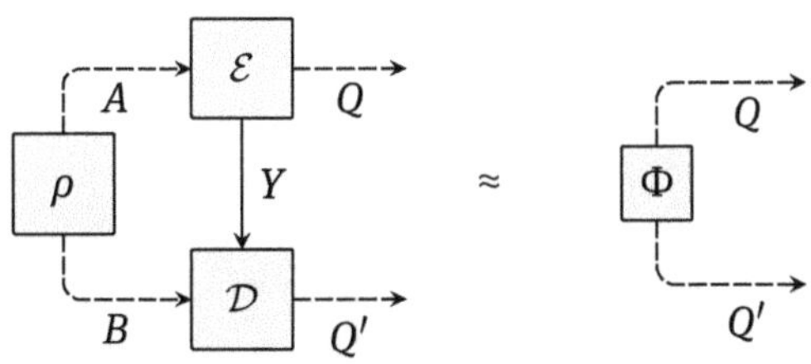

To quantify the entanglement of the output, we can again make use of the probability of obtaining the maximally entangled state $\mathrm{Tr}[\Phi_{QQ'}\, \mathcal{D}_{Q'|BY} \circ \mathcal{E}_{QY|A}[\rho_{AB}]]$. Then for all ρ_{AB}, instruments $\mathcal{E}_{QY|A}$, and channels $\mathcal{D}_{Q'|BY}$, we have

$$\mathrm{Tr}[\Phi_{QQ'}\mathcal{D}_{Q'|BY} \circ \mathcal{E}_{QY|A}[\rho_{AB}]] \leq \frac{1}{|Q|} \min(|A|, |B|)\,. \tag{8.32}$$

Therefore, if the output is precisely $\Phi_{QQ'}$, then the dimension of the output can never be larger than that of the input.

Note that the bound differs from the previous quantum bounds by a square root. This is due to the fact that Alice and Bob can create correlated states using classical communication. Indeed, the bound can be achieved for $|A| = |B| = 2^n$ and $|M| = 2^m$ with $m \geq n$ by choosing ρ_{AB} to be n maximally entangled states. The protocol leaves these states untouched while creating $m-n$ instances of $\tau = \frac{1}{2}(|00\rangle\langle 00| + |11\rangle\langle 11|)$. The state τ can be created by Alice randomly preparing $|0\rangle$ or $|1\rangle$ and informing Bob of her choice, so that he may prepare the same state. It is easy to confirm that $\mathrm{Tr}[\Phi\tau] = 1/2$, and so (8.32) can be achieved.

The statement follows more or less immediately from (8.14) and (8.11). Pinching the initial state ρ_{AB} results in a separable state times a system dimension, and the LOCC protocol will always result in another separable state at its output. Therefore (8.11) applies and gives the desired bound. Note that this argument goes through for arbitrarily many rounds of back and forth communication between Alice and Bob, i. e., all LOCC protocols.

8.7 Notes and further reading

Dense coding was introduced by Bennett and Wiesner [33] in 1992, teleportation by Bennett et al. [27] the following year. The general pinching inequality in (8.17) is due to Hayashi [124], who puts it to several uses in his book on quantum information theory [128]. The PPT bound is due to Rains [232].

9 Discriminating states and channels

It is difficult to make predictions, especially about the future.

unattributed remark of Danish origin[1]

Imagine that we are given one of two state preparation devices, one that always prepares ρ and another that always prepares σ. The states could be commuting operators, and then we could think of the devices as two different random number generators, one producing an output distributed according to P and the other to Q. Without looking at the internal operation of the device, how can we tell which device we have? The only option is to make a decision based on observing the outputs of the device.

The extent to which we can correctly determine which device we actually have—to discriminate between the two possibilities—defines the distinguishability of quantum states in a directly operational way. The setup fits naturally with the information processing context of this book, as an equivalent formulation is that the device is a channel with binary classical input and quantum output. The input specifies which state is generated on the output. We are then in the role of the decoder, attempting to determine the classical input from the quantum output. The issue is the same in the context of statistical modeling, though the language is different. There we obtain data from a physical system and would like to decide which model best describes the system. The two models correspond to the two states ρ and σ, and we speak of distinguishing between them as *hypothesis testing*.

Suppose that we make a decision by using the device just once, i. e., by observing only one output of the device. We need to measure the output in some way and then make a decision as to which device we have based on the outcome. Any procedure can be defined as a QC channel from the device output to the classical information representing our guess. Therefore the entire process may as well be described by a POVM with just two outputs, one output corresponding to a guess that the device produces ρ and the other to σ. Denoting Λ the POVM element corresponding to ρ, the POVM element corresponding to σ is just $\mathbb{1} - \Lambda$. When the device actually produces ρ, the probability of an incorrect guess is $\mathrm{Tr}[(\mathbb{1} - \Lambda)\rho]$, whereas $\mathrm{Tr}[\Lambda\sigma]$ is the probability of error when the device actually produces σ.

9.1 Two approaches

There are two main approaches to quantifying the how good a particular Λ is at distinguishing the states and, consequently, of defining the optimal measurement $\Lambda^\star$. (We

1 Det er vanskeligt at spå, især naar det gælder Fremtiden. [260, Section Niels Bohr]

https://doi.org/10.1515/9783110570250-009

will regularly use the superscript * to denote the optimal variable.) In the Bayesian approach to hypothesis testing, we suppose that ρ and σ occur with some prior probabilities p and $1-p$, respectively, and then consider the average probability of successfully guessing,

$$P_{\text{guess}} = p\,\mathrm{Tr}[\Lambda\rho] + (1-p)\,\mathrm{Tr}[(\mathbb{1}-\Lambda)\sigma] = (1-p) + \mathrm{Tr}[\Lambda(p\rho - (1-p)\sigma)]\,. \tag{9.1}$$

We can regard the setup of state and prior probability as defining a CQ state $\tau_{XB} = |0\rangle\langle 0|_X \otimes p\rho_B + |1\rangle\langle 1|_X \otimes (1-p)\sigma_B$. Let us write the guessing probability for a given measurement Λ as $P_{\text{guess}}(X|B)_{\tau,\Lambda}$. The dependence on the CQ state from the setup of the problem is indicated in the subscript, along with the dependence on the measurement. The optimal guessing probability, denoted $P_{\text{guess}}(X|B)_\tau$, is

$$P_{\text{guess}}(X|B)_\tau := \max_\Lambda \{P_{\text{guess}}(X|B)_{\tau,\Lambda} : 0 \le \Lambda \le \mathbb{1}\}\,. \tag{9.2}$$

Due to the form of the guessing probability in (9.1), we have

$$P_{\text{guess}}(X|B)_\tau = (1-p) + \max\{\mathrm{Tr}[\Lambda(p\rho - (1-p)\sigma)] : 0 \le \Lambda \le \mathbb{1}\}\,. \tag{9.3}$$

Strictly speaking, we should write the optimization with a supremum instead of a maximum, to account for the possibility that there is no measurement that precisely attains the optimal value. However, we will see momentarily that the optimum is always attained.

Meanwhile, in the Neyman[2]–Pearson[3] approach to hypothesis testing, we do not involve any prior probability and just take the two errors separately. Then the optimal measurement minimizes one of the errors for a fixed value of the other. Let us denote by $\beta_\alpha(\rho,\sigma)$ the smallest error for σ given a fixed error for ρ of $1-\alpha$. This parameterization is convenient since $\alpha = \mathrm{Tr}[\Lambda\rho]$, that is,

$$\beta_\alpha(\rho,\sigma) := \min_\Lambda \{\mathrm{Tr}[\Lambda\sigma] : \mathrm{Tr}[\Lambda\rho] = \alpha,\ 0 \le \Lambda \le \mathbb{1}\}\,. \tag{9.4}$$

An important feature of the operational approach employed here and of the associated variational definitions is that the crucial property of *monotonicity* follows more or less immediately. Monotonicity, which plays a crucial role throughout information theory, just says that the discrimination task cannot be made easier by first applying a channel to the state. The optimal guessing probability will not increase, neither will the smallest error in σ for fixed error in ρ decrease. The reason is that the possibility of applying a channel is already considered in the optimization.

---

2 Jerzy Neyman, born Jerzy Spława-Neyman, 1894–1981.

3 Egon Sharpe Pearson, 1895–1980.

To appreciate the monotonicity argument more formally, consider a channel $\mathcal{E}_{C|B}$ and the probability of distinguishing $\mathcal{E}_{C|B}[\rho_B]$ versus $\mathcal{E}_{C|B}[\sigma_B]$ for arbitrary prior distribution specified by p. By (9.3) we are interested in $\max\{\mathrm{Tr}[\Lambda(p\mathcal{E}[\rho] - (1-p)\mathcal{E}[\sigma])] : 0 \leq \Lambda \leq \mathbb{1}\}$. Using the adjoint, we can express the action of the channel on the POVM element Λ, moving from the Schrödinger to Heisenberg pictures. The objective function then reads $\mathrm{Tr}[\mathcal{E}^*[\Lambda](p\rho - (1-p)\sigma)]$. Meanwhile, since $\mathcal{E}^*$ is completely positive and unital (cf. Exercises 5.3 and 5.8), applying it to the constraints gives $0 \leq \mathcal{E}^*[\Lambda] \leq \mathbb{1}$. Therefore $\Lambda' = \mathcal{E}^*[\Lambda]$ is a potential POVM element for the optimization, usually referred to as being *feasible*. Hence $P_{\text{guess}}(X|B)_\tau \geq P_{\text{guess}}(X|C)_\omega$, where $\omega_{XC} = \mathcal{E}_{C|B}[\tau_{XC}]$. The same argument implies $\beta_\alpha(\rho,\sigma) \leq \beta_\alpha(\mathcal{E}[\rho],\mathcal{E}[\sigma])$.

Though the setup of the Bayesian and Neyman–Pearson approaches are different, it turns out that the form of the optimal measurement is essentially the same in both cases. To demonstrate this, we need to examine the optimizations more carefully. Both the maximization in (9.3) and the minimization in (9.4) are *semidefinite programs*, a specific kind of *convex optimization.*

A convex optimization is an optimization in which the set of possible optimization variables (called the *feasible set*) is convex, while the *objective function* to be optimized is suitably convex or concave: convex for minimization and concave for maximization. Semidefinite programs are convex optimizations whose objective function is linear and whose variables are subject to positive semidefinite and linear constraints. The constraint $\Lambda \leq \mathbb{1}$ for $\Lambda \in \mathrm{Lin}(\mathcal{H})$ is an example of a semidefinite constraint, and $\mathrm{Tr}[\Lambda\rho] = \alpha$ is a linear constraint. Since the objective function is both convex and concave, it is always possible to find an optimal variable among the extreme points of the feasible set (if an optimal variable exists at all). We will develop most of the concepts and properties needed to work with SDPs here in the main text, but see Appendix C for a more general presentation.

Convex optimization has the very nice property that its optimizers can be recognized locally. In a general optimization problem, there are many local optima, which makes it difficult to determine the global optimal value. This does not occur in convex optimization. Consider the minimization of a convex function $f : \mathbb{R}^n \to \mathbb{R}$ and a point x that is the minimum of f in some small region $R \subseteq \mathbb{R}^n$ containing x. If x were not the global optimum, we would run into a contradiction. Suppose $y \notin R$ is such that $f(y) < f(x)$ and consider the points $z = (1-\lambda)x + \lambda y$ for $\lambda \in [0,1]$, which interpolate between x and y. For suitably small λ, $z \in R$. However, by convexity of f, $f(z) \leq (1-\lambda)f(x) + \lambda f(y) < f(x)$. This contradicts the assumption that x is the local optimizer.

9.2 Bayesian hypothesis testing

Let us first consider the Bayesian approach to hypothesis testing. It is relatively straightforward to work out the optimal Λ^* in (9.3). Since $p\rho - (1-p)\sigma$ is Hermitian, a good guess for $\Lambda^\star$ is just the projection onto the subspace of positive eigenvalues of $p\rho - (1-p)\sigma$. The negative eigenvalues would only decrease the objective function, so it is sensible to leave them out. We are free to include the zero eigenvalues or indeed set the value of Λ on this subspace to any value between zero and one, since these eigenvalues do not contribute to the objective function. Let us denote the projection by $\{p\rho - (1-p)\sigma > 0\}$, i. e., for arbitrary Hermitian M, $\{M > 0\}$ is the projection onto the subspace of positive eigenvalues. Similarly, $\{M \geq 0\}$ is the projection onto the nonnegative subspace, $\{M < 0\}$ is the projection onto the negative, and so forth.

In the classical case of commuting ρ and σ, our guess for the optimal measurement is based on the *likelihood ratio*. Let $\rho = \sum_x P(x)|x\rangle\langle x|$ and $\sigma = \sum_x Q(x)|x\rangle\langle x|$. Then $p\rho-(1-p)\sigma > 0$ for all $|x\rangle$ such that $P(x)/Q(x) > (1-p)/p$. The quantity $L(x) = P(x)/Q(x)$ is the likelihood ratio, and the measurement decides for ρ whenever x is such that $L(x)$ exceeds $(1-p)/p$. This threshold is the prior likelihood of σ relative to ρ, meaning that the decision is based on whether the observed likelihood $L(x)$ is more biased toward ρ than the prior probability is biased away from it.

Let us define $\Lambda(\gamma) = \{\rho - \gamma\sigma > 0\}$ for arbitrary $\gamma \geq 0$. Note that $\Lambda(\frac{1-p}{p}) = \{p\rho - (1-p)\sigma > 0\}$ since rescaling does not affect the operator inequality defining the projection. Certainly, $\Lambda(\frac{1-p}{p})$ is a reasonable guess for $\Lambda^\star$, but it is not immediately clear that it is in fact optimal. For one thing, should $\Lambda^\star$ necessarily commute with $p\rho - (1-p)\sigma$? To see that this is in fact the case, we derive an upper bound on the maximization that matches the optimal value. This leads to the *dual optimization*. First, let us define the *primal optimization*

$$f(M) = \sup_{\Lambda}\{\mathrm{Tr}[\Lambda M] : 0 \leq \Lambda \leq \mathbb{1}, \Lambda \in \mathrm{Lin}(\mathcal{H})\}, \tag{9.5}$$

now in terms of an arbitrary Hermitian input $M \in \mathrm{Lin}(\mathcal{H})$. We revert to using the supremum instead of maximum to illustrate the general setup. Using the supremum in a generic optimization not only accounts for the possibility, already mentioned, that the optimal value is not obtained, but also handles the case that the feasible set is empty. Then the value of the optimization is $-\infty$.

Neither is an issue here, of course. The feasible set is not empty, and the optimum value will be attained: Since the objective function $\Lambda \mapsto \mathrm{Tr}[\Lambda M]$ is continuous and the feasible set is closed and bounded (see Section B.5), the set of values the objective function can take is also closed. This will be the case for all the optimizations we consider in this book. Our considerations so far amount to the lower bound $f(M) \geq \mathrm{Tr}[\{M \geq 0\}M]$. Note that if M has zero eigenvalues, then we can take $\Lambda = \{M \geq 0\} + c\{M = 0\}$ for any $c \in [0,1]$ and get the same lower bound.

To derive an upper bound, we could use the method of Lagrange[4] multipliers, but for a linear problem, it is more direct to follow a method that goes back to von

4 Joseph-Louis Lagrange, born Giuseppe Lodovico Lagrangia, 1736–1813.

Neumann. Consider the constraint $\Lambda \leq \mathbb{1}$ from the optimization and take the Hilbert–Schmidt inner product of each side with some operator θ. If $\theta \geq 0$, then we obtain the inequality $\mathrm{Tr}[\Lambda\theta] \leq \mathrm{Tr}[\theta]$, since $\theta^{1/2}(\mathbb{1} - \Lambda)\theta^{1/2} \geq 0$ and the trace of positive operators is positive. Next, suppose that the left-hand side is larger than $\mathrm{Tr}[\Lambda M]$ for all feasible Λ, i. e. $\mathrm{Tr}[\Lambda M] \leq \mathrm{Tr}[\Lambda\theta]$. Then $f(M) \leq \mathrm{Tr}[\theta]$. Since $\Lambda \geq 0$, it is sufficient to require $\theta \geq M$ for the second inequality to hold. Therefore the tightest upper bound is given by

$$f^{\dagger}(M) = \inf_{\theta}\{\mathrm{Tr}[\theta] : \theta \geq M, \theta \geq 0, \theta \in \mathrm{Lin}(\mathcal{H})\}, \tag{9.6}$$

and $f(M) \leq f^{\dagger}(M)$. The fact that the primal is bounded by the dual in this way is known as *weak duality*. The difference $f^{\dagger}(M) - f(M)$ is called the *duality gap*. Again, we only use the infimum to illustrate the general case; here infeasibility causes the infimum to take the value $+\infty$.

Just as there was a clear guess for the primal optimal variable, there is also a clear guess for the optimal dual variable $\theta^{\star}$. Since we can work in the eigenbasis of M, take $\theta^{\star}$ to be the positive part of M, which we denote by $\{M\}_+$. Then $f^{\dagger}(M) \leq \mathrm{Tr}[\{M\}_+]$. In fact, this matches the lower bound we already obtained from the primal optimization: Because M and $\{M \geq 0\}$ commute, $\{M \geq 0\}M = \{M\}_+$. Hence $f(M) = f^{\dagger}(M) = \mathrm{Tr}[\{M\}_+]$. Equality of the primal and dual optimizations is *strong duality*. We have thus shown that strong duality holds for this optimization, though we had to find the optimizer to do it.

A nice closed-form expression comes from using the trace norm, which for Hermitian M is just $\|M\|_1 = \mathrm{Tr}[\{M\}_+] - \mathrm{Tr}[\{M\}_-]$. Since $\mathrm{Tr}[M] = \mathrm{Tr}[\{M\}_+] + \mathrm{Tr}[\{M\}_-]$, it follows that

$$f(M) = f^{\dagger}(M) = \tfrac{1}{2}(\mathrm{Tr}[M] + \|M\|_1). \tag{9.7}$$

Applied to the original problem, we find

$$P_{\text{guess}}(X|B)_{\tau} = \tfrac{1}{2}(1 + \|p\rho - (1-p)\sigma\|_1). \tag{9.8}$$

Exercise 9.1. Compute the probability of correctly distinguishing the qubit states $|0\rangle\langle 0|$ and $|+\rangle\langle +|$, assuming uniform prior probability.

Exercise 9.2. Express the probability of correctly distinguishing between two arbitrary qubit states ρ and σ with arbitrary prior probabilities p and $1 - p$ in terms of the Bloch representation.

9.3 Neyman–Pearson hypothesis testing

9.3.1 Testing region

Now let us turn to the Neyman–Pearson approach. Consider all the possible outcome probabilities for a given pair of states ρ and σ when ranging over the set of POVM elements. Let us denote this the *testing region* $\mathscr{R}(\rho,\sigma)$. It is a subset of the unit square in $\mathbb{R}^2$ defined as

$$\mathscr{R}(\rho,\sigma) := \{(\alpha,\beta) : \alpha = \mathrm{Tr}[\Lambda\rho], \beta = \mathrm{Tr}[\Lambda\sigma], 0 \le \Lambda \le \mathbb{1}\}. \tag{9.9}$$

Moreover, $\mathscr{R}(\rho,\sigma)$ is a convex set, since the convex combination of effects is still an effect operator. We are naturally interested in the extreme points of the set or, more generally, its boundary. Observe that $\mathscr{R}(\rho,\sigma)$ necessarily contains the diagonal, all points (c,c) for $c \in [0,1]$, by choosing $\Lambda = c\mathbb{1}$. This implies $\beta_\alpha(\rho,\sigma) \le \alpha$. Additionally, $(1-\alpha, 1-\beta) \in \mathscr{R}(\rho,\sigma)$ when $(\alpha,\beta) = \mathscr{R}(\rho,\sigma)$ since $\mathbb{1} - \Lambda$ is an effect if Λ is. Thus the upper boundary is the image of the lower boundary under a rotation around the point $(1/2, 1/2)$ by the angle π. The function β_α defined in (9.4) therefore completely characterizes $\mathscr{R}(\rho,\sigma)$. Figure 9.1 depicts a testing region of two probability distributions.

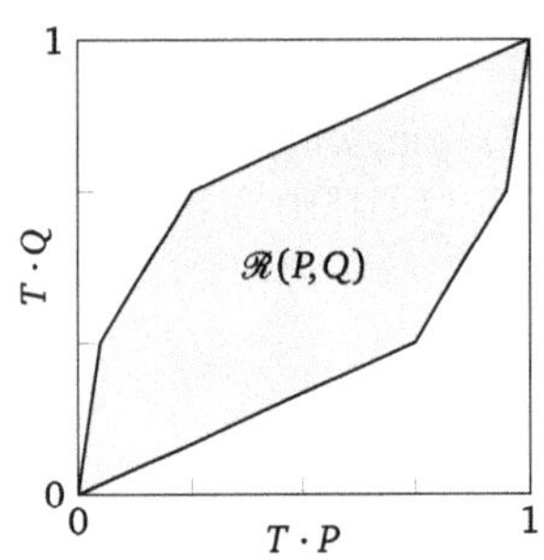

Figure 9.1: An example of a testing region $\mathscr{R}(P,Q)$. Here $P = (15,4,1)/20$ and $Q = (1,1,1)/3$.

Exercise 9.3. Determine $\mathscr{R}(P,Q)$ for $P = (p, 1-p)$ and $Q = (q, 1-q)$.

Exercise 9.4. Using (9.5), show that $\mathrm{Tr}[\Lambda(\gamma)\sigma] \le \frac{1}{\gamma}\mathrm{Tr}[\Lambda(\gamma)\rho]$ for $\Lambda(\gamma) = \{\rho - \gamma\sigma > 0\}$ as defined in Section 9.2.

As opposed to the Bayesian case, a good guess for the optimal measurement $\Lambda^\star$ for given α is not as clear. We can also try to get some insight from the dual optimization. In the previous example, there was only one constraint besides positivity of Λ, whereas here there are two. Hence we consider two dual variables m and θ associated with the constraints $\mathrm{Tr}[\Lambda\rho] = \alpha$ and $\Lambda \le \mathbb{1}$, respectively. The remaining constraint $\Lambda \ge 0$ will not need a dual variable, but will play a role in the derivation of the dual. Let us write the inequality constraint as $-\Lambda \ge -\mathbb{1}$ and take the inner product with $\theta \ge 0$ to obtain

$-\operatorname{Tr}[\Lambda\theta] \geq -\operatorname{Tr}[\theta]$. The reason for doing so will be evident momentarily. Adding m times the first constraint then yields

$$m\operatorname{Tr}[\Lambda\rho] - \operatorname{Tr}[\Lambda\theta] \geq m\alpha - \operatorname{Tr}[\theta]\,. \tag{9.10}$$

Observe that there is no requirement that m be positive, as the associated constraint is an equality. The primal optimization is a minimization for which upper bounds are easily obtained from feasible choices of Λ. To get a lower bound, we require

$$\operatorname{Tr}[\Lambda\sigma] \geq m\operatorname{Tr}[\Lambda\rho] - \operatorname{Tr}[\Lambda\theta]\,. \tag{9.11}$$

For this purpose, it is sufficient to require $\sigma \geq m\rho - \theta$ since $\Lambda \geq 0$. The tightest lower bound is thus $\beta_\alpha(\rho,\sigma) \geq \beta_\alpha^\dagger(\rho,\sigma)$ for

$$\beta_\alpha^\dagger(\rho,\sigma) := \max_{m,\theta}\{m\alpha - \operatorname{Tr}[\theta] : m\rho - \theta \leq \sigma, \theta \geq 0, \theta \in \operatorname{Lin}(\mathcal{H}), m \in \mathbb{R}\}\,. \tag{9.12}$$

Notice in the above construction that there is one dual variable for each constraint of the primal. This is the general pattern. Reviewing the derivation, it is also clear that we needed to express $\Lambda \leq \mathbb{1}$ as a lower bound so as to be consistent with obtaining a lower bound on $\beta_\alpha(\rho,\sigma)$. In general, it is convenient to write the inequality constraints of minimizations as lower bounds and maximizations as upper bounds.

We can construct the dual of the dual, but it is just the primal again.

Exercise 9.5. Confirm that the dual of $\beta_\alpha^\dagger(\rho,\sigma)$ is $\beta_\alpha(\rho,\sigma)$.

Exercise 9.6. Show that replacing $\operatorname{Tr}[\Lambda\rho] = \alpha$ with $\operatorname{Tr}[\Lambda\rho] \geq \alpha$ does not change the value of $\beta_\alpha(\rho,\sigma)$. What is the corresponding change in the dual?
Hint: Use the properties of $\mathscr{R}(\rho,\sigma)$.

Exercise 9.7. Using (9.12), show that for all $\alpha \in [0,1]$, $\gamma \geq 0$, and states ρ and σ,

$$\alpha - \gamma\beta_\alpha(\rho,\sigma) \leq \operatorname{Tr}[\Lambda(\gamma)\rho] - \gamma\operatorname{Tr}[\Lambda(\gamma)\sigma]\,. \tag{9.13}$$

Slater's[5] *condition* is an easily checked condition which implies strong duality. It states that if the primal (dual) is feasible and the dual (primal) is *strictly feasible*, then the duality gap is zero, and there exists a primal (dual) optimizer. Proposition C.2 in the appendix provides more details. In the present context, strict feasibility means that all equality constraints are satisfied and all inequality constraints are strictly satisfied. Thus, for instance, we can infer that $f(M)$ and $f^\dagger(M)$ from (9.5) and (9.6) must be equal by the fact that $\Lambda = \frac{1}{2}\mathbb{1}$ and $\theta = 2M + \mathbb{1}$ are strictly feasible for the respective optimizations. (The additional $\mathbb{1}$ handles rank-deficient M.)

5 Morton Lincoln Slater, 1921–2002.

Exercise 9.8. Show that strong duality holds for $\beta_\alpha(\rho,\sigma)$ and $\beta_\alpha^\dagger(\rho,\sigma)$ and that both primal and dual optimizers exist.

9.3.2 Optimal tests

Given that strong duality holds and the optimal values are both achieved, (9.10) and (9.11) must be satisfied with equality. This is known as *complementary slackness*; the reason for the name will become clear momentarily. Making use of the equality constraint $\mathrm{Tr}[\Lambda\rho] = \alpha$, equality in (9.10) just amounts to $\mathrm{Tr}[(\mathbb{1} - \Lambda^\star)\theta^\star] = 0$. Because this is the trace of the product of two positive operators, the product itself must be zero (see Lemma B.3), i. e., $(\mathbb{1} - \Lambda^\star)\theta^\star = 0$. Similarly, equality in (9.11) implies

$$\Lambda^\star(\sigma - m^\star\rho + \theta^\star) = 0\,. \tag{9.14}$$

Thus there cannot be "slack" in both the constraint and the associated dual variable. Either the constraint is satisfied with equality (called *binding*), or the dual variable is zero, or both. Since the conditions are formulated for operators, the constraint can be binding in a subspace and slack in its orthogonal complement; then the dual variable will necessarily be zero on the complement.

We can now try to determine the form of the optimizers by making use of the slackness conditions. First note that they are equivalent to

$$\Lambda^\star\theta^\star = \theta^\star \quad \text{and} \quad \Lambda^\star(m^\star\rho - \sigma) = \theta^\star\,. \tag{9.15}$$

Therefore, since $\Lambda^\star$, $\theta^\star$, and $m^\star\rho - \sigma$ are all Hermitian, they all must commute. Due to the positivity of $\theta^\star$ in the second equation, it follows that $\theta^\star$ is zero on the zero eigenspace of $m^\star\rho - \sigma$ and that $\Lambda^\star$ annihilates the negative part of $m^\star\rho - \sigma$. Using the second to substitute for $\theta^\star$ in the first gives $(\Lambda^\star)^2(m^\star\rho - \sigma) = \Lambda^\star(m^\star\rho - \sigma)$, meaning that $\Lambda^\star$ has eigenvalue 1 on the positive part of $m^\star\rho - \sigma$. Hence the slackness conditions imply that $\Lambda^\star = \{m^\star\rho - \sigma > 0\} + c\{m^\star\rho - \sigma = 0\}$ for some $c \in [0,1]$ and $\theta^\star = \{m^\star\rho - \sigma\}_+$. The remaining constraint $\mathrm{Tr}[\Lambda^\star\rho] = \alpha$ fixes the value of c. We have therefore shown that likelihood ratio tests are also optimal in $\beta_\alpha(\rho,\sigma)$. When the likelihood ratio is greater than $1/m^\star$, the test decides for ρ, whereas if the likelihood is exactly this value the test decides for ρ with probability c. This is precisely the *Neyman–Pearson lemma* of classical statistics.

Proposition 9.1 (Neyman–Pearson lemma). *For any states ρ, σ and $\alpha \in [0,1]$, there exist $m^\star > 0$ and $c \in [0,1]$ such that for $\Lambda^\star = \{m^\star\rho - \sigma > 0\} + c\{m^\star\rho - \sigma = 0\}$, it holds that* $\mathrm{Tr}[\Lambda^\star\rho] = \alpha$ *and* $\mathrm{Tr}[\Lambda^\star\sigma] = \beta_\alpha(\rho,\sigma)$.

For classical distributions P and Q, the $\{m^\star P - Q = 0\}$ portion of the optimal test arises from interpolation of likelihood tests $\Lambda(y)$. In general, the testing region is the convex hull of the points generated by the extreme points in the set of tests, i. e., projection operations. In the classical case, these projections form a discrete set, and therefore β_α must be a piecewise linear function joining the values obtained from such tests, in particular, from simple likelihood tests. However, the piecewise linear form of β_α does not generally hold in the quantum case.

Exercise 9.9. Determine $\beta_\alpha(|0\rangle\langle 0|, |+\rangle\langle +|)$ for $\alpha \in [0,1]$.

Exercise 9.10. Consider a tangent to $\mathscr{R}(\rho,\sigma)$ at the point (α, β_α) such that the testing region lies above the tangent line, and let $m(\alpha)$ be its slope. Show that $m(\alpha)\alpha - \beta_\alpha \geq m\,\mathrm{Tr}[\Lambda\rho] - \mathrm{Tr}[\Lambda\sigma]$ for all effects Λ. Combine this inequality with (9.7) to infer the Neyman–Pearson lemma. Conclude that the optimal m in $\beta^\dagger(\rho,\sigma)$ is $m(\alpha)$.

Exercise 9.11. Consider the distribution P_{XY} such that P_X is uniform and $X = Y$, along with the completely uniform distribution Q_{XY}. Show that $\beta_\alpha(P_{XY}, Q_{XY}) = \alpha$ for all $\alpha \in [0,1]$.

Exercise 9.12. Suppose X is a $\mathbb{Z}_2$-valued random variable with $P_X(1) = p < 1/2$. Determine $\beta_\alpha(P_{X^n}, Q_{X^n})$ for all $\alpha \in [0,1]$, where $P_{X^n} = P_X^{\times n}$, and Q_{X^n} is the uniform distribution. In particular, for $a_t = \sum_{j=0}^{t} \binom{n}{j}(1-p)^{n-j}p^j$ and $b_t = \frac{1}{2^n}\sum_{j=0}^{t}\binom{n}{j}$, along with $a_{-1} = b_{-1} = 0$, show that when $\alpha = (1-\lambda)a_{t-1} + \lambda a_t$ for some $t \in \{0,\dots,n\}$ and $\lambda \in (0,1)$,

$$\beta_\alpha(P_{X^n}, Q_{X^n}) = (1-\lambda)b_{t-1} + \lambda b_t\,. \tag{9.16}$$

Exercise 9.13. Show that $\beta_\alpha(\rho, \mathbb{1}) = \beta_\alpha(P, \mathbb{1})$ for all $\alpha \in [0,1]$ and all states ρ, where P is the distribution of eigenvalues of ρ.

Exercise 9.14. Show that if $0 \leq \sigma \leq \tau$, then $\beta_\alpha(\rho,\sigma) \leq \beta_\alpha(\rho,\tau)$ for any state ρ and $\alpha \in [0,1]$.

Exercise 9.15. Show that $\beta_\alpha(\rho,\sigma) = \beta_\alpha(\rho \oplus 0, \sigma \oplus \tau)$ for every $\tau \geq 0$. Here $\sigma \oplus \tau$ denotes the direct sum of operators, i. e., $|0\rangle\langle 0| \otimes \sigma + |1\rangle\langle 1| \otimes \tau$.

Exercise 9.16. Show that $\beta_\alpha(\tau \otimes \rho, \tau \otimes \sigma) = \beta_\alpha(\rho,\sigma)$ for all ρ, σ, τ.

Exercise 9.17. Show that for any CQ states $\tau_{XB} = \sum_{x\in\mathcal{X}} P_X(x)|x\rangle\langle x|_X \otimes \rho_B(x)$ and $\theta_{XB} = \sum_{x\in\mathcal{X}} P_X(x)|x\rangle\langle x|_X \otimes \sigma_B(x)$ sharing the same distribution P_X and all $\alpha \in [0,1]$,

$$\beta_\alpha(\tau_{XB}, \theta_{XB}) \leq \sum_{x\in\mathcal{X}} P_X(x)\beta_\alpha(\rho_B(x), \sigma_B(x))\,. \tag{9.17}$$

9.4 Distinguishability

In the case of $p = 1/2$ the guessing probability satisfies $P_{\text{guess}}(X|B)_\tau = \frac{1}{2}(1 + \frac{1}{2}\|\rho - \sigma\|_1)$. A sensible notion of the *distinguishability* of two quantum states ρ and σ is therefore captured by

$$\delta(\rho, \sigma) := 2P_{\text{guess}}(X|B)_\tau - 1 \tag{9.18}$$

for $\tau = \frac{1}{2}|0\rangle\langle 0|_X \otimes \rho_B + \frac{1}{2}|1\rangle\langle 1|_X \otimes \sigma_B$. The easier it is to determine which is the true state, the more distinguishable the pair, and vice versa.

By definition, $0 \leq \delta(\rho, \sigma) \leq 1$. The two limits are obtained by identical states, $\rho = \sigma$, and by completely disjoint states, $\rho\sigma = 0$, respectively. In the latter case the supports of the two states are orthogonal subspaces, so it is possible to make a projective measurement that can distinguish between them without error.

Using (9.8), the distinguishability has the closed form

$$\delta(\rho, \sigma) = \tfrac{1}{2}\|\rho - \sigma\|_1, \tag{9.19}$$

and via (9.7) the dual variational forms

$$\delta(\rho, \sigma) = \max_{\Lambda}\{\mathrm{Tr}[\Lambda(\rho - \sigma)] : 0 \leq \Lambda \leq \mathbb{1}\} \tag{9.20}$$

$$= \min_{\theta}\{\mathrm{Tr}[\theta] : \theta \geq \rho - \sigma, \theta \geq 0\}. \tag{9.21}$$

Moreover, the distinguishability measure is *faithful* in the sense that $\delta(\rho, \sigma) = 0$ if and only if $\rho = \sigma$. Any difference between the states leads to some bias in the guessing probability away from 1/2. This can be seen from the closed form or from (9.21): $\mathrm{Tr}[\theta] = 0$ for $\theta \geq 0$ if and only if $\theta = 0$, and therefore there is no positive part of $\rho - \sigma$. Interchanging ρ and σ, the same argument implies that there is no negative part, and hence $\rho = \sigma$.

Similarly, the case of disjoint ρ and σ is the only situation in which the distinguishability can achieve its maximal value. For the optimal $\Lambda^\star$ in (9.20), $\mathrm{Tr}[\Lambda^\star\rho] - \mathrm{Tr}[\Lambda^\star\sigma] = 1$; since each term is bounded between 0 and 1, the second must be zero, and the first must be unity. Hence $\Lambda^\star$ projects onto the support of ρ and annihilates that of σ, so the supports are disjoint.

Exercise 9.18. Show that for any three states $\rho, \sigma \in \mathrm{Lin}(\mathcal{H}_A)$ and $\tau \in \mathrm{Lin}(\mathcal{H}_B)$,

$$\delta(\rho_A \otimes \tau_B, \sigma_A \otimes \tau_B) = \delta(\rho_A, \sigma_A). \tag{9.22}$$

Exercise 9.19. Show that the distinguishability is invariant under unitary channels applied to both arguments.

Typically, (9.19) is taken as the definition of the distinguishability, though doing so is perhaps not quite right from a conceptual point of view. The guessing probability is the operationally meaningful quantity and therefore should be used in the definition. Had the guessing probability turned out to be related to some other norm, say the Hilbert–Schmidt norm, we would just as happily use that closed-form expression instead.

The distinguishability satisfies several important properties. Foremost is monotonicity under the action of an arbitrary quantum channel $\mathcal{E}$, as we saw in Section 9.1. Another important property is the *triangle inequality*. For any three states ρ, σ, and τ, we have

$$\delta(\rho,\sigma) \le \delta(\rho,\tau) + \delta(\tau,\sigma)\,. \tag{9.23}$$

To see this, suppose $\Lambda^\star$ is optimal in the variational form (9.20) for $\delta(\rho,\sigma)$. Then

$$\delta(\rho,\sigma) = \mathrm{Tr}\big[\Lambda^\star(\rho-\tau)\big] + \mathrm{Tr}\big[\Lambda^\star(\tau-\sigma)\big] \le \delta(\rho,\tau) + \delta(\tau,\sigma)\,. \tag{9.24}$$

Exercise 9.20. Prove the triangle inequality and monotonicity of the distinguishability using the variational form in (9.21).

Closely related to monotonicity is the *joint convexity* of the distinguishability, which means that the function δ is jointly convex in its arguments. Specifically, for every probability distribution P over $\mathcal{X}$ and collection of states $\{\rho(x)\}_{x\in\mathcal{X}}$ and $\{\sigma(x)\}_{x\in\mathcal{X}}$,

$$\delta\Big(\sum_{x\in\mathcal{X}} P(x)\rho(x), \sum_{x\in\mathcal{X}} P(x)\sigma(x)\Big) \le \sum_{x\in\mathcal{X}} P(x)\,\delta(\rho(x),\sigma(x))\,. \tag{9.25}$$

Monotonicity implies joint convexity for the following reason. Fix the probability distribution P and define the CQ states $\tau_{XB} = \sum_x P(x)|x\rangle\langle x|_X \otimes \rho_B(x)$ and $\theta_{XB} = \sum_x P(x)|x\rangle\langle x|_X \otimes \sigma_B(x)$. Then $\delta(\tau_{XB},\theta_{XB}) = \sum_x P(x)\,\delta(\rho(x),\sigma(x))$; this is especially easy to see using the closed-form expression for the distinguishability. Then (9.25) follows by monotonicity under the channel Tr_X.

In fact, joint convexity implies monotonicity as well. This requires a little more work to show. Let us first establish that joint convexity implies monotonicity under the partial trace. To do so, consider arbitrary states ρ_{BR} and σ_{BR}. Using (5.5), the partial trace over R can be written as $\rho_R \otimes \pi_R = \frac{1}{|R|^2}\sum_{j=1}^{|R|^2} W_R(j)\rho_{BR}W_R(j)^*$ for appropriate unitaries $W_R(j)$ and $\pi_R = \frac{1}{|R|}\mathbb{1}_R$. Using the shorthand $\mathcal{V}_R(j)[\rho_{BR}] = V_R(j)\rho_{BR}V_R(j)^*$, joint convexity implies

$$\delta(\rho_B\otimes\pi_R, \sigma_B\otimes\pi_R) \le \frac{1}{|R|^2}\sum_{j=1}^{|R|^2} \delta(\mathcal{V}_R(j)[\rho_{BR}], \mathcal{V}_R(j)[\sigma_{BR}])\,. \tag{9.26}$$

Since the distinguishability is invariant under unitaries, $\delta(\rho_B \otimes \pi_R, \sigma_B \otimes \pi_R) \le \delta(\rho_{BR},\sigma_{BR})$. Applying (9.22) gives the desired result, $\delta(\rho_B,\sigma_B) \le \delta(\rho_{BR},\sigma_{BR})$. The case of a general channel now follows from the above argument by first using the Stinespring representation $V_{BR|A}$ for a channel $\mathcal{E}_{B|A}$. Specifically, $\delta(\rho_A,\sigma_A) = \delta(\mathcal{V}_{BR|A}[\rho_A], \mathcal{V}_{BR|A}[\sigma_A]) \ge \delta(\mathrm{Tr}_R\circ\mathcal{V}_{BR|A}[\rho_A], \mathrm{Tr}_R\circ\mathcal{V}_{BR|A}[\sigma_A]) = \delta(\mathcal{E}_{B|A}[\rho_A], \mathcal{E}_{B|A}[\sigma_A])$.

Exercise 9.21. Show that $\delta(\rho,\sigma) \geq \alpha - \beta_\alpha(\rho,\sigma)$ for all $\alpha \in [0,1]$. For what value of α does equality hold?

Exercise 9.22. Extending the previous exercise, give a geometric interpretation of the lower boundary of $\mathscr{R}(\rho,\sigma)$ in terms of $\delta(\rho,\gamma\sigma)$ for $\gamma \in \mathbb{R}$.

Exercise 9.23. Show that if $\delta(\rho,\rho') \leq \varepsilon$, then $\beta_{\alpha+\varepsilon}(\rho,\sigma) \geq \beta_\alpha(\rho',\sigma)$ for all positive σ and $\alpha \in [0,1-\varepsilon]$.

9.5 Channel distinguishability

9.5.1 Definition

The notion of state distinguishability can be easily extended to the case of distinguishing channels. Suppose $\mathcal{E}_{B|A}$ and $\mathcal{F}_{B|A}$ are two channels from system A to system B. To distinguish them, we are free to choose the input to the channel and test the output as we like. Even more, we are free to generate an entangled state ρ_{AR} with arbitrary R, subject A to the channel, and then test the combined output system BR. Assuming a uniform prior probability, the average probability of correctly guessing which of $\mathcal{E}_{B|A}$ and $\mathcal{F}_{B|A}$ is the true channel is given by $\frac{1}{2}(1+\delta(\mathcal{E}_{B|A},\mathcal{F}_{B|A}))$ for

$$\begin{aligned}\delta(\mathcal{E}_{B|A},\mathcal{F}_{B|A}) := \underset{\rho_{AR},\Lambda_{BR}}{\text{supremum}} \quad & \mathrm{Tr}[\Lambda_{BR}(\mathcal{E}_{B|A}[\rho_{AR}] - \mathcal{F}_{B|A}[\rho_{AR}])] \\ \text{such that} \quad & \mathrm{Tr}[\rho_{AR}] = 1\,, \quad \Lambda_{BR} \leq \mathbb{1}_{BR}\,, \\ & \rho_{AR}, \Lambda_{BR} \geq 0\,. \end{aligned} \tag{9.27}$$

In the definition, we make use of the supremum since the dimension of the additional system R is not restricted. Just as in the case of distinguishing states, the channel distinguishability satisfies monotonicity and the triangle inequality.

Proposition 9.2 (Triangle inequality of distinguishability). *For any three channels $\mathcal{E}$, $\mathcal{F}$, and $\mathcal{G}$ with the same input and output spaces,*

$$\delta(\mathcal{E},\mathcal{G}) \leq \delta(\mathcal{E},\mathcal{F}) + \delta(\mathcal{F},\mathcal{G})\,. \tag{9.28}$$

Proposition 9.3 (Monotonicity of distinguishability). *For every two channels $\mathcal{F}_{C|B}$ and $\mathcal{F}'_{C|B}$ and any channels $\mathcal{E}_{B|A}$ and $\mathcal{G}_{D|C}$,*

$$\delta\big(\mathcal{G}\circ\mathcal{F}\circ\mathcal{E},\mathcal{G}\circ\mathcal{F}'\circ\mathcal{E}\big) \leq \delta\big(\mathcal{F},\mathcal{F}'\big)\,. \tag{9.29}$$

Note that these two statements also encompass the corresponding results for states, since we can regard states as channels with trivial input.

Exercise 9.24. Prove Propositions 9.2 and 9.3.

9.5.2 Composability

Due to these two properties, the notion of resource composability can be formalized in terms of the channel distinguishability. An actual resource $\mathcal{R}$ is said to simulate an ideal resource $\mathcal{R}'$ to within ε if $\delta(\mathcal{R},\mathcal{R}') \leq \varepsilon$. No experiment can distinguish them with probability better than $\frac{1}{2}(1+\varepsilon)$. Now suppose, as in the example of the Introduction, we are interested in composing two actual resources $\mathcal{R}$ and $\mathcal{S}$ (protocols for noisy channel coding and data compression, say) in the hopes that together they simulate the ideal resource (transmitting the output of the data source to the receiver), which is the composition of the ideal resources $\mathcal{R}'$ and $\mathcal{S}'$. By the triangle inequality and monotonicity the overall approximation error ε will be bounded by the sum of the approximation errors of either resource. Formally,

$$\begin{aligned}\delta(\mathcal{R}\circ\mathcal{S},\mathcal{R}'\circ\mathcal{S}') &\leq \delta(\mathcal{R}\circ\mathcal{S},\mathcal{R}'\circ\mathcal{S}) + \delta(\mathcal{R}'\circ\mathcal{S},\mathcal{R}'\circ\mathcal{S}') \\ &\leq \delta(\mathcal{R},\mathcal{R}') + \delta(\mathcal{S},\mathcal{S}')\,.\end{aligned} \tag{9.30}$$

9.5.3 SDP formulation

The optimization in definition (9.27) may look somewhat intractable, as it involves a potentially very large reference system R. However, it can be transformed into an SDP that makes no reference to R. To do so, first observe that without loss of generality we can restrict to pure ρ_{AR}. As the objective function is linear in ρ_{AR}, a mixed input will simply lead to a convex combination of objective functions with pure state inputs. We may simply select the largest term in the convex combination. We can then show the following:

Proposition 9.4 (Channel distinguishability SDP). *Let E_{BA} and E'_{BA} be the Choi operators of any two channels $\mathcal{E}_{B|A}$ and $\mathcal{E}'_{B|A}$, respectively. Then the channel distinguishability $\delta(\mathcal{E}_{B|A},\mathcal{E}'_{B|A})$ can be expressed as the following semidefinite program:*

$$\begin{aligned}\delta(\mathcal{E},\mathcal{E}') = \underset{\rho_A,\Gamma_{AB}}{\text{maximum}} \quad & \mathrm{Tr}\big[\Gamma_{BA}\big(E_{BA}-E'_{BA}\big)\big] \\ \text{such that} \quad & \mathrm{Tr}[\rho_A]=1\,,\quad \Gamma_{BA}\leq \mathbb{1}_B\otimes\rho_A^T\,, \\ & \rho_A\geq 0\,,\quad \Gamma_{BA}\geq 0\,.\end{aligned} \tag{9.31}$$

Proof. Since ρ_{AR} is pure, by (4.26) we can write, for some $K_{B|A}$,

$$\rho_{AR} = (\mathbb{1}_A\otimes K_{R|A'})\,\Omega_{AA'}\,(\mathbb{1}_A\otimes K^*_{R|A'})\,, \tag{9.32}$$

$$\rho_{AR}^{T_A} = (\mathbb{1}_A\otimes K_{R|A'})\,\Omega_{AA'}^{T_A}\,(\mathbb{1}_A\otimes K^*_{R|A'}) = (\mathbb{1}_A\otimes K_{R|A'})\,\Upsilon_{AA'}\,(\mathbb{1}_A\otimes K^*_{R|A'})\,. \tag{9.33}$$

Note that $\rho_A^T = \mathrm{Tr}_R[\rho_{AR}^{T_A}] = \mathrm{Tr}_R[K_{R|A'}\Upsilon_{AA'}(K_{R|A'})^*] = \mathrm{Tr}_{A'}[(K_{R|A'})^* K_{R|A'}\Upsilon_{AA'}]$. A useful property of the swap operator is that for any operators $O_{BA'}$ and O_{BA},

$$\mathrm{Tr}[O_{BA'}O'_{BA}\Upsilon_{AA'}] = \mathrm{Tr}[O_{BA}O'_{BA}]. \tag{9.34}$$

Therefore $\rho_A^T = (K_{R|A})^* K_{R|A}$. For the objective function, we now have $\mathrm{Tr}[\Lambda_{BR}(\mathcal{E}_{B|A}[\rho_{AR}] - \mathcal{E}'_{B|A}[\rho_{AR}])] = \mathrm{Tr}[(\mathbb{1}_B \otimes K^*_{R|A'})\Lambda_{BR}(\mathbb{1}_B \otimes K_{R|A'})(E_{BA} - E'_{BA})\Upsilon_{AA'}]$. If we define $\Gamma_{BA'} = (\mathbb{1}_B \otimes K^*_{R|A'})\Lambda_{BR}(\mathbb{1}_B \otimes K_{R|A'})$, then the objective function becomes simply $\mathrm{Tr}[\Gamma_{BA}(E_{BA} - E'_{BA})]$. The condition $0 \le \Lambda_{BR} \le \mathbb{1}$ translates to

$$0 \le \Gamma_{BA} = (\mathbb{1}_B \otimes K^*_{R|A})\Lambda_{BR}(\mathbb{1}_B \otimes K_{R|A}) \le \mathbb{1}_B \otimes (K_{R|A})^* K_{R|A} = \mathbb{1}_B \otimes \rho_A^T. \tag{9.35}$$

Hence we have

$$\begin{aligned} \delta(\mathcal{E}_{B|A}, \mathcal{E}'_{B|A}) \ge \underset{\rho_A, \Gamma_{AB}}{\text{maximum}} \quad & \mathrm{Tr}[\Gamma_{BA}(E_{BA} - E'_{BA})] \\ \text{such that} \quad & \mathrm{Tr}[\rho_A] = 1, \quad \Gamma_{BA} \le \mathbb{1}_B \otimes \rho_A^T, \\ & \rho_A \ge 0, \quad \Gamma_{BA} \ge 0. \end{aligned} \tag{9.36}$$

The transpose on ρ does not really serve any purpose in the optimization and can be safely omitted. However, by keeping it the purification of ρ is the optimal input test state. Moreover, $\rho_A \ge 0$ is redundant in light of the other two operator constraints.

We have an inequality in (9.36) because we relaxed the constraint on Λ_{BR} by conjugating with K: It could be that not every feasible Γ_{AB} comes from a Λ_{BR} in this way. However, equality holds. Given feasible ρ_A and Γ_{AR}, consider any "matrix square root" $K_{R|A}$ of ρ_A^T, i. e., $K_{R|A}$ such that $\rho_A^T = (K_{R|A})^* K_{R|A}$. The polar decomposition of $K_{R|A}$ is just $K_{R|A} = V_{R|A}\sqrt{\rho_A^T}$ for some isometry $V_{R|A}$. Now define $\Lambda_{BR} = V_{R|A}(\rho_A^T)^{-1/2}\Gamma_{AB}(\rho_A^T)^{-1/2}(V_{R|A})^*$. It is positive by construction, and the constraint $\Gamma_{AB} \le \rho_A^T \otimes \mathbb{1}_B$ becomes $\Lambda_{BR} \le V_{R|A}(V_{R|A})^* \otimes \mathbb{1}_B = \Pi_R \otimes \mathbb{1}_B \le \mathbb{1}_{BR}$, where Π_R is the projection onto the image of $V_{R|A}$. Therefore we have the desired statement. □

Exercise 9.25. Show that the dual form is as follows and that strong duality holds:

$$\begin{aligned} \delta(\mathcal{E}_{B|A}, \mathcal{E}'_{B|A}) = \underset{\lambda, T_{AB}}{\text{minimum}} \quad & \lambda \\ \text{such that} \quad & \lambda\mathbb{1}_A - \mathrm{Tr}_B[T_{BA}] \ge 0, \\ & T_{BA} \ge E_{BA} - E'_{BA}, \quad T_{BA} \ge 0. \end{aligned} \tag{9.37}$$

Of course, we expect that for a classical channel, the optimal experiment is to choose a fixed value of x at the input and then make a measurement to distinguish the output distributions, as with the statistical distance. Indeed, this works for all CQ channels.

Exercise 9.26. Show that for any two CQ channels $\mathcal{E}_{B|X}$ and $\mathcal{F}_{B|X}$,

$$\delta(\mathcal{E}_{B|X}, \mathcal{F}_{B|X}) = \max_{x \in \mathcal{X}} \delta(\mathcal{E}_{B|X=x}, \mathcal{F}_{B|X=x}) . \tag{9.38}$$

For $B = Y$ classical and $\mathcal{Y} = \mathcal{X}$, show that this leads to $\delta(W, I) = 1 - \min_{x \in \mathcal{X}} W_{Y|X=x}(x)$, where I is the identity channel.

Exercise 9.27. Consider an arbitrary classical channel $W_{Y|X}$ with $\mathcal{Y} = \mathcal{X}$, and let $P_{XY}(x,y) = P_{XX'}(x,y) = \frac{1}{|X|}\delta_{x,y}$. Show that $P_{\text{agree}}(W_{Y|X}) = 1 - \delta(P_{XY}, W_{Y|X'}P_{XX'})$.

9.5.4 Need for entanglement

Entanglement is useful for distinguishing quantum channels, even in the case that one of the channels is the identity, as shown in the following exercises. For certain channels, it is necessary for achieving the optimal distinguishability, though this is not always the case.

Exercise 9.28. Determine $\delta(\mathcal{E}, \mathcal{I})$ for an arbitrary Pauli channel $\mathcal{E}$. Give an optimal input state and measurement that works for all Pauli channels.

Exercise 9.29. Show that the optimal strategy for distinguishing the depolarizing channel $\rho \mapsto (1-q)\rho + q\,\mathrm{Tr}[\rho]\pi$ from the identity channel using a nonentangled input is $q/2$, whereas $3q/4$ is achievable using entangled states.

Exercise 9.30. Show that $\delta(\mathcal{N}_p, \mathcal{I}) = p$ for $\mathcal{N}_p$, the amplitude damping channel of (5.3) with damping parameter p.

A particularly dramatic example of the gap in distinguishability when using entangled versus unentangled inputs is given by the so-called Werner[6]–Holevo[7] channels in dimension d, which were considered in Exercise 5.23:

$$\mathcal{E}[\rho] = \frac{1}{d+1}(\mathrm{Tr}[\rho]\mathbb{1} + \rho^T) \quad \text{and} \quad \mathcal{F}[\rho] = \frac{1}{d-1}(\mathrm{Tr}[\rho]\mathbb{1} - \rho^T) . \tag{9.39}$$

Acting on system A of the maximally entangled state Φ_{AB}, the two channels produce $\mathcal{E}_A[\Phi_{AB}] = \frac{1}{d(d+1)}(\mathbb{1}_{AB} + \Upsilon_{AB})$ and $\mathcal{F}_A[\Phi_{AB}] = \frac{1}{d(d-1)}(\mathbb{1}_{AB} - \Upsilon_{AB})$. A direct calculation shows that the product of these two states is zero. (In fact, they are proportional to the projection operators on the symmetric and antisymmetric subspaces of $\mathcal{H}_A \otimes \mathcal{H}_B$, respectively.) Therefore $\delta(\mathcal{E}, \mathcal{F}) = 1$.

On the other hand, observe that $\mathcal{E}[\rho] - \mathcal{F}[\rho] = \frac{2}{d^2-1}(d\rho^T - \mathbb{1})$. Using the triangle inequality of the trace norm, it follows that $\delta(\mathcal{E}[\rho], \mathcal{F}[\rho]) = \frac{1}{d^2-1}\|d\rho^T - \mathbb{1}\|_1 \leq \frac{2d}{d^2-1}$.

6 Reinhard Frank Werner, born 1954.

7 Alexander Holevo, born 1943. Also transliterated as Kholevo.

This quantity tends to zero as $1/d$ for large d, and hence in principle there can be a dimension-dependent gap in the distinguishability when using entangled inputs and when not.

Exercise 9.31. Confirm that the distinguishability of the Z channel of Figure 3.1 to the identity channel is just r, the probability of an input 1 being output as 0. Now consider the equal convex combination of the Z channel with its bit-flipped version, in which the input 0 is output as 1 with probability r. Show that the resulting channel is just BSC($r/2$), and therefore the distinguishability has decreased to $r/2$. It is tempting to think that it should be possible to derandomize the latter scenario and restore the larger distinguishability. Why is this not the case?
Hint: What is the appropriate input state to use, given the choice of the two versions of the Z channel?

The channel distinguishability inherits joint convexity from the state distinguishability. This can be easily established using the dual formulation (9.37) and monotonicity applied to the heralded extension $\mathcal{E}_{XB|A}$ of an arbitrary convex combination $\sum_x P(x)\,\mathcal{E}_{B|A}(x)$ of channels, given by $\mathcal{E}_{XB|A} = \sum_x P(x)|x\rangle\langle x|_X \otimes \mathcal{E}_{B|A}(x)$.

Exercise 9.32. Show that for any collection of channels $\mathcal{E}_{B|A}(x)$ and $\mathcal{E}'_{B|A}(x)$ and probability distribution P,

$$\delta\Big(\sum_x P(x)\mathcal{E}_{B|A}(x), \sum_x P(x)\mathcal{E}'_{B|A}(x)\Big) \le \sum_x P(x)\,\delta(\mathcal{E}_{B|A}(x), \mathcal{E}'_{B|A}(x))\,. \tag{9.40}$$

Exercise 9.33. Show that the distinguishability of heralded version of the random choice of the two Z channels in Exercise 9.31 from the (heralded) identity channel is not increased over the unheralded case.

Exercise 9.34. Suppose $\mathcal{E}_{B|A}$ and $\mathcal{F}_{B|A}$ are any two channels, and let $\rho_{AB} = \mathcal{E}_{B|A'}[\Phi_{A'A}]$ and $\sigma_{AB} = \mathcal{F}_{B|A'}[\Phi_{A'A}]$ be the normalized versions of their associated Choi representatives. Show that, for the dimension d_A of the input system A,

$$\delta(\rho_{AB}, \sigma_{AB}) \le \delta(\mathcal{E}_{B|A}, \mathcal{F}_{B|A}) \le d_A \delta(\rho_{AB}, \sigma_{AB})\,. \tag{9.41}$$

Hint: Relax the constraints in (9.31) *by using* $\rho \le \mathbb{1}$.

9.6 Notes and further reading

Bayesian and Neyman–Pearson hypothesis testing of quantum states was first considered by Helstrom for projective measurements [137], who showed (9.8) in this case, and later extended by Holevo [143] to arbitrary POVMs. The classical Neyman–Pearson lemma is from [210] and was shown by Holevo in full generality in [143]. Here we follow an SDP approach taken in [86]; the more geometric approach of Exercise 9.10 follows

Holevo (see also Helstrom [138]). For more on semidefinite programming, including proof of Slater's condition; see Boyd and Vandenberghe, either their review [289] or their book on convex optimization [44], or Barvinok [13].

Channel distinguishability was considered by Kitaev [161] and Aharonov, Nisan, and Kitaev [1] in terms of the *completely bounded trace norm* on the one hand (often called the "diamond" norm as the notation is $\|\cdot\|_{\diamond}$), and on the other in terms of the completely bounded operator norm on the adjoint of the channel (working in the Heisenberg picture instead of the Schrödinger picture as here), as in Werner [299]. These norms are equivalent. For more, see Paulsen [217]. Using the trace norm formulation of the distinguishability in our definition leads to the characterization in terms of the completely bounded trace norm, i. e., the trace norm distinguishability extended by entangled inputs. Gilchrist, Langford, Nielsen [109] showed that the channel distinguishability is a convex optimization, and Watrous [297] showed that it can be formulated as a semidefinite program.

10 Fidelity

FIDELITY, n. A virtue peculiar to those who are about to be betrayed.

from *The Cynic's Word Book* by Ambrose Bierce

Continuing with the Bayesian approach to discrimination introduced in the previous chapter, we may ask what happens if the purifying system is included in the state or channel distinguishability task. This leads to the quantity known as fidelity, which is often much simpler to work with than the distinguishability itself. This chapter takes up the definition of the fidelity and explores many of its properties.

10.1 Definition

For states ρ and σ on $\mathcal{H}_A$, let $|\psi_\rho\rangle_{AR}$ and $|\psi_\sigma\rangle_{AR}$ denote purifications of the states to $\mathcal{H}_R$, respectively. It is not difficult to show that the distinguishability of the purifications $|\psi_\rho\rangle$ and $|\psi_\sigma\rangle$ is given by

$$\delta(\psi_\rho, \psi_\sigma)^2 = 1 - \left|\langle\psi_\rho|\psi_\sigma\rangle\right|^2 . \tag{10.1}$$

Since purifications are not unique, the best case distinguishability of the purifications of two density operators ρ and σ is in fact unity, as the supports of the purification states could be disjoint. For instance, $|\psi_\sigma\rangle_{AR}|0\rangle_{R'}$ and $|\psi_\sigma\rangle_{AR}|1\rangle_{R'}$ are legitimate purifications. Thus, to have a meaningful measure, we should consider the worst-case distinguishability when using the purification. In light of (10.1), we define the fidelity $F(\rho, \sigma)$ of two density operators ρ and σ on $\mathcal{H}_A$ to be the largest possible overlap of their purifications,

$$F(\rho, \sigma) := \sup_R \max_{|\psi_\rho\rangle, |\psi_\sigma\rangle} \left|\langle\psi_\rho|\psi_\sigma\rangle_{AR}\right| . \tag{10.2}$$

The supremum is taken over purifying systems, which can in principle be of arbitrary dimension. There is little difference in optimizing the overlap versus the square of the overlap, and the reader is advised that both choices are made in the literature. We will need to have $|R|$ larger than the rank of either state to be sure there are purifications for both states on $\mathcal{H}_A \otimes \mathcal{H}_R$.

From the definition it is clear that $F(\rho, \sigma) \in [0, 1]$. The most important property of the fidelity, monotonicity under channel action, follows immediately from the definition.

https://doi.org/10.1515/9783110570250-010

Proposition 10.1 (Monotonicity of the fidelity). *For any density operators ρ_A, σ_A and a channel $\mathcal{E}_{B|A}$,*

$$F(\mathcal{E}_{B|A}[\rho_A], \mathcal{E}_{B|A}[\sigma_A]) \geq F(\rho_A, \sigma_A) . \tag{10.3}$$

Equality holds if $\mathcal{E}_{B|A}$ is an isometry, i. e., $\mathcal{E}_{B|A}[\tau_A] = V_{B|A}\tau_A V^*_{B|A}$ for isometry $V_{B|A}$.

Proof. For convenience, let $\rho'_B = \mathcal{E}_{B|A}[\rho_A]$ and $\sigma'_B = \mathcal{E}_{B|A}[\sigma_A]$. Take the equality case first and observe that for every purification $|\psi\rangle_{AE}$ of ρ_A and $|\varphi\rangle_{AE}$ of σ_A, $V_{B|A}|\psi\rangle_{AE}$ and $V_{B|A}|\varphi\rangle_{AE}$ are purifications of ρ'_B and σ'_B, respectively. Conversely, since ρ'_B and σ'_B are supported in the image of $V_{B|A}$ by construction, for every purification $|\psi'\rangle_{BE}$ of ρ'_B and $|\varphi'\rangle_{BE}$ of σ'_B, the states $V^*_{B|A}|\psi'\rangle_{BE}$ and $V^*_{B|A}|\varphi'\rangle_{BE}$ are purifications of ρ_A and σ_A, respectively. Therefore $\langle\psi|\varphi\rangle = \langle\psi|V^*V|\varphi\rangle$ and $\langle\psi'|\varphi'\rangle = \langle\psi'|VV^*|\varphi'\rangle$ imply that the fidelities are equal.

For a general channel $\mathcal{E}_{B|A}$, let $V_{BR|A}$ be a Stinespring isometry and define $|\psi'\rangle_{BRE} = V_{BR|A}|\psi\rangle_{AE}$ and $|\varphi'\rangle_{BRE} = V_{BR|A}|\varphi\rangle_{AE}$. The equality condition implies that $F(\rho_A, \sigma_A) = F(\psi'_{BR}, \varphi'_{BR})$. Thus it only remains to show that the fidelity cannot decrease under partial trace, so that $F(\rho'_B, \sigma'_B) = F(\psi'_B, \varphi'_B) \geq F(\psi'_{BR}, \varphi'_{BR})$. However, this follows at once since every purification of a joint state is a purification of its marginals. □

10.2 Closed-form expression

Despite the optimization in the definition, it turns out that the fidelity has a relatively simple closed-form expression. Using the Schmidt decomposition, we express purifications of arbitrary ρ and σ as $|\psi_\rho\rangle = \sum_{k=1}^{|A|} \sqrt{\lambda_k}|\lambda_k\rangle_A \otimes |\xi_k\rangle_R$ and $|\psi_\sigma\rangle = \sum_{k=1}^{|A|} \sqrt{\theta_k}|\theta_k\rangle_A \otimes |\eta_k\rangle_R$, where now we include zero eigenvalues into the expression so that the summations run over complete orthonormal bases in system A. The overlap is therefore

$$\langle\psi_\rho|\psi_\sigma\rangle_{AR} = \sum_{k=1}^{|A|}\sum_{k'=1}^{|A|} \sqrt{\lambda_k\theta_{k'}}\,\langle\lambda_k|\theta_{k'}\rangle\langle\xi_k|\eta_{k'}\rangle . \tag{10.4}$$

Note that this expression implies

$$F(\rho, \sigma) = F(\Pi_\sigma\rho\Pi_\sigma, \sigma) , \tag{10.5}$$

where Π_σ is the projector onto the support of σ, since only those $|\theta_k\rangle$ in the support of σ contribute to the sum. By the same argument, $F(\rho, \sigma) = F(\rho, \Pi_\rho\sigma\Pi_\rho)$.

Rewriting the overlap expression a little gives

$$\begin{aligned}\langle\psi_\rho|\psi_\sigma\rangle_{AR} &= \mathrm{Tr}\left[\sqrt{\sigma}\sum_{k=1}^{d}\sum_{k'=1}^{d} |\theta_{k'}\rangle\langle\xi_k|\eta_{k'}\rangle\langle\lambda_k|\sqrt{\rho}\right] \\ &= \mathrm{Tr}\left[\sqrt{\sigma}\sum_{k=1}^{d}\sum_{k'=1}^{d} |\theta_{k'}\rangle\langle\overline{\eta}_{k'}|\overline{\xi}_k\rangle\langle\lambda_k|\sqrt{\rho}\right] = \mathrm{Tr}[\sqrt{\rho}\sqrt{\sigma}U] ,\end{aligned} \tag{10.6}$$

where $U = \sum_{k,k'=1}^{d} |\theta_{k'}\rangle\langle\overline{\eta}_{k'}|\overline{\xi}_k\rangle\langle\lambda_k|$. Until now we have not placed any constraints on the states $\{|\xi_k\rangle_R\}$ and $\{|\eta_k\rangle_R\}$, apart from the fact that they are both orthogonal sets, in accordance with the Schmidt decomposition. However, to maximize the overlap, these sets should span the same subspace of R. Hence U must be a unitary operator since it is the product of two maps, each of which takes an orthonormal set to another orthonormal set of the same size, namely $\sum_k^d |\overline{\xi}_k\rangle\langle\lambda_k|$ and $\sum_{k'=1}^d |\theta_{k'}\rangle\langle\overline{\eta}_{k'}|$. Therefore we have $F(\rho,\sigma) = \max_U |\operatorname{Tr}[\sqrt{\rho}\sqrt{\sigma}U]|$. By Lemma B.5 this is just $\|\sqrt{\rho}\sqrt{\sigma}\|_1$. Thus we have shown the following:

Proposition 10.2 (Uhlmann's theorem). *For any two density operators $\rho, \sigma \in \operatorname{Lin}(\mathcal{H})$,*

$$F(\rho,\sigma) = \|\sqrt{\rho}\sqrt{\sigma}\|_1 . \tag{10.7}$$

Furthermore, there exist optimal purifications of ρ and σ in $\mathcal{H}\otimes\mathcal{H}$. In particular, for U such that $\sqrt{\rho}\sqrt{\sigma}U = |\sqrt{\rho}\sqrt{\sigma}|$, the optimal purifications are $\sqrt{\rho}\otimes\mathbb{1}|\Omega\rangle$ and $\sqrt{\sigma}\otimes U^T|\Omega\rangle$, respectively.

The trace norm expression is often written $F(\rho,\sigma) = \operatorname{Tr}[\sqrt{\sqrt{\sigma}\rho\sqrt{\sigma}}]$, which is a little more explicit but less obviously symmetric in the arguments.

Exercise 10.1. Using this form, show that

$$F(\rho, |\psi\rangle\langle\psi|)^2 = \langle\psi|\rho|\psi\rangle . \tag{10.8}$$

When ρ and σ commute, so that their eigenvalues define the probability distributions P and Q, respectively, the expression simplifies to $F(\rho,\sigma) = \sum_k \sqrt{P(k)\,Q(k)}$. In classical information theory, this is known as the *Bhattacharyya*[1] *coefficient.*

Exercise 10.2. Show that the fidelity is multiplicative, that is, $F(\rho\otimes\rho', \sigma\otimes\sigma') = F(\rho,\sigma)\,F(\rho',\sigma')$ for all positive ρ, ρ', σ, and σ',.

Exercise 10.3. Following the argument for joint convexity of the distinguishability in (9.25), show that in this case monotonicity implies joint concavity of the fidelity.

10.3 SDP formulation

Perhaps surprisingly, the fidelity $F(\rho,\sigma)$ can be formulated as the optimum value of an SDP in which the states ρ and σ appear linearly in the optimization. Consider a purification of the operator

$$\Theta = \begin{pmatrix} \rho & X \\ X^* & \sigma \end{pmatrix} , \tag{10.9}$$

1 Anil Kumar Bhattacharyya, 1915–1996.

where we restrict attention to those X such that $\Theta \geq 0$. We can regard Θ as a (non-normalized) state on AB, where ρ and σ are states on B, and A is a qubit, i. e., $\Theta_{AB} = |0\rangle\langle 0|_A \otimes \rho_B + |0\rangle\langle 1|_A \otimes X_B + |1\rangle\langle 0|_A \otimes X_B^* + |1\rangle\langle 1|_A \otimes \sigma_B$. Since Θ_{AB} is positive, it has a purification $|\Theta\rangle_{ABR}$, and this can be written

$$|\Theta\rangle_{ABR} = |0\rangle_A \otimes |\psi\rangle_{BR} + |1\rangle_A \otimes |\varphi\rangle_{BR} \tag{10.10}$$

for some vectors $|\psi\rangle_{BR}$ and $|\varphi\rangle_{BR}$. By the form of Θ_{AB} these vectors must be purifications of ρ_B and σ_B, respectively. Therefore the trace of X_B is the overlap of the purifications, since $\mathrm{Tr}[X] = \langle\varphi|\psi\rangle$ The overlap need not be real-valued, but taking the real part leads to the following SDP:

$$\begin{aligned} F(\rho,\sigma) = \underset{X}{\text{maximum}} \quad & \tfrac{1}{2}\,\mathrm{Tr}[X + X^*] \\ \text{such that} \quad & \begin{pmatrix} \rho & X \\ X^* & \sigma \end{pmatrix} \geq 0\,. \end{aligned} \tag{10.11}$$

To see that this equality statement is correct, i. e., that the optimal value of the SDP does deliver the fidelity, note that the overlap of any two purifications can be made real by adjusting the global phase of one of the pure states. Hence the maximization is effectively over the absolute value of the overlap.

Exercise 10.4. Prove monotonicity of the fidelity using the SDP formulation. Show that, in fact, monotonicity holds for all positive arguments ρ and σ (not necessarily of unit trace) and superoperators $\mathcal{E}$, which do not decrease the trace and for which $\mathcal{E} \otimes \mathcal{I}_2$ is positive, with $\mathcal{I}_2$ the identity superoperator on a two-dimensional system.

Exercise 10.5. Confirm that for σ a pure state, $X = \rho\sigma / F(\rho,\sigma)$ is the optimizer in (10.11). *Hint: Make use of* (10.5).

The optimization is not in our usual SDP form, but this is easily done. Instead of X as the variable and the stated block-operator constraint, take the block operator $W = \left(\begin{smallmatrix} U & X \\ X^* & V \end{smallmatrix}\right)$ as the variable subject to the constraints $W = 0$, $U = \rho$, and $V = \sigma$. Using the composite system AB from above to express the block form, the objective function is $\mathrm{Tr}[\frac{1}{2}((\sigma_x)_A \otimes \mathbb{1}_B)W_{AB}]$, and the constraints involving the states can be expressed as $\mathrm{Tr}_A[|0\rangle\langle 0|_A W_{AB}] = \rho_B$ and $\mathrm{Tr}_A[|1\rangle\langle 1|_A W_{AB}] = \sigma_B$.

Now we can more easily derive the dual. It has two variables from the two constraints, denote them $\frac{1}{2}Y$ and $\frac{1}{2}Z$. The prefactors will be useful momentarily. Taking the inner product with the corresponding constraints and adding the two results gives

$$\mathrm{Tr}[\tfrac{1}{2}(|0\rangle\langle 0|_A \otimes Y_B + |1\rangle\langle 1|_A \otimes Z_B)W_{AB}] = \tfrac{1}{2}\,\mathrm{Tr}[Y\rho] + \tfrac{1}{2}\,\mathrm{Tr}[Z\sigma]\,. \tag{10.12}$$

Here we require only Hermiticity of Y and Z for the equality to hold. To ensure that the left-hand side is larger than the objective function, it suffices to require

$$(\sigma_x)_A \otimes \mathbb{1}_B \leq |0\rangle\langle 0|_A \otimes Y_B + |1\rangle\langle 1|_A \otimes Z_B\,. \tag{10.13}$$

Rewriting in block form yields the upper bound $F(\rho,\sigma) \leq F^\dagger(\rho,\sigma)$ with

$$\begin{aligned} F^\dagger(\rho,\sigma) = \underset{Y,Z}{\text{minimum}} \quad & \tfrac{1}{2}(\mathrm{Tr}[Y\rho] + \mathrm{Tr}[Z\sigma]) \\ \text{such that} \quad & \begin{pmatrix} Y & -\mathbb{1} \\ -\mathbb{1} & Z \end{pmatrix} \geq 0\,. \end{aligned} \tag{10.14}$$

In fact, the Slater criterion implies that $F^\dagger(\rho,\sigma) = F(\rho,\sigma)$, and the infimum is attained by some optimal $Y^\star$ and $Z^\star$. Simply take $W = |0\rangle\langle 0|_A \otimes \rho_B + |1\rangle\langle 1|_A \otimes \sigma_B$ (i. e., $X = 0$) as strictly feasible in the primal and $Y = Z = 2\mathbb{1}$ in the dual. Having established strong duality, we can then learn some interesting properties of the optimal solution from the complementary slackness conditions. Equality of the primal and dual implies, from (10.13), that

$$\mathrm{Tr}\left[\begin{pmatrix} \rho & X \\ X^* & \sigma \end{pmatrix}\begin{pmatrix} Y & -\mathbb{1} \\ -\mathbb{1} & Z \end{pmatrix}\right] = 0\,. \tag{10.15}$$

Since this is the trace of the product of two positive operators, the product itself must be zero (see Lemma B.3). Therefore the optimal X, Y, and Z satisfy $\rho = XZ$, $\sigma = X^*Y$, $X = \rho Y$, and $X^* = \sigma Z$. This tells us that $\mathrm{Tr}[Y\rho] = \mathrm{Tr}[Z\sigma]$ as well as $\rho = \rho YZ$ and $\sigma = \sigma ZY$. Thus $Y = Z^{-1}$ at least on the support of ρ and the support of σ.

Exercise 10.6. Show that for $\rho = |\varphi\rangle\langle\varphi|$ and $\sigma = |\psi\rangle\langle\psi|$, with phases chosen such that $F(\rho,\sigma) = \langle\varphi|\psi\rangle$, the choice $Y = \sigma/F$ and $Z = \rho/F$ is optimal.

10.4 Further properties

10.4.1 Achievability by measurement

The slackness conditions allow us to prove Alberti's[2] characterization of the fidelity,

Proposition 10.3 (Alberti's theorem). *For any two density operators ρ and σ,*

$$\begin{aligned} F(\rho,\sigma)^2 = \underset{Z}{\text{minimum}} \quad & \mathrm{Tr}[Z\sigma]\ \mathrm{Tr}\big[Z^{-1}\rho\big] \\ \textit{such that} \quad & Z > 0 \textit{ on the support of } \rho\,. \end{aligned} \tag{10.16}$$

2 Peter M. Alberti.

Proof. By (10.5) we can restrict σ to the support of ρ and then compute the fidelity in this restricted subspace. Then, by the slackness conditions, we can replace Y by Z^{-1} in (10.14). The constraint reduces to $Z > 0$ by the properties of the *Schur complement* from Section B.8. The objective function becomes $\frac{1}{2}(\mathrm{Tr}[Z\sigma] + \mathrm{Tr}[Z^{-1}\rho])$, and the inequality of arithmetic and geometric means implies the squared fidelity is larger than the optimal value in (10.16). However, note that it is always possible to find an optimal Z in (10.16) such that the two factors are equal simply by rescaling Z. Then the arithmetic and geometric means are equal, so a feasible pair (Y, Z) for (10.14) can be constructed from the optimal Z in (10.16) such that the objective functions are equal. Therefore the two optimizations have the same optimal value. □

Alberti's theorem implies that there exists a projective measurement with projectors $\Pi(x)$ such that $F(\rho, \sigma) = F(P, Q)$ for $P(x) = \mathrm{Tr}[\Pi(x)\rho]$ and $Q(x) = \mathrm{Tr}[\Pi(x)\sigma]$. To see this, first, observe that by monotonicity it follows that $F(\rho, \sigma) \le F(P, Q)$ for any measurement $\Pi(x)$. On the other hand, consider the eigenbasis of an optimal Z. For $Z = \sum_x v_x \Pi(x)$, we have

$$F(\rho, \sigma)^2 = \sum_x v_x \mathrm{Tr}[\Pi(x)\sigma] \sum_{x'} v_{x'}^{-1} \mathrm{Tr}[\Pi(x)\rho]\,. \tag{10.17}$$

Defining $u_x = \sqrt{v_x Q(x)}$ and $v_x = \sqrt{v_x^{-1} P(x)}$, this is $F(\rho, \sigma)^2 = (u \cdot u)(v \cdot v)$, and by Cauchy[3]–Schwarz[4] we have

$$F(\rho, \sigma) \ge |u \cdot v| = \left| \sum_x \sqrt{P(x)\,Q(x)} \right| = F(P, Q)\,. \tag{10.18}$$

Hence both inequalities hold, meaning that $F(\rho, \sigma) = F(P, Q)$ for measurement in the eigenbasis of the optimal Z.

Exercise 10.7. Using the singular value decomposition, show that the operator $Z = \sigma^{-1/2}(\sigma^{1/2}\rho\sigma^{1/2})^{1/2}\sigma^{-1/2}$ satisfies $Z^{-1} = \rho^{-1/2}(\rho^{1/2}\sigma\rho^{1/2})^{1/2}\rho^{-1/2}$. Therefore this Z is optimal in Alberti's theorem, and measurement in its eigenbasis does not change the fidelity.

10.4.2 Bounds between fidelity and distinguishability

It turns out that the worst-case distinguishability when using the purification is not terribly different from the distinguishability when not using the purification at all.

3 Augustin-Louis Cauchy, 1789–1857.

4 Karl Hermann Amandus Schwarz, 1843–1921.

Proposition 10.4. *For any two density operators ρ and σ,*

$$\delta(\rho,\sigma) + F(\rho,\sigma) \geq 1, \tag{10.19}$$

$$\delta(\rho,\sigma)^2 + F(\rho,\sigma)^2 \leq 1. \tag{10.20}$$

Proof. Suppose P and Q are the classical distributions resulting from the optimal measurement for the fidelity as above. Then we have $\delta(\rho,\sigma) \geq \delta(P,Q)$. But $\delta(P,Q) \geq 1 - F(P,Q)$ by the following argument:

$$\begin{aligned} \delta(P,Q) &= \tfrac{1}{2}\sum_x |P_x - Q_x| = \tfrac{1}{2}\sum_x \left|\sqrt{P_x} - \sqrt{Q_x}\right|\left(\sqrt{P_x} + \sqrt{Q_x}\right) \\ &\geq \tfrac{1}{2}\sum_x \left(\sqrt{P_x} - \sqrt{Q_x}\right)^2 = 1 - F(P,Q)\,. \end{aligned} \tag{10.21}$$

To show the second, suppose ψ and φ are the optimal purifications of ρ and σ from Uhlmann's theorem. By (10.1), $\delta(\psi,\varphi)^2 + F(\rho,\sigma)^2 = 1$. Monotonicity of the distinguishability finishes the argument. □

The case of pure states saturates the quadratic bound, as in (10.1). Meanwhile, the linear bound is saturated by the classical distributions $P = (1-t, t, 0)$ and $Q = (1-t, 0, t)$ for any $t \in [0,1]$. Observe that the linear bound is obtained by applying monotonicity of a quantity (distinguishability) to the measurement channel achieving another quantity (fidelity). This is a useful technique we will often employ. Indeed, it is one option for proving the bound in Exercise 9.21. Here are some other examples.

Exercise 10.8. Show that if one of ρ and σ is pure, then $\delta(\rho,\sigma) + F(\rho,\sigma)^2 \geq 1$. Furthermore, show that the same bound holds for any two states of dimension two. *Hint: Use monotonicity of the distinguishability under the measurement achieving the fidelity. Optimize over the free parameters.*

Exercise 10.9. Follow the strategy of the previous exercise but use the optimal distinguishing measurement in monotonicity of the fidelity. Show that this leads back to (10.20).

Exercise 10.10. Suppose a state ρ is measured with a POVM and the outcome corresponding to POVM element Λ occurs, such that the postmeasurement state is $\rho' = \sqrt{\Lambda}\rho\sqrt{\Lambda}/\operatorname{Tr}[\Lambda\rho]$. Show that the state does not change much, provided that the probability $\operatorname{Tr}[\Lambda\rho]$ is large: $F(\rho,\rho')^2 \geq \operatorname{Tr}[\Lambda\rho]$. *Hint: Use $\sqrt{\Lambda} \geq \Lambda$ for $\Lambda \geq 0$.*

10.4.3 Triangle inequality

Although the fidelity is in some sense derived from the distinguishability, due to its form, it does not satisfy the triangle inequality. However, if we interpret the overlap as defining an angle between vectors, then it is not too difficult to show that the angle itself does. In particular, we have the following:

Proposition 10.5 (Triangle inequality of the fidelity). *For all quantum states ρ, σ, and τ,*

$$\arccos F(\rho,\sigma) \leq \arccos F(\rho,\tau) + \arccos F(\tau,\sigma)\,. \tag{10.22}$$

Equivalently, with $x = F(\rho,\tau)$ and $y = F(\tau,\sigma)$,

$$F(\rho,\sigma) \geq xy - \sqrt{1-x^2}\sqrt{1-y^2}\,. \tag{10.23}$$

Equivalence of the two forms is easily seen by setting $x = \cos\alpha$ and $y = \cos\beta$ in the latter expression and then substituting $\alpha = \arccos F(\rho,\tau)$ and $\beta = \arccos F(\tau,\sigma)$. Observe that the bound is only useful when the right-hand side of the latter expression is positive, which amounts to $F(\rho,\tau)^2 + F(\tau,\sigma)^2 > 1$.

Proof. For a given purification $|\theta\rangle$ of τ, we can choose $|\varphi\rangle$ to be a purification of ρ such that $F(\rho,\tau) = \langle\varphi|\theta\rangle$. Continuing, we may choose $|\psi\rangle$ to be a purification of σ such that $F(\tau,\sigma) = \langle\theta|\psi\rangle$. Finally, by adjusting the phases of all three states we can ensure that $\langle\varphi|\psi\rangle \geq 0$. Therefore the three vectors can be represented in $\mathbb{R}^3$ and $F(\rho,\sigma) \geq \langle\varphi|\psi\rangle$. We are interested in finding the smallest $\langle\varphi|\psi\rangle$ given this setup, so that the resulting bound on $F(\rho,\sigma)$ holds for all states.

Geometrically, the bound is the intuitive statement that for three vectors in $\mathbb{R}^3$ such that the angles between two pairs are fixed, the maximal angle for the third pair occurs when the vectors lie in a plane. Though it is perhaps unnecessary to give a more explicit proof, let us take the opportunity to further illustrate the technique of working with SDPs. Consider the matrix formed by the overlaps of these three vectors, their *Gram*[5] *matrix*,

$$G = \begin{pmatrix} 1 & x & z \\ x & 1 & y \\ z & y & 1 \end{pmatrix} \tag{10.24}$$

for $x = \langle\varphi|\theta\rangle$, $y = \langle\theta|\psi\rangle$, and $z = \langle\varphi|\psi\rangle$. Every Gram matrix of a collection of vectors $|v_k\rangle$ is positive semidefinite, since it is just $G = T^*T$ for $T = \sum_k |v_k\rangle\langle k|$. Thus, we are looking for $f(x,y) = \inf_z\{z : G \geq 0\}$, which is an SDP.

It is a good guess that the minimal z will occur when there is one zero eigenvalue and the other two are positive. This also accords with the geometric intuition that in the optimal configuration the vectors lie in plane: The rank of a Gram matrix is the dimension of the span of the associated vectors, which follows because T^*T and TT^* have the same nonzero eigenvalues (see Appendix B.6).

5 Jørgen Pedersen Gram, 1850–1916.

Again, we make use of the Schur complement. The upper left 2×2 block $A = \left(\begin{smallmatrix} 1 & x \\ x & 1 \end{smallmatrix}\right)$ is positive when $x \le 1$, and its inverse is simply $A^{-1} = \frac{1}{1-x^2}\left(\begin{smallmatrix} 1 & -x \\ -x & 1 \end{smallmatrix}\right)$. By Lemma B.6 positivity of G is therefore equivalent to $1 - (z, y)^T A^{-1}(z, y) \ge 0$, which is just $x^2 + y^2 + z^2 \ge 1 + 2xyz$. Setting this equal to zero and using the quadratic formula yields $z = xy \pm \sqrt{x^2y^2 + (1 - x^2 - y^2)}$. The choice $z = xy - \sqrt{1 - x^2}\sqrt{1 - y^2}$ would therefore seem to be the minimal value.

To be certain, we turn to the dual. It is given by

$$
\begin{aligned}
f^\dagger(x, y) = \underset{J}{\text{supremum}} \quad & -\mathrm{Tr}\left[J \begin{pmatrix} 1 & x & 0 \\ x & 1 & y \\ 0 & y & 1 \end{pmatrix}\right] \\
\text{such that} \quad & \mathrm{Tr}\left[J \begin{pmatrix} 0 & 0 & 1 \\ 0 & 0 & 0 \\ 1 & 0 & 0 \end{pmatrix}\right] \le 1, \quad J \ge 0 .
\end{aligned}
\tag{10.25}
$$

Using the slackness conditions as a guide, we can construct a feasible dual variable J with the same value of the objective function, thereby establishing strong duality. The slackness conditions are $JG = 0$ and equality in the trace constraint of the dual. Combining these two implies equality of the objective functions, so everything is consistent. Observe that for the putatively optimal value of z, G has two positive eigenvalues $\frac{1}{2}(3 \pm \sqrt{1 + 8xyz})$. Therefore we can choose J to be proportional to the zero eigenvector of G, implying $J \ge 0$, scaled so that the trace constraint is satisfied with equality. This choice of z and J is both feasible and fulfills the slackness conditions and hence is optimal. □

We should expect this form for the optimal J: the positivity constraint for G is only "barely" satisfied, and so the constraint is binding only on one dimension. By slackness the dual variable can be zero on the orthogonal subspace.

10.5 Channel fidelity

10.5.1 Definition and SDP formulation

We can define a useful fidelity measure for two channels by optimizing the fidelity of their outputs over all possible input states, including entangled states:

$$
F(\mathcal{E}_{B|A}, \mathcal{E}'_{B|A}) := \inf_{\rho_{AR}} F(\mathcal{E}_{B|A}[\rho_{AR}], \mathcal{E}'_{B|A}[\rho_{AR}]) . \tag{10.26}
$$

Joint concavity of the fidelity (see Exercise 10.3) implies that we may as well take ρ_{AR} in the optimization to be pure.

Due to the properties of the fidelity, it is tempting to think that this quantity is unchanged if the system R is omitted. After all, the fidelity will anyway involve purification. However, the fidelity involves separate purifications for each argument, whereas here we demand the input state to both channels be identical.

As with the channel distinguishability, it is possible to cast this optimization as a semidefinite program, this time building on the SDP in (10.14).

Proposition 10.6 (Channel fidelity SDP). *Let E_{BA} and E'_{BA} be the Choi operators of any two quantum channels $\mathcal{E}_{B|A}$ and $\mathcal{E}'_{B|A}$. Then the channel fidelity $F(\mathcal{E}_{B|A}, \mathcal{E}'_{B|A})$ can be expressed as the following semidefinite program:*

$$
\begin{aligned}
F(\mathcal{E}_{B|A}, \mathcal{E}'_{B|A}) = \underset{\rho,\Lambda,\Gamma}{\text{minimum}} \quad & \tfrac{1}{2}\left(\mathrm{Tr}[E_{BA}\Gamma_{BA}] + \mathrm{Tr}\left[E'_{BA}\Lambda_{BA}\right]\right) \\
\text{such that} \quad & \begin{pmatrix} \Gamma_{BA} & -\mathbb{1}_B \otimes \rho_A^T \\ -\mathbb{1}_B \otimes \rho_A^T & \Lambda_{BA} \end{pmatrix} \geq 0\,, \\
& \mathrm{Tr}\left[\rho_A^T\right] = 1\,, \\
& \rho_A, \Gamma_{BA}, \Lambda_{BA} \geq 0\,.
\end{aligned}
\tag{10.27}
$$

We leave the transpose in this expression, because then (the purification of) ρ_A itself is the optimal input to distinguish the channels.

Proof. First, use (10.14) to write

$$
\begin{aligned}
F(\mathcal{E}, \mathcal{E}') = \text{minimum} \quad & \tfrac{1}{2}(\mathrm{Tr}[Y\mathcal{E}[\rho]] + \mathrm{Tr}[Z\mathcal{E}'[\rho]]) \\
\text{such that} \quad & \begin{pmatrix} Y & -\mathbb{1} \\ -\mathbb{1} & Z \end{pmatrix} \geq 0\,, \quad \mathrm{Tr}[\rho^T] = 1\,, \quad \rho_{AR}, Y_{BR}, Z_{BR} \geq 0\,.
\end{aligned}
\tag{10.28}
$$

We can restrict to pure ρ_{AR} since the objective is linear in ρ_{AR} and proceed as in the channel distinguishability proof. Using the Choi representation and (9.33), $F(\mathcal{E}, \mathcal{E}')$ can be written as

$$
\begin{aligned}
F(\mathcal{E}, \mathcal{E}') = \text{minimum} \quad & \tfrac{1}{2}(\mathrm{Tr}[Y_{BR}E_{BA}K_{R|A'}\Upsilon_{AA'}(K_{R|A'})^*] \\
& + \mathrm{Tr}[Z_{BR}E'_{BA}K_{R|A'}\Upsilon_{AA'}(K_{R|A'})^*]) \\
\text{such that} \quad & \begin{pmatrix} Y & -\mathbb{1} \\ -\mathbb{1} & Z \end{pmatrix} \geq 0\,, \quad \mathrm{Tr}[(K_{R|A})^*K_{R|A}] = 1\,, \\
& Y_{BR}, Z_{BR} \geq 0\,.
\end{aligned}
\tag{10.29}
$$

Next, we show that any choice of feasible variables in (10.29) can be converted into a feasible set for (10.27) having the same value of the objective function and vice versa. Turning the former into the latter is simple. Define $\Gamma_{BA'} = (K_{R|A'})^* Y_{BR} K_{R|A'}$ and $\Lambda_{BA'} = (K_{R|A'})^* Z_{BR} K_{R|A'}$ so that the objective is simply $\frac{1}{2}(\mathrm{Tr}[\Gamma_{BA}E_{BA}]+\mathrm{Tr}[\Lambda_{BA}E'_{BA}])$. Conjugating the block matrix by M^* on the left and $M = \mathrm{diag}(K_{B|A}, K_{B|A})$ on the right yields the form in (10.27). This implies that the channel fidelity $F(\mathcal{E},\mathcal{E}')$ is at least as large as the optimal value in (10.27), as conjugation by M relaxes the constraint.

The other direction is complicated by the possibility that ρ_A does not have full rank. If it does have full rank, then a simple choice is $K_{R|A} = V_{R|A}\sqrt{\rho_A^T}$ for a density operator ρ_A and an arbitrary isometry $V_{R|A}$, along with

$$Y_{BR} = V_{R|A}(\rho_A^T)^{-1/2}\Gamma_{BA}(\rho_A^T)^{-1/2}(V_{R|A})^* \tag{10.30}$$

and Z_{BR} defined similarly from Λ_{BA} instead of Γ_{BA}. Direct calculation shows that $\mathrm{Tr}[Y_{BR}\mathcal{E}_{B|A}[\rho_A]] = \mathrm{Tr}[\Pi_A\Gamma_{BA}\Pi_A E_{BA}]$ for the projection Π_A onto the support of ρ_A^T. Without loss of generality the optimal Γ will be such that $\Pi_A\Gamma_{BA}\Pi_A = \Gamma_{BA}$ since replacing any feasible Γ_{BA} by $\Pi_A\Gamma_{BA}\Pi_A$ will maintain the constraints and only potentially lower the objective. As much holds for Λ. Hence (10.27) and (10.29) are equivalent for ρ_A of full rank.

The case of ρ_A of lower rank can presumably be dealt with by continuity, but here is a more direct solution. Keep the same form of $K_{R|A}$ but set $Y_{BR} = Y'_{BR}+\mathbb{1}_B\otimes(\mathbb{1}_R-\Pi'_R)$ for Y'_{BR} defined from (10.30) and $\Pi'_R = V_{R|A}\Pi_A V_{R|A}^*$, and similarly for Z_{BR}. These extra terms will not contribute to the objective function due to the presence of ρ_A^T. Conjugation of the block matrix constraint in (10.27) by $M = \mathrm{diag}(L,L)$ for $L_{R|A} = V_{R|A}(\rho_A^T)^{-1/2}$ gives

$$\begin{pmatrix} Y'_{BR} & -\mathbb{1}_B\otimes\Pi'_R \\ -\mathbb{1}_B\otimes\Pi'_R & Z'_{BR}\end{pmatrix} \geq 0\,. \tag{10.31}$$

Adding $\left(\begin{smallmatrix} \mathbb{1}_B & -\mathbb{1}_B \\ -\mathbb{1}_B & \mathbb{1}_B\end{smallmatrix}\right)\otimes(\mathbb{1}_R-\Pi'_R)\geq 0$ gives the block matrix constraint of (10.29). □

Exercise 10.11. Show that the dual is given by the following and that strong duality holds:

$$\begin{aligned} F(\mathcal{E},\mathcal{E}') = \underset{\mu,X_{AB}}{\text{maximum}} \quad & \mu \\ \text{such that} \quad & \mu\mathbb{1} - \tfrac{1}{2}\mathrm{Tr}_B[(X_{AB}+X_{AB}^*)] \leq 0\,, \\ & \begin{pmatrix} E_{AB} & X_{AB} \\ X_{AB}^* & E'_{AB}\end{pmatrix} \geq 0\,. \end{aligned} \tag{10.32}$$

Exercise 10.12. Show that for any two channels $\mathcal{E}_{B|A}$ and $\mathcal{E}'_{B|A}$,

$$\delta(\mathcal{E},\mathcal{E}') + F(\mathcal{E},\mathcal{E}') \geq 1 \quad \text{and} \tag{10.33}$$

$$\delta(\mathcal{E},\mathcal{E}')^2 + F(\mathcal{E},\mathcal{E}')^2 \leq 1\,. \tag{10.34}$$

Exercise 10.13. Show that $F(\mathcal{E},\mathcal{I}) = \sqrt{1-p}$ for every Pauli channel $\mathcal{E}$ whose probability of the identity operation is $1-p$.

Exercise 10.14. Show that $F(\mathcal{N}_\gamma, \mathcal{I}) = \sqrt{1-\gamma}$, where $\mathcal{N}_\gamma$ is the amplitude damping channel with parameter γ.

10.5.2 Comparing a channel to the identity

In Section 9.5, we observed that entangled inputs can provide a large advantage in the task of distinguishing two channels. However, when one of the channels is the identity channel, this advantage mostly disappears. This fact is easier to establish using fidelity than distinguishability.

When comparing a channel $\mathcal{E}_A$ to the identity $\mathcal{I}_A$, the channel fidelity can be expressed in terms of what is often called the *entanglement fidelity* $F_{\text{ent}}(\rho, \mathcal{E})$ of a state ρ_A and a channel $\mathcal{E}_A$. For an arbitrary purification ψ_{AR} of ρ_A,

$$F_{\text{ent}}(\rho, \mathcal{E}) := F(\mathcal{E}_A[\psi_{AR}], \psi_{AR}) . \tag{10.35}$$

Since any other purification differs only by action on R, which does not affect $\mathcal{E}_A$, this expression is independent of the choice of purification. Thus $F(\mathcal{E}, \mathcal{I}) = \min_\rho F_{\text{ent}}(\rho, \mathcal{E})$. Note that $P_{\text{agree}}(\mathcal{E}) = F_{\text{ent}}(\pi, \mathcal{E})^2$.

Exercise 10.15. Show that $F_{\text{ent}}(\rho, \mathcal{E})^2 = \sum_j |\operatorname{Tr}[K(j)\rho]|^2$ for a set $K(j)$ of Kraus operators of the channel. Using this form, argue that it is also independent of the choice of Kraus operators.

Exercise 10.16. Using (10.27), show that $F_{\text{ent}}(\rho, \mathcal{E})$ is convex in ρ for any channel $\mathcal{E}$; indeed, the fidelity $F(\rho_A, \mathcal{E}_{B|A}, \mathcal{E}'_{B|A})$ between $\mathcal{E}_{B|A}[\psi_{AR}]$ and $\mathcal{E}'_{B|A}[\psi_{AR}]$ for a purification ψ_{AR} of ρ_A is convex in ρ_A.

Now define the quantity $F_{\text{pure}}(\mathcal{E}) := \min_{|\psi\rangle} F(\mathcal{E}[|\psi\rangle\langle\psi|], |\psi\rangle\langle\psi|)$, which is the channel fidelity between $\mathcal{E}$ and $\mathcal{I}$, but restricted to channel inputs that are pure on A. We can show that

$$F(\mathcal{E}, \mathcal{I})^2 \geq 2F_{\text{pure}}(\mathcal{E})^2 - 1 . \tag{10.36}$$

Proof. Suppose that ρ is optimal in the $F_{\text{ent}}(\rho, \mathcal{E})$ optimization, and define $\{|k\rangle\}$ to be its eigenbasis and λ_k^2 its eigenvalues. Defining $|\psi\rangle_{AA'} = \sum_k \lambda_k |k\rangle_A \otimes |k\rangle_{A'}$, a straightforward calculation gives $F(\mathcal{E}, \mathcal{I})^2 = \sum_{jk} \lambda_j^2 \lambda_k^2 \langle j|\mathcal{E}[|j\rangle\langle k|]|k\rangle$. Next, define $|\varphi\rangle_A = \sum_k \lambda_k e^{i\theta_k} |k\rangle$ for arbitrary phases $\theta_k \in \mathbb{R}$ for all k. Then

$$\begin{aligned} F_{\text{pure}}(\mathcal{E})^2 &\leq \langle\varphi|\mathcal{E}[|\varphi\rangle\langle\varphi|]|\varphi\rangle \\ &= \sum_{jk\ell m} \lambda_j \lambda_k \lambda_\ell \lambda_m e^{i(-\theta_j+\theta_k-\theta_\ell+\theta_m)} \langle j|\mathcal{E}[|k\rangle\langle\ell|]|m\rangle . \end{aligned} \tag{10.37}$$

The phase factor vanishes when $j = k$, $\ell = m$ or $j = m$, $k = \ell$. Integrating the inequality over the four phases, each in the interval $(0, 2\pi]$, leaves only these two cases.

Attributing the contribution $j = k = \ell = m$ to the first case, we have

$$\begin{aligned} F_{\text{pure}}(\mathcal{E})^2 &\leq \sum_{j\ell} \lambda_j^2\lambda_\ell^2 \langle j|\mathcal{E}[|j\rangle\langle \ell|]|\ell\rangle + \sum_{k,j\neq k} \lambda_j^2\lambda_k^2 \langle j|\mathcal{E}[|k\rangle\langle k|]|j\rangle \\ &= F(\mathcal{E},\mathcal{I})^2 + \sum_{k,j\neq k} \lambda_j^2\lambda_k^2 \langle j|\mathcal{E}[|k\rangle\langle k|]|j\rangle \\ &\leq F(\mathcal{E},\mathcal{I})^2 + \sum_k \lambda_k^2 \sum_{j\neq k} \langle j|\mathcal{E}[|k\rangle\langle k|]|j\rangle \\ &\leq F(\mathcal{E},\mathcal{I})^2 + 1 - F_{\text{pure}}(\mathcal{E})^2 . \end{aligned} \tag{10.38}$$

The penultimate inequality is just $\lambda_j^2 \leq 1$ for all j. In the final inequality, we use the fact that $\sum_{j\neq k} \langle j|\mathcal{E}[|k\rangle\langle k|]|j\rangle \leq 1 - F_{\text{pure}}(\mathcal{E})^2$, since $\sum_j \langle j|\mathcal{E}[|k\rangle\langle k|]|j\rangle = 1$ and the $j = k$ term is at least $F_{\text{pure}}(\mathcal{E})^2$. This proves (10.36). □

10.5.3 Channel fidelity of unitary channels

It turns out that distinguishing a channel that applies a unitary operation from the identity channel does not require an entangled input. To see this, denote the unitary operator by U and its eigenvalues by $e^{i\theta_k}$ for $\theta_k \in \mathbb{R}$. Use the eigenbasis to define $|\Omega\rangle_{BA}$ and choose $X_{BA} = e^{-i\varphi} U_B |\Omega\rangle\langle\Omega|_{AB}$ for some $\varphi \in \mathbb{R}$ in the dual optimization (10.32). The Choi operator of the unitary channel is simply $U_B \Omega_{BA} U_B^*$, and so the block matrix in the constraint becomes

$$\begin{pmatrix} U\Omega U^* & e^{-i\varphi} U\Omega \\ e^{i\varphi}\Omega U^* & \Omega \end{pmatrix} = \begin{pmatrix} e^{-i\varphi} U & 0 \\ 0 & \mathbb{1} \end{pmatrix} \begin{pmatrix} \Omega & \Omega \\ \Omega & \Omega \end{pmatrix} \begin{pmatrix} e^{-i\varphi} U & 0 \\ 0 & \mathbb{1} \end{pmatrix}^* . \tag{10.39}$$

The right-hand side is positive if the middle operator is, by Exercise B.3. Conjugating the middle operator by $\frac{1}{\sqrt{2}}\left(\begin{smallmatrix} \mathbb{1} & \mathbb{1} \\ \mathbb{1} & -\mathbb{1} \end{smallmatrix}\right)$ gives $\operatorname{diag}(2\Omega, 0)$, and therefore the constraint is satisfied.

In the first constraint, this choice of X_{AB} gives $\operatorname{Tr}_B[X_{BA}] = e^{-i\varphi} U_A^T = e^{-i\varphi} U_A$, since we are working in the eigenbasis of U, as well as $\operatorname{Tr}_B[X_{BA}^*] = e^{i\varphi}\overline{U}_A$. Thus $\mu = \max_{\varphi\in\mathbb{R}} \min_k \cos(\theta_k - \varphi)$ is feasible, and $F(\mathcal{U},\mathcal{I}) \geq \max_{\varphi\in\mathbb{R}} \min_k \cos(\theta_k - \varphi)$. We can appreciate what this optimization is looking for geometrically. The eigenvalues $e^{i\theta_k}$ specify points on the unit circle in the complex plane. This forms a convex set inside the unit circle, and μ is the shortest distance from the origin to this set. Consider the edge closest to the origin. By rotating the entire set this edge can be oriented vertically along the imaginary axis, so that the minimal distance lies along the real axis. Performing this rotation is the accomplished by varying φ, while the distance along the real axis is precisely the cosine of the angle.

The closest edge is specified by two eigenvalues, denote them $e^{i\theta_j}$ and $e^{i\theta_k}$. The optimal φ must be $\frac{1}{2}(\theta_j + \theta_k)$, and therefore the optimal μ for this choice of X_{BA} is $\mu = \cos(\frac{1}{2}(\theta_j - \theta_k))$. Hence $F(\mathcal{U},\mathcal{I}) \geq \max_{j,k} \cos(\frac{1}{2}(\theta_j - \theta_k))$.

This lower bound can be achieved by the input state $|\psi\rangle = \frac{1}{\sqrt{2}}(|b_j\rangle + |b_k\rangle)$, where $|b_j\rangle$ is an eigenvector associated with eigenvalue $e^{i\theta_j}$ and similarly for $|b_k\rangle$. Therefore we have established that entanglement is not needed to optimally distinguish these channels, and

$$F(\mathcal{U},\mathcal{I}) = \min_{|\psi\rangle} |\langle\psi|U|\psi\rangle| = \max_{j,k} \cos(\tfrac{1}{2}(\theta_j - \theta_k)) . \tag{10.40}$$

Moreover, since the output states under the two channels are pure, the same conclusion regarding entanglement carries over to the channel distinguishability by (10.1).

10.6 Notes and further reading

The fidelity was originally introduced by Uhlmann in a more general setting as the square of our definition, which then has the interpretation as a transition probability from one pure state to the other [285]. The name “fidelity” is due to Jozsa [157]. As with channel distinguishability, in the literature, fidelity is often defined by the expression in (10.7), and then Uhlmann's theorem reads the other way. Again, we prefer the operationally motivated variational definition. Monotonicity was shown by Alberti and Uhlmann [3]. The SDP characterization was independently found by Killoran [160] and Watrous [298]. Alberti's theorem appears in [2]; in his proof, he uses the achievability by measurement shown earlier by Araki and Raggio [8], rather than the other way around here. The specific form of the measurement achieving the fidelity was determined by Fuchs and Caves [104]. The bounds in Proposition 10.4 are due to Fuchs and van de Graaf [103]. Exercise 10.10 is known as the “gentle measurement lemma”, originally due to Winter [305]. The SDP characterization of the channel fidelity was developed in writing this book and also independently by Katariya and Wilde [158, Proposition 50]. The entanglement fidelity was introduced by Schumacher [252] using the square as in Exercise 10.15. Convexity of the squared quantity was shown in [11]. The bound (10.36) is a simplification of Barnum, Knill, and Nielsen [11, Theorem 2]. That entangled inputs are not needed to distinguish unitaries from the identity (or each other) was shown by Aharonov, Nisan, and Kitaev [1], and a similar result appears in Childs, Preskill, and Renes [58].

11 Optimal and pretty good receivers

> Failing the possibility of measuring that which you desire, the lust for measurement may, for example, merely result in your measuring something else—and perhaps forgetting the difference—or in your ignoring some things because they cannot be measured.
>
> George Udny Yule

The decoder or receiver in a classical communication protocol is faced with the task of distinguishing between the states produced by different inputs to the encoder and channel, just as in the setup of Chapter 9, but it will generally need to distinguish between more than two possibilities. For quantum communication, the decoder has to restore the input state. In this chapter, we take up the question of how well this can be done by the optimal receivers and show that their "pretty good" cousins are indeed worthy of the name. In Part III, we will use the pretty good measurement to construct decoders for both classical and quantum communication tasks.

11.1 Optimal recovery of classical information

11.1.1 Definition

Given a CQ state $\rho_{XB} = \sum_{x\in\mathcal{X}} P_X(x)|x\rangle\langle x|_X \otimes \varphi_B(x)$ with arbitrary quantum states $\varphi_B(x)$ and probability distribution $P_X(x)$, suppose we would like to determine the value of X by making a measurement on B. For any POVM on B with elements $\Lambda_B(x)$, the average probability of guessing correctly is, generalizing (9.1), $P_{\text{guess}}(X|B)_{\rho,\Lambda} := \sum_{x\in\mathcal{X}} P_X(x)\ \text{Tr}[\Lambda_B(x)\varphi_B(x)]$. In terms of the QC measurement channel $\mathcal{M}_{X'|B}$ associated with the POVM Λ_B, this can be expressed more compactly as

$$P_{\text{guess}}(X|B)_{\rho,\mathcal{M}} = \text{Tr}[\Pi_{XX'}\mathcal{M}_{X'|B}[\rho_{XB}]]\,, \tag{11.1}$$

where $\Pi_{XX'} = \sum_x |x\rangle\langle x|_X \otimes |x\rangle\langle x|_{X'}$. This form emphasizes that the measurement would ideally transform the input state ρ_{XB} into the completely correlated joint distribution $P_{XX'}(x,x') = P_X(x)\delta_{x,x'}$. Then it is a straightforward exercise to show that

$$P_{\text{guess}}(X|B)_{\rho,\mathcal{M}} = 1 - \delta(\mathcal{M}_{X'|B}[\rho_{XB}], P_{XX'})\,. \tag{11.2}$$

The optimal guessing probability is therefore

$$P_{\text{guess}}(X|B)_\rho := \max_{\mathcal{M}_{Y|B}} \text{Tr}[\Pi_{XY}\mathcal{M}_{Y|B}[\rho_{XB}]] = \max_{\Lambda}\sum_x P_X(x)\ \text{Tr}[\Lambda_B(x)\varphi_B(x)]\,. \tag{11.3}$$

Before examining its properties, it is worth stressing that the optimal average guessing probability should not be confused with the optimal worst-case guessing probability. For a given measurement Λ, the latter is simply $\min_{x\in\mathcal{X}} \text{Tr}[\Lambda(x)\varphi(x)]$. Observe

https://doi.org/10.1515/9783110570250-011

that this quantity is independent of the prior probability $P_X(x)$ of the states. Therefore we should think of the worst-case guessing probability not as function of the state ρ_{XB}, but as a function of the CQ channel $\mathcal{E}_{B|X}$ that takes x to $\varphi(x)$. Using the best possible measurement gives $P^{\mathrm{wc}}_{\mathrm{guess}}(X|B)_{\mathcal{E}} := \max_\Lambda \min_x \mathrm{Tr}[\Lambda(x)\varphi(x)]$. We will not make very much use of this quantity as the average guessing probability will be sufficient for our purposes.

11.1.2 Classical case

Returning to $P_{\mathrm{guess}}(X|B)_\rho$, let us examine some of its properties. In the classical case of commuting $\varphi_B(x)$, the optimal measurement is deterministic. Here $\varphi_B(x) = \sum_y P_{Y|X=x}(y)|y\rangle\langle y|_B$ for some conditional probability distributions $P_{Y|X}$. The receiver observes the value of y, and so it is the relevant conditional distribution $P_{X|Y=y}$. Intuitively, the optimal guess for x simply maximizes $P_{X|Y=y}$. To establish this more formally, write the POVM elements as $\Lambda_B(x) = \sum_y W(x,y)|y\rangle\langle y|_B$ for some function W taking values in $[0,1]$. The POVM constraint $\sum_x \Lambda_B(x) = \mathbb{1}_B$ implies $\sum_x W(x,y) = 1$ for all y, so that $W(x,y)$ is a conditional probability distribution over x given the value y. For the guessing probability, we then have

$$\begin{aligned} P_{\mathrm{guess}}(X|B)_\rho &= \sum_x P_X(x) \sum_y W(x,y) P_{Y|X=x}(y) \\ &= \sum_y P_Y(y) \sum_x W(x,y) P_{X|Y=y}(x). \end{aligned} \tag{11.4}$$

For each y, the summation over x can be interpreted as a convex combination, with weights $W(x,y)$, of the values of the conditional distribution $P_{X|Y=y}$. Then the largest value that can be obtained from such a convex combination is indeed the largest value of $P_{X|Y=y}$.

Exercise 11.1. Assuming a uniform input distribution, confirm that the optimal guessing probabilities for BSC(p), BEC(q), and Z(r) are $1-p$, $1-q/2$, and $(2-r)/2$, respectively.

Exercise 11.2. Recall the pure state channel PSC(f) from Section 5.2. For uniform input distribution, what measurement optimizes the average guessing probability? Show that the optimal guessing probability is $\frac{1}{2}(1+\sqrt{1-f^2})$.

11.1.3 SDP formulation

The optimal measurement can be found by semidefinite programming.

Proposition 11.1 (Optimal guessing probability). *For every CQ state $\rho_{XB} \in \mathrm{Stat}(\mathcal{H}_X \otimes \mathcal{H}_B)$ with classical X,*

$$\begin{aligned} P_{\text{guess}}(X|B)_\rho = \underset{\Lambda_{XB}}{\text{maximum}} \quad & \mathrm{Tr}[\Lambda_{XB}\rho_{XB}] \\ \text{such that} \quad & \mathrm{Tr}_X[\Lambda_{XB}] = \mathbb{1}_B\,, \\ & \Lambda_{XB} \geq 0\,. \end{aligned} \tag{11.5}$$

Proof. Since ρ_{XB} is a CQ state, it is invariant under the pinch map: $\mathcal{P}_X[\rho_{XB}] = \rho_{XB}$. Observe that for any feasible Λ_{XB} in the optimization, $\mathcal{P}_X[\Lambda_{XB}]$ is also feasible. Only the diagonal part of Λ_{XB} is relevant in the objective function and equality constraint. Then we may as well consider CQ Λ_{XB} and write $\Lambda_{XB} = \sum_x |x\rangle\langle x|_X \otimes \Lambda_B(x)$. Hence the optimization finds the POVM with the largest guessing probability. □

Exercise 11.3. Show that strong duality holds with the dual SDP

$$\begin{aligned} P_{\text{guess}}(X|B)_\rho = \underset{\sigma_B}{\text{minimum}} \quad & \mathrm{Tr}[\sigma_B] \\ \text{such that} \quad & \mathbb{1}_X \otimes \sigma_B \geq \rho_{XB}\,. \end{aligned} \tag{11.6}$$

Exercise 11.4. Show that for any CQ channel $\mathcal{E}_{B|X} : x \mapsto \varphi_B(x)$,

$$\begin{aligned} P^{\text{wc}}_{\text{guess}}(X|B)_{\mathcal{E}} = \underset{\Lambda_B,\lambda}{\text{maximum}} \quad & \lambda \\ \text{such that} \quad & \lambda - \mathrm{Tr}[\Lambda_B(x)\,\varphi_B(x)] \leq 0 \quad \forall x \\ & \sum_x \Lambda_B(x) \leq \mathbb{1}_B\,, \\ & \Lambda_B(x) \geq 0 \quad \forall x\,. \end{aligned} \tag{11.7}$$

11.1.4 Largest and smallest values

The guessing probability is of course bounded between $1/|X|$ and 1. The former occurs if $\rho_{XB} = \pi_X \otimes \rho_B$, whereas the latter occurs if all of the conditional states $\varphi_B(x)$ are disjoint. In fact, these are the only cases in which the extreme values can occur, i. e., both "ifs" in the previous sentence can be replaced with "if and only if".

The former case is the most straightforward using the dual optimization. When the optimal value of $\mathrm{Tr}[\sigma] = 1/|X|$, the operator $\mathbb{1}_X \otimes \sigma_B$ is normalized, but we easily see that the operator inequality implies $\rho_{XB} = \mathbb{1}_X \otimes \sigma_B$, and therefore $\rho_{XB} = \pi_X \otimes \rho_B$.

Exercise 11.5. Suppose ρ and σ are positive operators such that $\mathrm{Tr}[\rho] = \mathrm{Tr}[\sigma]$ and $\rho \geq \sigma$. Show that $\rho = \sigma$. *Hint: Appeal to Lemma* B.3.

For the latter case, observe that each term $\mathrm{Tr}[\Lambda_B(x)\varphi_B(x)]$ in (11.3) must be equal to 1, since the summation is a convex combination and 1 is the largest possible value for each term. Similarly, since 1 is the largest eigenvalue of $\Lambda_B(x)$, $\mathrm{Tr}[\Lambda_B(x)\varphi_B(x)] = 1$

implies that, for each x, $\Lambda_B(x)$ is a projector on the support of $\varphi_B(x)$. Each $\Lambda_B(x)$ may be zero elsewhere, since this will not affect the optimal value. Finally, the normalization condition $\sum_x \Lambda_B(x) = \mathbb{1}_B$ implies that the $\Lambda_B(x)$ must be disjoint. For consider the possibility that $\Lambda_B(x)$ and $\Lambda_B(x')$ are not disjoint. Then there exists a vector $|\psi\rangle$ in the intersection of their supports, and therefore $\langle\psi|\Lambda_B(x)|\psi\rangle + \langle\psi|\Lambda_B(x')|\psi\rangle = 2$. This contradicts normalization, and hence the supports of $\varphi_B(x)$ must all be disjoint.

11.1.5 Monotonicity and chain rules

Clearly, the optimal guessing probability is monotonic under channels acting on B, since such channels can be regarded as part of $\mathcal{M}_{X'|B}$. The dual optimization implies a stronger monotonicity property. Namely, for an arbitrary quantum channel $\mathcal{E}_{C|B}$ and an arbitrary unital classical channel $\mathcal{F}_{Y|X}$,

$$P_{\text{guess}}(Y|C)_\tau \le P_{\text{guess}}(X|B)_\rho\,, \tag{11.8}$$

where $\tau_{YC} = \mathcal{F}_{Y|X} \circ \mathcal{E}_{C|B}[\rho_{XB}]$. The proof proceeds by constructing a feasible variable in the dual for the left-hand side from the optimal variable of the right-hand side. Suppose $\sigma_B^\star$ is optimal in $P_{\text{guess}}(X|B)_\rho$, so that $\mathrm{Tr}[\sigma_B^\star] = P_{\text{guess}}(X|B)_\rho$ and $\mathbb{1}_X \otimes \sigma_B^\star \ge \rho_{XB}$. Applying $\mathcal{F}_{Y|X} \circ \mathcal{E}_{C|B}$ to the operator inequality gives $\mathbb{1}_Y \otimes \theta_C \ge \tau_{YC}$ for $\theta_C = \mathcal{E}_{C|B}[\sigma_B^\star]$. Then θ_C is feasible in the dual SDP for $P_{\text{guess}}(Y|C)_\tau$, and $\mathrm{Tr}[\theta_C] = \mathrm{Tr}[\sigma_B^\star]$, which implies (11.8).

Exercise 11.6. Consider an arbitrary CQ state $\rho_{XB} = \sum_x |x\rangle\langle x|_X \otimes P(x)\varphi_B(x)$ and an arbitrary function $f : X \to Y$, and define $\sigma_{YB} = \sum_y |y\rangle\langle y| \otimes \sum_{x:f(x)=y} P(x)\varphi_B(x)$. Show that it is always easier to guess the output of a function than the input, i. e.,

$$P_{\text{guess}}(X|B)_\rho \le P_{\text{guess}}(Y|B)_\sigma\,. \tag{11.9}$$

Exercise 11.7. Show the following crude but useful chain rule for the guessing probability: For classical X, Y and quantum B,

$$P_{\text{guess}}(X|YB) \le |Y| P_{\text{guess}}(X|B)\,. \tag{11.10}$$

Give an example that saturates the bound. On the other hand, replacing classical Y by quantum C, show that the bound becomes

$$P_{\text{guess}}(X|BC) \le |C|^2 P_{\text{guess}}(X|B)\,. \tag{11.11}$$

Give an example ρ_{XBC} that saturates the bound.

Exercise 11.8. Show that the optimal probability of guessing X and Y by measuring B and C for a product state $\rho_{XYBC} = \sigma_{XB} \otimes \theta_{YC}$ is just given by $P_{\text{guess}}(XY|BC)_\rho =$

$P_{\text{guess}}(X|B)_\sigma P_{\text{guess}}(Y|C)_\theta$. *Hint: Use the channel* $S \mapsto S \otimes T$ *for positive* T *to infer that* $S \otimes T \geq S' \otimes T'$ *for* $S \geq S'$ *and* $T \geq T'$.

11.1.6 Conditions on the optimal measurement

The complementary slackness conditions give a fairly simple optimality condition on the measurement Λ_{XB}. Recall that the slackness conditions come from taking the Hilbert–Schmidt inner product of each inequality constraint with its dual variable and demanding the result be an equality. Here this procedure produces $\text{Tr}[\Lambda_{XB}(\mathbb{1}_X \otimes \sigma_B)] = \text{Tr}[\Lambda_{XB}\rho_{XB}]$. Since Λ_{XB} and $\mathbb{1}_X \otimes \sigma_B - \rho_{XB}$ are positive, it must be that

$$\Lambda_{XB}(\mathbb{1}_X \otimes \sigma_B) = \Lambda_{XB}\,\rho_{XB}\,. \tag{11.12}$$

Therefore the optimal POVM elements $\Lambda_B(x)$ and optimal σ_B satisfy $\Lambda_B(x)\sigma_B = P(x)\Lambda_B(x)\varphi_B(x)$ for all x. Summing over x (equivalently, taking the trace over X in the previous expression) gives $L_B := \sum_x P(x)\Lambda_B(x)\varphi_B(x) = \sigma_B$. The optimal $\Lambda_B(x)$ must therefore be such that L_B is Hermitian. The inequality constraint on σ_B implies that it must also satisfy

$$L_B \geq P(x)\varphi_B(x) \quad \forall x\,. \tag{11.13}$$

Exercise 11.9. Consider an ensemble of n pure qubit states $|\varphi_k\rangle$ uniformly separated along an equator of the Bloch sphere. Use (11.13) to show that $\Lambda(k) = \frac{2}{n}|\varphi_k\rangle\langle\varphi_k|$ is optimal for the ensemble with uniform probability.

Exercise 11.10. Derive the optimality of the optimal measurement in the classical case, which selects $\operatorname{argmax} P_{X|Y=y}(x)$, from the SDP formulation. What happens in the case of degeneracy?

11.2 Pretty good recovery of classical information

Instead of the optimal measurement to determine X from B, we could use the pretty good measurement constructed from ρ_{XB}, as defined in (6.11). The resulting guessing probability is simply

$$P_{\text{guess}}^{\text{PGM}}(X|B)_\rho := \text{Tr}\big[(\mathbb{1}_X \otimes \rho_B^{-1/2})\rho_{XB}(\mathbb{1}_X \otimes \rho_B^{-1/2})\rho_{XB}\big]\,. \tag{11.14}$$

The pretty good measurement has an appealing interpretation in the classical setting when all states $\varphi_B(x)$ commute. In this case, we have

$$\varphi_B(x) = \sum_y P_{Y|X=x}(y)|y\rangle\langle y|_B \tag{11.15}$$

for some orthonormal basis $\{|y\rangle\}$, and therefore the POVM elements of the pretty good measurement are given by $\Lambda_B(x) = \sum_y P_{X|Y=y}(x)|y\rangle\langle y|_B$. We may regard the entire POVM as consisting of two parts. The first part is a measurement in the common eigenbasis $\{|y\rangle\}$, which delivers a particular value of y. The second part is the generation of a guess from the measurement result; here the guess is generated by picking an x according to the conditional distribution $P_{X|Y=y}$, i. e., by *randomly sampling* from the distribution $P_{X|Y=y}$. The optimal strategy, as we saw above, is to deterministically pick the x that maximizes $P_{X|Y=y}$.

For example, consider the conditional distributions on Y generated by BSC(p) for uniformly random X. In this case, $P_{X|Y=y}(y) = 1 - p$, so the pretty good measurement in this case is to apply BSC(p) again.

Exercise 11.11. Show that the pretty good measurement for Y generated from uniform X by Z(r) is again a Z channel but with parameter $r/(1+r)$ and such that the input 1 is perfectly delivered to the output, not 0 (a "Σ" channel, as it were). What is the resulting guessing probability?

Exercise 11.12. Show that the pretty good measurements for BEC(q) and PSC(f) are each optimal, assuming uniform inputs. Furthermore, show that the measurement in Exercise 11.9 is the pretty good measurement.

Importantly, the pretty good measurement is guaranteed to be pretty good.

Proposition 11.2 (Quality of the pretty good measurement). *For every CQ state $\rho_{XB} \in \mathrm{Stat}(\mathcal{H}_X \otimes \mathcal{H}_B)$ with classical X,*

$$P_{\text{guess}}^{\text{PGM}}(X|B)_\rho \geq P_{\text{guess}}(X|B)_\rho^2 \,. \tag{11.16}$$

Proof. The proof is basically an application of the Cauchy–Schwarz inequality and (8.15). Suppose that $\Lambda_B(x)$ are the optimal POVM elements and define $\Lambda_{XB} = \sum_x |x\rangle\langle x|_X \otimes \Lambda_B(x)$. Then the guessing probability is $P_{\text{guess}}(X|B)_\rho = \mathrm{Tr}[\Lambda_{XB}\rho_{XB}]$. Let $\Gamma_{XB} = \rho_B^{1/4}\Lambda_{XB}\rho_B^{1/4}$ and then use Cauchy–Schwarz for the Hilbert–Schmidt inner product:

$$\begin{aligned} P_{\text{guess}}(X|B)_\rho^2 &= \mathrm{Tr}[\rho_B^{-1/4}\rho_{XB}\,\rho_B^{-1/4}\Gamma_{XB}]^2 \leq \mathrm{Tr}[(\rho_B^{-1/4}\rho_{XB}\,\rho_B^{-1/4})^2]\ \mathrm{Tr}[\Gamma_{XB}^2] \\ &= P_{\text{guess}}^{\text{PGM}}(X|B)_\rho\,\mathrm{Tr}[\Lambda_{XB}\,\rho_B^{1/2}\Lambda_{XB}\,\rho_B^{1/2}] \,. \end{aligned} \tag{11.17}$$

We can upper bound the second factor by 1 as follows. From (8.15) we have $\Lambda_{XB} \leq \mathbb{1}_X \otimes \Lambda_B$. The right-hand side is just $\mathbb{1}_{XB}$ since $\Lambda_B = \sum_x \Lambda_B(x) = \mathbb{1}_B$. Using $\mathrm{Tr}[S_{XB}(\mathbb{1}_{XB} - \Lambda_{XB})] \geq 0$ for $S_{XB} = \rho_B^{1/2}\Lambda_{XB}\,\rho_B^{1/2} \geq 0$, we can remove one factor of Λ_{XB} from the second factor. Tracing the remainder over X just leaves $\mathrm{Tr}[\rho_B] = 1$. □

Exercise 11.13. Show that $P^{\mathrm{PGM}}_{\mathrm{guess}}(Y|B)_\rho \geq P^{\mathrm{PGM}}_{\mathrm{guess}}(X|B)_\sigma$ for every CQ state ρ_{XB} and function $f : X \to Y$ generating the CQ state σ_{YB}.

11.3 Optimal entanglement recovery

11.3.1 Definition

In the guessing probability the task is to recover uniformly distributed classical information from quantum information to which it is correlated. A sensible quantum analog, as we examined in Chapter 8, is to recover maximal entanglement from a bipartite state by operations acting on one part alone. For a state ρ_{AB} and operation $\mathcal{E}_{A'|B}$, we define the *recoverable entanglement* as

$$R_{\mathrm{ent}}(A|B)_{\rho,\mathcal{E}} := \mathrm{Tr}[\Phi_{AA'}\mathcal{E}_{A'|B}[\rho_{AB}]] . \tag{11.18}$$

Then the *optimal recoverable entanglement* is

$$R_{\mathrm{ent}}(A|B)_\rho := \max_{\mathcal{E}_{A'|B}} \mathrm{Tr}[\Phi_{AA'}\mathcal{E}_{A'|B}[\rho_{AB}]] . \tag{11.19}$$

By (10.8) this is the optimal squared fidelity with the maximally entangled state. As in the classical case, the recovery operation is only applied one part of the state. Observe the similarities between (11.19) and (11.1). The recovery channel $\mathcal{E}_{A'|B}$ is the analog of the measurement $\mathcal{M}_{Y|B}$, and the maximally entangled projector $\Phi_{AA'}$ is the analog of the correlation projector Π_{XY}.

Exercise 11.14. Show that for any bipartite state ρ_{XB} and measurement $\Lambda_{X'|B}$,

$$P_{\mathrm{guess}}(X|B)_{\rho,\Lambda} = |X|\, R_{\mathrm{ent}}(X|B)_{\rho,\Lambda} . \tag{11.20}$$

The optimal $\mathcal{E}_{A'|B}$ can also be found via semidefinite programming.

Proposition 11.3 (Optimal recoverable entanglement). *For any bipartite state ρ_{AB}, the optimal recoverable entanglement is given by*

$$\begin{aligned} R_{\mathrm{ent}}(A|B)_\rho = \underset{E_{AB}}{\text{maximum}} \quad & \frac{1}{|A|}\mathrm{Tr}[E_{AB}\rho_{AB}] \\ \textit{such that} \quad & \mathrm{Tr}_A[E_{AB}] = \mathbb{1}_B , \quad E_{AB} \geq 0 . \end{aligned} \tag{11.21}$$

Proof. Using the Choi isomorphism, we have the SDP

$$\begin{aligned} R_{\mathrm{ent}}(A|B)_\rho = \text{maximum} \quad & \mathrm{Tr}[\Phi_{AA'}E_{A'B}\rho_{AB}^{T_B}] \\ \text{such that} \quad & \mathrm{Tr}_{A'}[E_{A'B}] = \mathbb{1}_B , \quad E_{A'B} \geq 0 . \end{aligned} \tag{11.22}$$

Here the superscript T_B denotes partial transposition on B, i. e., $\mathcal{T}_B$. We are free to transpose any systems inside the trace (see Exercise 5.16). Choose A' and B to obtain

$$\begin{aligned} R_{\text{ent}}(A|B)_\rho = \text{maximum} \quad & \frac{1}{|A|}\operatorname{Tr}[\Upsilon_{AA'}E^T_{A'B}\rho_{AB}] \\ \text{such that} \quad & \operatorname{Tr}_{A'}[E_{A'B}] = \mathbb{1}_B\,, \quad E_{A'B} \geq 0. \end{aligned} \tag{11.23}$$

Now $E^T \geq 0$ iff $E \geq 0$, so we just replace the variable. Then the trace over the swap operator just links A and A' as in (9.34), giving the desired form. □

Exercise 11.15. Show that the dual is given by the following and that strong duality holds:

$$\begin{aligned} R_{\text{ent}}(A|B)_\rho = \text{minimum} \quad & \operatorname{Tr}[\sigma] \\ \text{such that} \quad & |A|\mathbb{1}_A \otimes \sigma_B \geq \rho_{AB}\,. \end{aligned} \tag{11.24}$$

The dual form is nearly identical to that of the guessing probability, (11.6). Therefore the stronger monotonicity property also holds for the recoverable entanglement. That is, $R_{\text{ent}}(C|D)_\tau \leq R_{\text{ent}}(A|B)_\rho$ for arbitrary quantum channel $\mathcal{E}_{D|B}$, arbitrary unital quantum channel $\mathcal{F}_{C|A}$, and $\tau_{CD} = \mathcal{F}_{C|A} \circ \mathcal{E}_{D|B}[\rho_{AB}]$.

Exercise 11.16. Show that doing nothing is the optimal entanglement recovery channel for a state that is diagonal in the Bell basis, i. e., the output of a Pauli channel acting on the maximally entangled state.

Exercise 11.17. Show that in general $R_{\text{ent}}(A|BC)_\rho \leq |C|^2 R_{\text{ent}}(A|B)_\rho$ and, for classical X, $R_{\text{ent}}(A|BX)_\rho \leq |X| R_{\text{ent}}(A|B)_\rho$.

11.4 Pretty good entanglement recovery

Recalling the form of the pretty good measurement, we can see that it essentially uses the CQ state ρ_{XB} itself as the Choi map for the measurement, though this has to be distorted by $\rho_B^{-1/2}$ to ensure that the Choi map is trace-preserving. For entanglement recovery, we can do the same. It turns out that we need to use the transpose ρ^T_{AB}, so to keep the notation simple, let $\theta_{AB} = \rho^T_{AB}$. Then define the channel $\mathcal{E}^{\text{PGR}}_{A'|B}$ by the Choi operator $E_{A'B} = \theta_B^{-1/2}\theta_{A'B}\theta_B^{-1/2}$. The channel is completely positive by the Choi isomorphism and at least trace nonincreasing because $\operatorname{Tr}_{A'}[E_{A'B}] = \theta_B^{-1/2}\theta_B\theta_B^{-1/2} = \Pi_B$, where Π_B is the projector onto the support of θ_B. Just as for the pretty good measurement, here we will need to augment the channel action to make it trace-preserving in general. For simplicity, let us just consider the case that θ_B has full rank.

For example, consider any Pauli channel acting on the maximally entangled state $\Phi_{AA'}$. The result ρ_{AB} is a mixture of Bell states, whose matrix representation in the standard basis is real-valued, and hence $\theta_{AB} = \rho_{AB}$. Since $\theta_B = \frac{1}{2}\mathbb{1}_B$, the Choi operator of the pretty good recovery channel is just $2\rho_{AB}$, precisely the Choi operator of the original Pauli channel, that is, the pretty good recovery channel is again the original Pauli channel. Recall that the same occurred with the pretty good measurement for the BSC.

Exercise 11.18. Determine the pretty good recovery channel for the case of amplitude damping applied to $\Phi_{AA'}$.

The pretty good recoverable entanglement is simply the recoverable entanglement using this channel, $R_{\text{ent}}^{\text{PGR}}(A|B)_\rho := \text{Tr}[\Phi_{AA'}\mathcal{E}_{A'|B}^{\text{PGR}}(\rho_{AB})]$. Simplifying the expression for $R_{\text{ent}}^{\text{PGR}}(A|B)_\rho$ reveals a form entirely analogous to (11.14):

$$\begin{aligned} R_{\text{ent}}^{\text{PGR}}(A|B)_\rho &= \text{Tr}[\rho_{AB}^{T_B}\Phi_{AA'}E_{A'B}] = \frac{1}{|A|}\text{Tr}[\rho_{AB}^{T_B}E_{A'B}^{T_{A'}}\Upsilon_{AA'}] \\ &= \frac{1}{|A|}\text{Tr}[E_{A'B}^{T}\rho_{AB}\Upsilon_{AA'}] = \frac{1}{|A|}\text{Tr}[\rho_B^{-1/2}\rho_{A'B}\rho_B^{-1/2}\rho_{AB}\Upsilon_{AA'}] \\ &= \frac{1}{|A|}\text{Tr}[\rho_{AB}\rho_B^{-1/2}\rho_{AB}\rho_B^{-1/2}]. \end{aligned} \tag{11.25}$$

In the first step, partial transposition of A' reverses the order of the operators $\Phi_{AA'}$ and $E_{A'B}$, and $\Phi_{AA'}$ becomes $\frac{1}{|A|}\Upsilon_{AA'}$. The next step is transposition of B, which reverses the order of ρ_{AB} and $E_{A'B}$. We should confirm that $E_{A'B}^T = \rho_B^{-1/2}\rho_{A'B}\rho_B^{-1/2}$. To do so, first note that $(S^{-1})^T = (S^T)^{-1}$ for $S > 0$, since $(S^{-1})^T S^T = (SS^{-1})^T = \mathbb{1}$. Then it must be that transpose commutes with the inverse square root: $(S^{-1/2})^T = (S^T)^{-1/2}$, as $(S^{-1/2})^T(S^{-1/2})^T = (S^{-1/2}S^{-1/2})^T = (S^{-1})^T = (S^T)^{-1}$. Therefore

$$\begin{aligned} E_{A'B}^T &= (\theta_B^{-1/2}\theta_{A'B}\theta_B^{-1/2})^T = (\theta_B^{-1/2})^T\theta_{A'B}^T(\theta_B^{-1/2})^T \\ &= (\theta_B^T)^{-1/2}\theta_{A'B}^T(\theta_B^T)^{-1/2} = \rho_B^{-1/2}\rho_{A'B}\rho_B^{-1/2}. \end{aligned} \tag{11.26}$$

Exercise 11.19. What is the pretty good recoverable entanglement for the Pauli channel example above?

Just as for the pretty good measurement, we can show that

$$R_{\text{ent}}^{\text{PGR}}(A|B)_\rho \geq R_{\text{ent}}(A|B)_\rho^2. \tag{11.27}$$

Let E_{AB} be the optimizer in (11.21), and let $E'_{AB} = \frac{1}{|A|}\rho_B^{1/4}E_{AB}\rho_B^{1/4}$. The Cauchy–Schwarz inequality yields

$$\begin{aligned} R_{\text{ent}}(A|B)_\rho^2 &= \text{Tr}[E'_{AB}\rho_B^{-1/4}\rho_{AB}\rho_B^{-1/4}]^2 \leq |A|R_{\text{ent}}^{\text{PGR}}(A|B)_\rho\,\text{Tr}[(E'_{AB})^2] \\ &= \frac{1}{|A|}R_{\text{ent}}^{\text{PGR}}(A|B)_\rho\,\text{Tr}[E_{AB}\rho_B^{1/2}E_{AB}\rho_B^{1/2}]. \end{aligned} \tag{11.28}$$

By (8.16), $E_{AB} \leq |A| \mathbb{1}_A \otimes E_B$, and $E_B = \mathbb{1}_B$ by construction. Therefore we have

$$\frac{1}{|A|} \operatorname{Tr}[E_{AB} \rho_B^{1/2} E_{AB} \rho_B^{1/2}] \leq \operatorname{Tr}[\rho_B^{1/2} E_{AB} \rho_B^{1/2}] = 1, \tag{11.29}$$

which completes the proof.

Like the average guessing probability, the recoverable entanglement is a property of a bipartite state, not a channel. Though it is not our main focus here, it is worth mentioning that a channel property like the worst-case guessing probability can just as well be defined for quantum channels. For a fixed quantum channel $\mathcal{N}_{B|A}$, we could simply ask for the recovery operation $\mathcal{R}_{A|B}$ that minimizes $\delta(\mathcal{R} \circ \mathcal{N}, \mathcal{I})$ or maximizes $F(\mathcal{R} \circ \mathcal{N}, \mathcal{I})$. Using (9.37) or (10.32), each of these can even be formulated as an SDP.

11.5 Monotonicity of pretty good recoveries

Monotonicity of the optimal guessing probability and optimal recoverable entanglement under channels is immediate from their respective definitions. Not so, however, for the pretty good measurement and pretty good recoverable entanglement. We take this opportunity to delve more deeply into matrix analysis to establish monotonicity results for both quantities, and because it will be of use later coding theorems for classical communication over CQ channels and information reconciliation with quantum side information in Chapters 15 and 16.

Notice that if we define the quantity

$$Q(\rho, \sigma) := \operatorname{Tr}[\rho \sigma^{-1/2} \rho \sigma^{-1/2}], \tag{11.30}$$

then the pretty good guessing probability takes the form $P_{\text{guess}}^{\text{PGM}}(X|B)_\rho = Q(\rho_{XB}, \mathbb{1}_X \otimes \rho_B)$, while $R_{\text{ent}}^{\text{PGR}}(A|B)_\rho = \frac{1}{|A|} Q(\rho_{AB}, \mathbb{1}_A \otimes \rho_B)$. Hence, if $Q(\rho, \sigma) \geq Q(\mathcal{E}[\rho], \mathcal{E}[\sigma])$, then $P_{\text{guess}}^{\text{PGM}}$ and $R_{\text{ent}}^{\text{PGR}}$ cannot increase under general channels acting on B and unital channels acting on A or X.

To establish monotonicity, we proceed by first establishing joint convexity of Q. Then we follow the approach taken in Section 9.4 to show that joint convexity implies monotonicity. Note that for the two cases of interest, the particular inputs to Q are such that the support of the first argument is contained in that of the second. The proof below is specific to this condition, though presumably it can be generalized (e. g., by a continuity argument).

Proposition 11.4 (Joint convexity of Q). *For an arbitrary probability distribution $P(x)$ and arbitrary collections of states $\{\rho(x)\}_{x \in \mathcal{X}}$ and $\{\sigma(x)\}_{x \in \mathcal{X}}$ such that the support of $\rho(x)$ is contained in the support of $\sigma(x)$ for all $x \in \mathcal{X}$,*

$$Q\Big(\sum_x P(x)\rho(x), \sum_x P(x)\sigma(x)\Big) \leq \sum_x P(x) Q\big(\rho(x), \sigma(x)\big). \tag{11.31}$$

Proof. First, note that

$$Q(\rho,\sigma) = \sup_{\Lambda\geq 0}(2\,\mathrm{Tr}[\Lambda\rho] - \mathrm{Tr}[\Lambda\sigma^{1/2}\Lambda\sigma^{1/2}])\,. \tag{11.32}$$

The proof is simple. Observe that for positive K and L, we have $\mathrm{Tr}[(K-L)^2] \geq 0$, and therefore $2\,\mathrm{Tr}[KL] \leq \mathrm{Tr}[K^2] + \mathrm{Tr}[L^2]$. Using $K = \sigma^{-1/4}\rho\sigma^{-1/4}$ and $L = \sigma^{1/4}\Lambda\sigma^{1/4}$ gives the above variational expression. Note that this step makes use of the support condition on ρ and σ.

The usefulness of this form is that joint convexity of Q follows once we show that the map $g : \sigma \mapsto \mathrm{Tr}[\Lambda\sigma^{1/2}\Lambda\sigma^{1/2}]$ is *operator concave* for any fixed $\Lambda \geq 0$, that is, $g(\sum_x P(x)\sigma(x)) \geq \sum_x P(x)g(\sigma(x))$ for any probability distribution P. Operator concavity is much more subtle than concavity of scalar functions and is discussed in more detail in Section B.9. Writing

$$\mathrm{Tr}\big[\Lambda\sigma^{1/2}\Lambda\sigma^{1/2}\big] = \langle\Omega|(\Lambda\otimes\mathbb{1})(\sqrt{\sigma}\otimes\sqrt{\sigma}^{T})(\Lambda\otimes\mathbb{1})|\Omega\rangle\,, \tag{11.33}$$

it is apparent that the concavity of g follows from the operator concavity of $\sigma \mapsto \sqrt{\sigma}\otimes\sqrt{\sigma}^{T}$. Note that $\sqrt{\sigma}^{T}$ must be equal to $\sqrt{\sigma^T}$ since its square is σ^T (just as we argued in the previous section). Therefore we are interested in showing the operator concavity of $\sigma \mapsto \sqrt{\sigma}\otimes\sqrt{\sigma^T}$.

To do this, we make use of the *geometric mean* of two positive definite operators A and B, denoted $A\#B$. The operator geometric mean can be defined by extending the extremal property of the usual geometric mean of positive numbers a and b, namely that it is the largest c such that $ab - c^2 \geq 0$ or, in terms of matrices, $\left(\begin{smallmatrix} a & c \\ c & b\end{smallmatrix}\right) \geq 0$. This variational definition, detailed in Section B.9, has the advantage that joint concavity of $(A,B) \mapsto A\#B$ follows immediately. Moreover, as described there in (B.35), the geometric mean also has the closed-form expression $A\#B = A^{1/2}(A^{-1/2}BA^{-1/2})^{1/2}A^{1/2}$. Hence we can write $\sqrt{\sigma}\otimes\sqrt{\sigma^T}$ as $(\sigma\otimes\mathbb{1})\#(\mathbb{1}\otimes\sigma^T)$. Note that we can restrict attention to the support of σ to ensure the diagonal elements in the definition of the geometric mean are positive definite. Since the geometric mean is jointly concave, the operator concavity of $\sigma \mapsto \sqrt{\sigma}\otimes\sqrt{\sigma^T}$ therefore follows. □

From here it is a short step back to monotonicity following the method in Section 9.4. For this, we need to establish two simple properties of Q. First, Q is invariant under isometries, so that for $\rho,\sigma \in \mathrm{Stat}(\mathcal{H})$ and $V \in \mathrm{Lin}(\mathcal{H},\mathcal{H}')$,

$$Q(V\rho V^*, V\sigma V^*) = Q(\rho,\sigma)\,. \tag{11.34}$$

This follows by straightforward calculation using the fact that $(V\sigma V^*)^{1/2} = V\sigma^{1/2}V^*$. Second, appending an additional system in the same state to both arguments does not change the value of Q: for any state τ,

$$Q(\rho\otimes\tau, \sigma\otimes\tau) = Q(\rho,\sigma)\,. \tag{11.35}$$

Again, this follows by straightforward calculation, this time using the fact that $(\sigma \otimes \tau)^{-1/2} = \sigma^{-1/2} \otimes \tau^{-1/2}$. Then we proceed just as in the case of the distinguishability.

Proposition 11.5 (Monotonicity of Q). *For any quantum channel $\mathcal{E}$ and any quantum states ρ and σ for which the support of ρ is contained in that of σ,*

$$Q(\mathcal{E}[\rho], \mathcal{E}[\sigma]) \le Q(\rho, \sigma) . \tag{11.36}$$

With monotonicity of Q in hand, we can sharpen the PGM and PGR bounds (11.16) and (11.27) by considering a measurement related to the optimal guessing probability. This is the general method we mentioned in (10.19). Consider the optimal measurement $\mathcal{M}_{X'|B}$ in $P_{\text{guess}}(X|B)_\rho$ from (11.1), and the POVM element $\Pi_{XX'}$, which checks for equality. Then we have a binary-output measurement, which applied to ρ_{XB} gives the distribution $P = (P_{\text{guess}}(X|B)_\rho, 1 - P_{\text{guess}}(X|B)_\rho)$. For $\mathbb{1}_X \otimes \rho_B$, the same operation produces $P' = (1, |X| - 1)$. Then by the monotonicity of Q we have

$$P_{\text{guess}}^{\text{PGM}}(X|B)_\rho \ge Q(P, P') = P_{\text{guess}}(X|B)_\rho^2 + \frac{(1 - P_{\text{guess}}(X|B)_\rho)^2}{|X| - 1} . \tag{11.37}$$

Observe that this bound is saturated by the example of Y generated from uniform X via BSC(p). Moreover, unlike (11.16), the bound is nontrivial for $P_{\text{guess}}(X|B)_\rho \in [\frac{1}{|X|}, \frac{1}{\sqrt{|X|}}]$.

Exercise 11.20. Adapt this argument to the case of recoverable entanglement and show that

$$R_{\text{ent}}^{\text{PGR}}(A|B)_\rho \ge R_{\text{ent}}(A|B)_\rho^2 + \frac{(1 - R_{\text{ent}}(A|B)_\rho)^2}{|A|^2 - 1} . \tag{11.38}$$

Confirm that the bound is saturated by the result of the depolarizing channel acting on the maximally entangled state.

11.6 Notes and further reading

The quote from Udny Yule appears in [313]. Yuen, Kennedy, and Lax [311, 312] investigated the optimal guessing probability and stated the optimizations in (11.3) and (11.6). Later Eldar, Megretski, and Verghese [92] recognized this as an SDP. The optimality conditions were found by Holevo [143, 144] and Yuen, Kennedy, and Lax [311, 312]. The pretty good measurement deserves its name by Proposition 11.2 due to Barnum and Knill [10].

Entanglement recovery as formulated here was studied by Reimpell and Werner [235], Fletcher, Shor, and Win [101], and Kosut and Lidar [170]. The latter two make use of the SDP characterization. Later it was realized by König, Renner, and Schaffner [176] that the min-entropy, a quantity utilized by Renner in conjunction with QKD [241], is related to the recoverable entanglement. The pretty good recovery map was introduced by Berta, Coles, and Wehner [35] and shown to be pretty good by Dupuis, Fawzi, and Wehner [85].

The quantity Q is the logarithm of the Rényi entropy of order 2, which is sometimes called the collision entropy. For more on Rényi entropies, see Tomamichel [280]. The variational form of Q is adapted from Frank and Lieb [102], while our proof of concavity of the geometric mean follows methods of Ando [6], itself based on methods of Uhlmann [286]. See [54] for an excellent overview. The geometric mean itself was introduced by Pusz and Woronowicz [231].

12 Entropy

You should call it entropy, for two reasons. In the first place your uncertainty function has been used in statistical mechanics under that name, so it already has a name. In the second place, and more important, no one knows what entropy really is, so in a debate you will always have the advantage.

John von Neumann, discussion with Claude Shannon

I don't understand everything I know about entropy.

John T. "Ted" Norgord, lecture to engineering students at the University of Idaho

One of the major breakthroughs of Shannon's 1948 paper was the quantification of the amount information in a random variable via the *entropy*. He defined the entropy $H(X)$ of a random variable X with distribution P_X as

$$H(X)_P := \sum_x P_X(x) \log \frac{1}{P_X(x)} . \tag{12.1}$$

As mentioned obliquely in the first quote above, the Shannon entropy has the same form as the Gibbs entropy in classical statistical mechanics. Heuristically, X carries a lot of information if learning the actual value is "surprising". Conversely, X carries little information if we are already fairly certain of its value; the value itself does not tell us anything new. The entropy $H(X)_P$, called the *Shannon entropy*, is the expected value (under P_X) of the *surprisal* $\log \frac{1}{P_X(x)}$, which can be regarded as quantifying the amount of surprise. For small $P_X(x)$, the surprisal is large, and nears zero as $P_X(x)$ tends to 1. Here and throughout this book, we use the binary logarithm, the logarithm base 2. Then both the surprisal and the entropy are measured in *bits*.

To be sure, the quantity $\log \frac{1}{P_X(x)}$ is not the only function of $P_X(x)$ with this property. There exist several axiomatic derivations of the particular functional form of the entropy as a measure of information content, starting from "reasonable" axioms. However, from our operational viewpoint, the ultimate justification is that the entropy plays a crucial role in characterizing information-processing tasks, e. g., as Shannon showed for source and channel coding in his 1948 paper, and which we will encounter in Chapters 14 and 15.

The operational viewpoint is especially relevant in the quantum setting, where an axiomatic approach is not so straightforward, and the quantum analog of the Shannon entropy as an information measure is not so apparent. For one thing, there is no immediate quantum counterpart to the surprisal. Although von Neumann translated the Gibbs entropy to quantum statistical mechanics and found that the analogous quantity is given by

$$H(A)_\rho := -\operatorname{Tr}[\rho_A \log \rho_A] , \tag{12.2}$$

https://doi.org/10.1515/9783110570250-012

his argument by itself does not by imply the usefulness of the *von Neumann entropy* in quantum information processing. We will see the relevance of the von Neumann entropy in this setting in Chapter 14.

More immediately, we can relate the Shannon and von Neumann entropies to the *relative entropy* or *Kullback*[1]*–Leibler*[2] *divergence*, which has an operational meaning in the setting of Neyman–Pearson hypothesis testing. The relative entropy $D(\rho, \sigma)$ of two states ρ and σ is defined as

$$D(\rho, \sigma) := \mathrm{Tr}[\rho(\log \rho - \log \sigma)] \tag{12.3}$$

for σ whose support is contained in the support of ρ. If this support condition is not satisfied, then $D(\rho, \sigma) = \infty$. We will also consider $D(\rho, \tau)$ with τ not necessarily normalized. Later in this chapter, we will prove *Stein's*[3] *lemma*, which states that in the task of discriminating the two states $\rho^{\otimes n}$ and $\sigma^{\otimes n}$, the optimal measurement at fixed error for $\rho^{\otimes n}$ will have an error for $\sigma^{\otimes n}$ that decays exponentially in n at a rate given by $D(\rho, \sigma)$. Here $\rho^{\otimes n}$ denotes the n-fold tensor product of ρ with itself. This fact will enable us to prove a crucial property of the entropy, the data processing inequality, and will also be critical in the proofs of the coding theorems of Part III.

12.1 Entropy and relative entropy

First, let us formulate the entropy, conditional entropy, and mutual information quantities and explore their more basic properties. We start with the von Neumann entropy. In terms of the relative entropy,

$$H(A)_\rho = -D(\rho_A, \mathbb{1}_A) = \log |A| - D(\rho_A, \pi_A)\,, \tag{12.4}$$

where π_A is the maximally mixed state $\pi_A = \frac{1}{|A|}\mathbb{1}_A$. Note that the von Neumann entropy is simply the Shannon entropy of the probability distribution formed by the eigenvalues of the state ρ. The definition already covers the case of multiple systems, the *joint entropy*

$$H(AB)_\rho = -D(\rho_{AB}, \mathbb{1}_{AB})\,. \tag{12.5}$$

Several important properties of the entropy can be derived just from the positivity of the relative entropy, a statement known as Klein's[4] inequality in the quantum case and Gibbs' inequality in the classical case.

1 Solomon Kullback, 1907–1994.

2 Richard A. Leibler, 1914–2003.

3 Charles Max Stein, 1920–2016.

4 Oskar Benjamin Klein, 1894–1977.

Proposition 12.1 (Klein's inequality). *For all $\rho, \sigma \in \mathrm{Stat}(\mathcal{H})$, $D(\rho, \sigma) \geq 0$ with equality if and only if $\rho = \sigma$.*

Of course, nonnegativity of the relative entropy must follow from Stein's lemma mentioned above; otherwise, the error probability would be *increasing* with increasing n, which is nonsensical. However, appealing to Stein's lemma in this way does not give the equality condition, as the precise statement involves a limit.

To prove Klein's inequality directly, it is useful to first note that the quantum relative entropy is equal to a classical relative entropy for a specific pair of probability distributions. Therefore nonnegativity of the quantum relative entropy follows from Gibbs' inequality. Expressing ρ and σ in their respective eigenbases as $\rho = \sum_x R(x)|u_x\rangle\langle u_x|$ and $\sigma = \sum_y S(y)|v_y\rangle\langle v_y|$, and defining $W(x,y) = |\langle u_x|v_y\rangle|^2$, the distributions in question are

$$P_{XY}(x,y) = R(x)W(x,y)\,, \tag{12.6}$$

$$Q_{XY}(x,y) = S(y)W(x,y)\,. \tag{12.7}$$

Note that $W(x,y)$ are the entries of a doubly stochastic matrix, i. e., $\sum_x W(x,y) = 1$ for all y and similarly for x. Then we have

$$\begin{aligned}
D(\rho,\sigma) &= \mathrm{Tr}\Big[\rho\Big(\log\rho - \sum_y \log S(y)|v_y\rangle\langle v_y|\Big)\Big] && (12.8a)\\
&= \sum_x R(x)\log R(x) - \sum_{x,y} R(x)W(x,y)\log S(y) && (12.8b)\\
&= \sum_{x,y} R(x)W(x,y)\log\frac{R(x)W(x,y)}{S(y)W(x,y)} && (12.8c)\\
&= D(P_{XY}, Q_{XY})\,. && (12.8d)
\end{aligned}$$

Despite this equivalence, there is no channel mapping ρ to P_{XY} and σ to Q_{XY}, as is the case with the distinguishability and fidelity.

Proof of Klein's inequality. By (12.8) positivity of $D(\rho,\sigma)$ is implied by positivity of $D(P,Q)$ for all probability distributions P and Q. We can restrict attention to the case that $P(x) = 0$ implies $Q(x) = 0$, as otherwise $D(P,Q) = \infty$, and there is nothing to prove.

Now observe that $\ln y \leq y - 1$ for all $y > 0$. At $y = 1$, $\ln y$ is tangent to $y - 1$ as is seen from their first derivatives, and the inequality holds because $\ln y$ is strictly concave, as seen from the negativity of its second derivative. Equality thus only holds at $y = 1$. Therefore

$$\begin{aligned}
D(P,Q) &= -\frac{1}{\ln 2}\sum_x P(x)\ln\Big(\frac{Q(x)}{P(x)}\Big)\\
&\geq -\frac{1}{\ln 2}\sum_x P(x)\Big(\frac{Q(x)}{P(x)} - 1\Big) = \frac{1}{\ln 2}\sum_x P(x) - Q(x) = 0\,.
\end{aligned} \tag{12.9}$$

Equality holds if and only if $P(x) = Q(x)$ for all x, i. e., $P = Q$. This also covers the case of commuting ρ and σ.

To establish the equality condition in the quantum case, start from the inequality in (12.8b) with $D(\rho, \sigma) = 0$. Since the $W(x, y)$ form a probability distribution over y and the logarithm is concave, Jensen's inequality (2.18) implies

$$\sum_y W(x,y) \log S(y) \leq \log\Big(\sum_y W(x,y)S(y)\Big) = \log Q_X(x) , \tag{12.10}$$

using (12.7). Hence $D(\rho, \sigma) \geq D(P_X, Q_X) \geq 0$, meaning that equality must hold in (12.10) in order to satisfy $D(\rho, \sigma) = 0$.

As stated in the discussion of Jensen's inequality, equality only holds for strictly concave functions such as log if the points in the convex combination (here, the $S(y)$) are all identical or the convex combination is trivial. The former case only occurs when the eigenvalues of σ are uniform on the support of ρ; then $[\rho, \sigma] = 0$, which is already covered by the classical case above. Thus, equality only holds in the generic case when $W(x, y)$ represents a deterministic reversible classical channel, i. e., a permutation. Because $W(x, y)$ is the matrix of overlaps of the eigenvectors of ρ and σ, this implies that these sets of eigenvectors are identical. Then, returning to (12.8b), it is clear that this expression is only zero if the eigenvalues are also equal. Hence $\rho = \sigma$ if and only if $D(\rho, \sigma) = 0$, as claimed. □

Now we can enumerate several useful properties following from Klein's inequality.

Proposition 12.2 (Properties of the entropy of a single system).

1. *Positivity:* $H(A)_\rho \geq 0$ *for all* ρ *with equality iff* ρ *is a pure state,*
2. *Unitary invariance:* $H(A)_{U\rho U^*} = H(A)_\rho$ *for unitary* U,
3. *Upper bound:* $H(A)_\rho \leq \log |A|$ *with equality iff* $\rho = \pi$,
4. *Concavity:* $H(A)_\rho \geq \sum_{x\in\mathcal{X}} P_X(x)\, H(A)_{\rho(x)}$ *for* $\rho = \sum_{x\in\mathcal{X}} P_X(x)\rho(x)$, *and*
5. *Increase under pinching: For any complete set of projectors* $\Pi(k)$ *(not necessarily rank-one) and* $\sigma = \sum_k \Pi(k)\,\rho\,\Pi(k)$, $H(A)_\sigma \geq H(A)_\rho$.

The last property states that measurements increase entropy when we forget the measurement outcome. Of course, entropy hopefully decreases given the measurement result. One clear example of the latter is a measurement with rank-one POVM elements. In this case the postmeasurement state is pure, and its entropy is therefore zero.

To show the final property, it is useful to note that

$$\log \sigma = \mathcal{P}[\log \sigma] \tag{12.11}$$

for any $\sigma = \mathcal{P}[\sigma]$ and any pinching operation $\mathcal{P}$. This is most easily understood by regarding $\mathcal{P}[\sigma]$ as a block-diagonal matrix, from which it is clear that its logarithm is the logarithm of the blocks. More algebraically, the pinch map will restrict the eigenvectors to the subspaces of the projections in the pinch (i. e., the blocks in the block-diagonal matrix).

Exercise 12.1. Prove the properties in Proposition 12.2 using Klein's inequality.
Hint: make use of Exercise 5.11.

Proposition 12.3 (Properties of the entropy of several systems).
1. *Duality:* $H(A)_\rho = H(B)_\rho$ *for pure* ρ_{AB},
2. *Subadditivity:* $H(AB)_\rho \leq H(A)_\rho + H(B)_\rho$ *with equality iff* $\rho_{AB} = \rho_A \otimes \rho_B$,
3. *Triangle inequality:* $H(AB)_\rho \geq |H(A)_\rho - H(B)_\rho|$.

Proof. Since the entropy is a function only of the eigenvalues of the reduced state, duality follows from the form of the purification in (6.1).

For subadditivity, a simple computation shows that

$$D(\rho_{AB}, \rho_A \otimes \rho_B) = H(A)_\rho + H(B)_\rho - H(AB)_\rho\,, \tag{12.12}$$

where as usual $\rho_A = \mathrm{Tr}_B[\rho_{AB}]$, and similarly for ρ_B. Thus positivity of the relative entropy implies subadditivity of the entropy. The equality condition follows from the equality condition in Klein's inequality.

The triangle equality follows from subadditivity by making use of duality. Let R be a purifying reference system so that $|\psi\rangle_{RAB}$ is a purification of ρ_{AB}. Then

$$H(B)_\psi = H(RA)_\psi \leq H(A)_\psi + H(R)_\psi = H(A)_\psi + H(AB)_\psi, \tag{12.13}$$

which implies that $H(AB)_\rho \geq H(B)_\rho - H(A)_\rho$. Interchanging A and B in the argument gives the bound $H(A)_\rho - H(B)_\rho$. □

12.2 Conditional entropy and mutual information

Again using the relative entropy, we define the *conditional entropy*

$$H(A|B)_\rho := -D(\rho_{AB}, \mathbb{1}_A \otimes \rho_B) = \log|A| - D(\rho_{AB}, \pi_A \otimes \rho_B) \tag{12.14}$$

and the *mutual information*

$$I(A:B)_\rho := D(\rho_{AB}, \rho_A \otimes \rho_B)\,. \tag{12.15}$$

By straightforward calculation we can see that the conditional entropy of a CQ state $\rho_{YA} = \sum_y P(y)|y\rangle\langle y|_Y \otimes \rho_A(y)$ is the average of the entropy of the conditional state,

i. e.,

$$H(A|Y)_\rho = \sum_y P(y)\, H(A)_{\rho_A(y)} \,. \tag{12.16}$$

In this sense the conditional entropy is the entropy of A conditioned on knowing the value of Y. Note that concavity of entropy can be expressed as $H(A|X) \le H(A)$.

Meanwhile, the classical mutual information $I(X{:}Y)_P$ is heuristically the amount of information Y has about X (or X about Y, since $I(X{:}Y)$ is symmetric in X and Y). Again, a straightforward calculation reveals that $I(X{:}Y)$ is the average, under P_{XY}, of the surprisal $\log \frac{1}{P_X(x)}$ minus the surprisal $\log \frac{1}{P_{X|Y=y}(x)}$. This difference is the amount we learn about X by learning the precise value of Y.

However, although definitions (12.14) and (12.15) carry over to the case of quantum states, the interpretation of the mutual information and the conditional entropy is not at all clear, since $H(A|B)_\rho$ can be negative and $I(A:B)$ can be larger than $\log |A|$. Unsurprisingly, the maximally entangled state Φ_{AB} gives an immediate example of both.

Exercise 12.2. Show that $H(A|B)_\Phi = -\log |A|$ and $I(A:B)_\Phi = 2\log |A|$.

Nonetheless, the quantum conditional entropy and quantum mutual information also show up in the characterization of information processing tasks.

Conditioning as averaging does continue to hold for classical information in the context of conditional quantum information.

Exercise 12.3. Consider an arbitrary CQQ state, i. e., $\rho_{XAB} = \sum_x P_X(x) \otimes \sigma_{AB}(x)$ for arbitrary distribution P_X and normalized states $\sigma_{AB}(x)$. Show that

$$H(A|BX)_\rho = \sum_{x \in \mathcal{X}} P_X(x)\, H(A|B)_{\sigma(x)} \,. \tag{12.17}$$

The conditional entropy and mutual information satisfy the following *chain rules*:

$$H(A|B)_\rho = H(AB)_\rho - H(B)_\rho \quad \text{and} \tag{12.18}$$

$$I(A:B)_\rho = H(A)_\rho + H(B)_\rho - H(AB)_\rho \tag{12.19}$$

$$= H(A) - H(A|B) \,. \tag{12.20}$$

The chain rules give intuitive interpretations of the conditional entropy and mutual information, and will prove very useful in the analysis of information processing protocols. The chain rule for conditional entropy essentially says that the uncertainty of a joint system AB is given by the uncertainty of one (B), plus the uncertainty of the second (A) conditioned on the first ($H(A|B)$). The mutual information of a joint system is simply the uncertainty of one part (A) minus the uncertainty of that part conditioned on the other ($H(A|B)$).

The first mutual information chain rule (12.19) is just (12.12), and the second follows from the first and (12.18). So it remains to show the conditional entropy chain rule. It follows from properties of the relative entropy, in particular, (12.22) in the following:

Proposition 12.4 (Properties of the relative entropy).

1. *Additivity: For any* $\rho, \sigma \in \mathrm{Stat}(\mathcal{H})$ *and* $\theta, \tau \in \mathrm{Stat}(\mathcal{H}')$,
$$D(\rho \otimes \theta, \sigma \otimes \tau) = D(\rho, \sigma) + D(\theta, \tau), \tag{12.21}$$
2. *Chain rules: For any* $\rho \in \mathrm{Stat}(\mathcal{H}_A \otimes \mathcal{H}_B)$ *and an arbitrary pinch map* $\mathcal{P}_A$ *on* $\mathcal{H}_A$,
$$D(\rho_{AB}, \pi_{AB}) = D(\rho_{AB}, \pi_A \otimes \rho_B) + D(\rho_B, \pi_B) \quad \text{and} \tag{12.22}$$
$$D(\rho_{AB}, \pi_A \otimes \rho_B) = D\big(\rho_{AB}, \mathcal{P}_A[\rho_{AB}]\big) + D\big(\mathcal{P}_A[\rho_{AB}], \pi_A \otimes \rho_B\big), \tag{12.23}$$
where π_A *is the maximally mixed state on* $\mathcal{H}_A$, *and similarly for* π_{AB}.

Proof. Additivity of the relative entropy follows from the fact that $\log(\rho \otimes \theta) = \log \rho \otimes \mathbb{1} + \mathbb{1} \otimes \log \theta$. This can be seen by using the spectral decomposition and observing that the eigenvectors of $\rho \otimes \theta$ are the tensor products of eigenvectors of ρ and θ. Then we have

$$\begin{aligned} D(\rho \otimes \theta, \sigma \otimes \tau) &= \mathrm{Tr}\big[(\rho \otimes \theta)(\log \rho \otimes \mathbb{1} + \mathbb{1} \otimes \log \theta - \log \sigma \otimes \mathbb{1} - \mathbb{1} \otimes \log \tau)\big] \\ &= \mathrm{Tr}\big[\rho(\log \rho - \log \sigma) \otimes \theta\big] + \mathrm{Tr}\big[\rho \otimes \theta(\log \theta - \log \tau)\big] \\ &= D(\rho, \sigma) + D(\theta, \tau). \end{aligned} \tag{12.24}$$

For the first chain rule, simply add and subtract $\log(\pi_A \otimes \rho_B)$ to obtain

$$\begin{aligned} D(\rho_{AB}, \pi_{AB}) &= \mathrm{Tr}\big[\rho_{AB}(\log \rho_{AB} - \log(\pi_A \otimes \rho_B) + \log(\pi_A \otimes \rho_B) - \log \pi_{AB})\big] \\ &= D(\rho_{AB}, \pi_A \otimes \rho_B) + \mathrm{Tr}\big[\rho_{AB}(\log(\pi_A \otimes \rho_B) - \log(\pi_A \otimes \pi_B))\big]. \end{aligned} \tag{12.25}$$

Using the previously mentioned property of the logarithm, in the second term, we have

$$\log(\pi_A \otimes \rho_B) - \log(\pi_A \otimes \pi_B) = \mathbb{1}_A \otimes \log \rho_B - \mathbb{1}_A \otimes \log \pi_B. \tag{12.26}$$

Therefore the second term simplifies to $D(\rho_B, \pi_B)$.

The second chain rule is similar but relies on (12.11). Let $\bar{\rho}_{AB} = \mathcal{P}_A[\rho_{AB}]$ for ease of notation. Adding and subtracting $\log \bar{\rho}_{AB}$ as in the first chain rule gives

$$D(\rho_{AB}, \pi_A \otimes \rho_B) = D(\rho_{AB}, \bar{\rho}_{AB}) + \mathrm{Tr}\big[\rho_{AB}(\log \bar{\rho}_{AB} - \log(\pi_A \otimes \rho_B))\big]. \tag{12.27}$$

Both $\bar{\rho}_{AB}$ and $\pi_A \otimes \rho_B$ are CQ states in the same basis on A, meaning that so are their logarithms. By the results of Exercise 5.11 the operator ρ_{AB} in the trace can therefore be replaced with $\bar{\rho}_{AB}$, giving the desired statement. □

Proposition 12.5 (Properties of the conditional entropy).

1. *Duality: For ρ_{ABC} pure, $H(A|B)_\rho = -H(A|C)_\rho$,*
2. *Bounds: $-\log|A| \leq H(A|B)_\rho \leq \log|A|$ with equality in the upper bound iff $\rho_{AB} = \pi_A \otimes \rho_B$ and equality in the lower bound iff $R_{\text{ent}}(A|B)_\rho = 1$,*
3. *Bound for classical systems: For a CQ state ρ_{XB} with classical X,*

$$H(X|B)_\rho \geq 0\,. \tag{12.28}$$

Equality holds iff $P_{\text{guess}}(X|B)_\rho = 1$.

Proof. By the chain rule the first statement is equivalent to $H(AB)_\rho - H(B)_\rho = -H(AC)_\rho + H(C)_\rho$, whence the previously established duality is equivalent to the conditional version.

The upper bound on the conditional entropy follows from the positivity of the relative entropy and the expression $H(A|B)_\rho = \log|A| - D(\rho_{AB}, \pi_A \otimes \rho_B)$. By Klein's inequality the bound is attained if and only if $\rho_{AB} = \pi_A \otimes \rho_B$.

The lower bound follows from the upper bound by duality since $H(A|B)_\rho = -H(A|C)_\rho \geq -\log d$ for a purification ρ_{ABC} of ρ_{AB}. Then ρ_{AC} must equal $\rho_{AC} = \pi_A \otimes \rho_C$, whose purification is $V_{B|B_1B_2}(\Phi_{AB_1} \otimes \sigma_{B_2C})V^*_{B|B_1B_2}$ for a purification σ_{B_2C} of ρ_C, the maximally entangled state Φ_{AB_1}, and an isometry $V_{B|B_1B_2}$. Therefore $V^*\rho_{AB}V = \Phi_{AB_1} \otimes \sigma_{B_2}$, and hence $R_{\text{ent}}(A|B)_\rho = 1$. On the other hand, if $R_{\text{ent}}(A|B)_\rho = 1$, then the purification ρ_{ABC} must take the form $\pi_A \otimes \rho_C$ and $H(A|C)_\rho = \log d$. By duality the lower bound is therefore saturated.

To establish the positivity of the conditional entropy for CQ states, consider the purification of a generic CQ state $\rho_{XB} = \sum_x P_X(x)|x\rangle\langle x|_X \otimes \varphi_B(x)$:

$$|\psi\rangle_{XABR} = \sum_x \sqrt{P_X(x)}|x\rangle_X|x\rangle_A|\varphi(x)\rangle_{BR}\,. \tag{12.29}$$

Here A is an additional system which purifies X, whereas R purifies B. We can regard this state as the result of measuring A in the standard basis, i. e., applying the Stinespring isometry $V_{XA|A}|x\rangle_A = |x\rangle_X|x\rangle_A$ to

$$|\psi'\rangle_{ABR} = \sum_x \sqrt{P_X(x)}|x\rangle_A|\varphi(x)\rangle_{BR}\,. \tag{12.30}$$

Since projective measurement increases entropy, it follows that $H(AR)_\psi \geq H(AR)_{\psi'}$. Moreover, since entropy is invariant under unitaries and, by the same reasoning, under isometries, it follows that $H(XAR)_\psi = H(AR)_{\psi'}$. Then we have

$$\begin{aligned} H(X|B)_\rho &= H(X|B)_\psi = H(XB)_\psi - H(B)_\psi \\ &= H(AR)_\psi - H(XAR)_\psi \geq H(AR)_{\psi'} - H(AR)_{\psi'} = 0\,. \end{aligned} \tag{12.31}$$

The first equality is the chain rule, and the second is entropy duality. □

Exercise 12.4. Show that $H(Y|B) \leq H(X|B)$ when $Y = f(X)$ for any function f.
Hint: Consider $H(XY|B)$.

From the chain rules the properties of the conditional entropy translate into the following properties of the mutual information.

Proposition 12.6 (Properties of the mutual information).

1. *Duality:* $I(A:B)_\rho + I(A:C)_\rho = 2H(A)_\rho$ *if* ρ_{ABC} *is a pure state.*
2. *Bounds:* $0 \leq I(A:B)_\rho \leq 2\min(\log|A|, \log|B|)$.
3. *Bound for classical systems:* $I(X:B)_\rho \leq \log|X|$ *for a CQ state* ρ_{XB}.

The conditional entropy chain rule extends to the case of conditioning every term in the expression on an additional system C: $H(A|BC)_\rho = H(AB|C)_\rho - H(B|C)_\rho$ for any tripartite state ρ_{ABC}. Using the chain rule, it is also possible to define a *conditional mutual information* of an arbitrary tripartite state ρ_{ABC}:

$$I(A:B|C)_\rho := H(A|C)_\rho - H(A|BC)_\rho . \tag{12.32}$$

Exercise 12.5. Show that $I(A:B|C) = I(A:BC) - I(A:C)$ and the equivalent of (12.17), $I(A:B|X)_\rho = \sum_{x\in\mathcal{X}} P_X(x)\, I(A:B)_{\sigma(x)}$.

Exercise 12.6. Show that $I(A:B|C)_\rho = I(A:B)_\rho$ if ρ_{ABC} is a pure state.

12.3 Stein's lemma

Having exhausted all the properties we can obtain about the entropy from Klein's inequality, we turn to *Stein's lemma.*

Proposition 12.7 (Stein's lemma). *For all* $\rho, \sigma \in \mathrm{Stat}(\mathcal{H})$ *and* $\alpha \in (0,1)$,

$$\lim_{n\to\infty} \frac{-1}{n} \log \beta_\alpha\left(\rho^{\otimes n}, \sigma^{\otimes n}\right) = D(\rho,\sigma). \tag{12.33}$$

To prove Stein's lemma in the quantum case, we will make use of the result in the classical case. In both cases, we proceed by showing matching upper and lower bounds. The lower bound is the *achievability bound*, as we must construct an appropriate POVM with the desired exponential decay of error. The upper bound is the *converse bound.*

Proof of Stein's lemma for commuting ρ *and* σ. Let $\rho = \sum_x P_X(x)|x\rangle\langle x|$ and $\sigma = \sum_x Q_X(x)|x\rangle\langle x|$ for probability distributions P_X and Q_X. For the achievability bound, set $\delta > 0$ and pick Λ to be the projector $\Lambda = \{\rho^{\otimes n} \geq \gamma\sigma^{\otimes n}\}$ for $\gamma = 2^{n(D(\rho,\sigma)-\delta)}$. It is of course diagonal in the basis $|x^n\rangle = |x_1\rangle \otimes |x_2\rangle \otimes \cdots \otimes |x_n\rangle$, where x^n denotes the sequence

$x_1, x_2, \ldots, x_n$. The x^n component of Λ, denote it $\lambda(x^n) = \langle x^n|\Lambda|x^n\rangle$, is

$$\begin{aligned}\lambda(x^n) &= 1\left[\log\frac{P_{X^n}(x^n)}{Q_{X^n}(x^n)} \geq n(D(\rho,\sigma)-\delta)\right]\\ &= 1\left[\frac{1}{n}\sum_{j=1}^{n}\log\frac{P_X(x_j)}{Q_X(x_j)} \geq D(\rho,\sigma)-\delta\right].\end{aligned} \tag{12.34}$$

Observe that the classical relative entropy is the average, under P, of the log-likelihood ratio $\log P(x)/Q(x)$. Since the log-likelihood for i. i. d. distributions is a sum of the individual parts, we can appeal to the weak law of large numbers for a lower bound on $\mathrm{Tr}[\Lambda\rho^{\otimes n}]$. In particular,

$$\mathrm{Tr}[\Lambda\rho^{\otimes n}] = \sum_{x^n} P_{X^n}(x^n)\,1\left[\frac{1}{n}\sum_{j=1}^{n}\log\frac{P(x_j)}{Q(x_j)} \geq D(\rho,\sigma)-\delta\right]. \tag{12.35}$$

By (2.21) for any $\alpha \in (0,1)$, there is an n such that $\mathrm{Tr}[\Lambda\rho^{\otimes n}]$ is greater than α. Thus, for large enough n, Λ is feasible for $\beta_\alpha(\rho^{\otimes n},\sigma^{\otimes n})$.

On the other hand, using Exercise 9.4 immediately yields $\mathrm{Tr}[\Lambda\sigma^{\otimes n}] \leq 2^{-n(D(\rho,\sigma)-\delta)}$. Taking the negative logarithm, the limit as $n \to \infty$, and then the limit as $\delta \to 0$ gives the lower bound

$$\lim_{n\to\infty}\frac{-1}{n}\beta_\alpha(\rho^{\otimes n},\sigma^{\otimes n}) \geq D(\rho,\sigma). \tag{12.36}$$

For the upper bound, choose $\gamma = 2^{n(D(\rho,\sigma)+\delta)}$ for $\delta > 0$ and use (9.13) to obtain

$$\alpha - \gamma\beta_\alpha(\rho^{\otimes n},\sigma^{\otimes n}) \leq \sum_{x^n} P(x^n)\,1\left[\log\frac{P(x^n)}{Q(x^n)} \geq n(D(\rho,\sigma)+\delta)\right]. \tag{12.37}$$

The right-hand side vanishes as $n \to \infty$, again by the weak law of large numbers. Therefore $\beta_\alpha(\rho^{\otimes n},\sigma^{\otimes n}) \geq \alpha\,2^{-n(D(\rho,\sigma)+\delta)}$. Now the negative logarithm, the limit as $n \to \infty$, and then the limit as $\delta \to 0$ give the upper bound

$$\lim_{n\to\infty}\frac{-1}{n}\beta_\alpha(\rho^{\otimes n},\sigma^{\otimes n}) \leq D(\rho,\sigma), \tag{12.38}$$

and the proof is complete. □

Stein's lemma is stated for fixed α (independent of n) not equal to zero or one but using Hoeffding's inequality instead of the weak law in the proof means that the conclusion holds even when α approaches these limits exponentially fast in n.

Exercise 12.7. Show that for some $c > 0$, using either the sequence $\alpha_n = 1 - e^{-cn}$ or $\alpha_n = e^{-cn}$, we still have

$$\lim_{n\to\infty}\frac{-1}{n}\log\beta_{\alpha_n}(P^{\times n},Q^{\times n}) = D(P,Q). \tag{12.39}$$

12.3.1 Achievability in the quantum case

Now we turn to the achievability bound in the quantum case. The proof strategy is to reduce to the classical case as follows. First, given any two states ρ and σ, suppose we construct a POVM element Λ such that

$$\mathrm{Tr}[\Lambda\rho] \geq \mathrm{Pr}_{P_{XY}}\left[\frac{P_{XY}(x,y)}{Q_{XY}(x,y)} \geq \gamma\right] \tag{12.40}$$

and

$$\mathrm{Tr}[\Lambda\sigma] \leq \frac{1}{\gamma} \tag{12.41}$$

for P_{XY} and Q_{XY} from (12.6) and (12.7), respectively. This implies

$$\beta_\alpha(\rho,\sigma) \leq \frac{1}{\gamma} \tag{12.42}$$

for all $\alpha \leq \mathrm{Pr}_{P_{XY}}[\frac{P_{XY}(x,y)}{Q_{XY}(x,y)} \geq \gamma]$. Then making the replacements $\rho \leftarrow \rho^{\otimes n}$, $\sigma \leftarrow \sigma^{\otimes n}$, $P_{XY} \leftarrow P_{X^nY^n}$, and $Q_{XY} \leftarrow Q_{X^nY^n}$, setting $\gamma = 2^{n(D(P_{XY},Q_{XY})-\delta)}$, and following the argument for the classical achievability bound yields

$$\lim_{n\to\infty} \frac{-1}{n} \log \beta_\alpha(\rho^{\otimes n}, \sigma^{\otimes n}) \geq D(P_{XY}, Q_{XY}) \tag{12.43}$$

for all $\alpha < 1$. Finally, by (12.8) the right-hand side is equal to $D(\rho,\sigma)$.

Thus we need to construct Λ satisfying (12.40) and (12.41) for generic states ρ and σ. First, fix the indexing of the eigenvalues $R(x)$ of ρ so that $R(x)$ is increasing with x. Next, fix $\gamma \geq 0$ and define the sets $T(x) = \{y : \gamma S(y) \leq R(x)\}$ and the projectors $\Pi(x) = \sum_{y\in T(x,y)} |v_y\rangle\langle v_y|$. By the above convention, $T(x') \subseteq T(x)$ for all $x' \leq x$, meaning that the $\Pi(x)$ are similarly nested and project onto ever-larger eigensubspaces of σ as x increases. Then set $|\xi_x\rangle = \Pi(x)|u_x\rangle$ and let Λ be the projector onto the span of $\{|\xi_x\rangle\}_x$.

Establishing (12.40) is fairly straightforward. Let $|\bar{\xi}_x\rangle$ be the normalized version of $|\xi_x\rangle$ (or zero if $|\xi_x\rangle = 0$). It follows that $|\bar{\xi}_x\rangle\langle\bar{\xi}_x| \leq \Lambda$. Therefore we have

$$\begin{aligned}
\mathrm{Tr}[\Lambda\rho] &\geq \sum_x R(x)|\langle u_x|\bar{\xi}_x\rangle|^2 = \sum_x R(x)\frac{\langle u_x|\Pi(x)|u_x\rangle^2}{\langle \xi_x|\xi_x\rangle} \\
&= \sum_x R(x)\langle \xi_x|\xi_x\rangle = \sum_x \sum_{y\in T(x)} R(x)W(x,y) \\
&= \sum_{x,y} P_{XY}(x,y)1[\gamma S(y) \leq R(x)]\,,
\end{aligned} \tag{12.44}$$

which is (12.40).

Establishing (12.41) takes a little more effort. Let $\{|\hat{\xi}_x\rangle\}_x$ be the vectors constructed by the Gram–Schmidt procedure applied to $\{|\xi_x\rangle\}_x$ in increasing order in x. This set forms an eigendecomposition of Λ. Note that some of the $|\hat{\xi}_x\rangle$ could be zero if the set $|\xi_x\rangle$ is not linearly independent. Nevertheless, they satisfy

$$|\hat{\xi}_x\rangle = \sum_{x'\leq x} c_{xx'}|\xi_{x'}\rangle = \sum_{x'\leq x} c_{xx'}\Pi(x')|u_{x'}\rangle = \sum_{x'\leq x} c_{xx'} \sum_{y\in T(x')} \langle v_y|u_{x'}\rangle\, |v_y\rangle \tag{12.45}$$

for some complex coefficients $c_{xx'}$. Due to the nested properties of the $T(x)$, it follows that $|\hat{\xi}_x\rangle$ is supported on the span of the $|v_y\rangle$ with $y \in T(x)$, i. e., $|\hat{\xi}_x\rangle = \sum_{y\in T(x)} t_{xy}|v_y\rangle$ for some complex coefficients t_{xy}. Since both $\{|v_y\rangle\}_y$ and the nontrivial $\{|\hat{\xi}_x\rangle\}_x$ are orthonormal sets, it must be that $\sum_{y\in T(x)} |t_{xy}|^2 = 1$ for all x such that $|\hat{\xi}_x\rangle \neq 0$. Otherwise, $t_{xy} = 0$. Now we have

$$\begin{aligned}\mathrm{Tr}[\Lambda\sigma] &= \sum_x \langle\hat{\xi}_x|\sigma|\hat{\xi}_x\rangle = \sum_x \sum_{y\in T(x)} S(y)|t_{xy}|^2 \\ &\leq \frac{1}{\gamma}\sum_x \sum_{y\in T(x)} R(x)|t_{xy}|^2 \leq \frac{1}{\gamma},\end{aligned} \tag{12.46}$$

and the reduction to the classical case is complete.

12.3.2 Converse in the quantum case

The proof of the converse bound in the quantum case also proceeds by reduction to the classical case, again via P_{XY} and Q_{XY}. The main idea is to pinch ρ in the eigenbasis of σ to obtain commuting operators whose eigenvalue distributions are the marginals P_Y and Q_Y. Doing so creates additional terms in the bound, which vanish in the asymptotic limit by the strengthened version of Stein's lemma in (12.39).

Let us start with the pinching inequality (8.17) and express it here as $\rho \leq \nu(\sigma)\, \mathcal{P}_\sigma[\rho]$ for the pinching map $\mathcal{P}_\sigma$ composed of the projection operators onto the eigensubspaces of σ and the number $\nu(\sigma)$ of distinct eigenvalues of σ. For the optimal test $\Lambda^\star$ in $\beta_\alpha(\rho,\sigma)$ for any $\alpha \in (0,1)$, the pinching inequality implies that $\Lambda^\star$ is feasible in $\beta_{\alpha/\nu(\sigma)}(\mathcal{P}_\sigma[\rho],\sigma)$ with the same value of the objective as $\beta_\alpha(\rho,\sigma)$. Therefore we have the first step in the proof,

$$\beta_{\alpha/\nu(\sigma)}(\mathcal{P}_\sigma[\rho],\sigma) \leq \beta_\alpha(\rho,\sigma)\,. \tag{12.47}$$

In the next step, we relate $\beta_\alpha(\mathcal{P}_\sigma[\rho],\sigma)$ to $\beta_\alpha(P_{XY},Q_{XY})$. For ρ and σ with degenerate eigenvalues, which will assuredly be the case in the i. i. d. setting, there is some ambiguity in the definition of P_{XY} and Q_{XY} corresponding to the choice of eigenbasis. A suitable choice for the eigenbasis of σ will ensure that the eigenvalues of $\mathcal{P}_\sigma[\rho]$ are given by P_Y, the marginal of the associated P_{XY}. Consider the matrix whose (y,y') entry is $\langle v_y|\rho|v_{y'}\rangle$ for an eigenbasis $|v_y\rangle$ of σ. The pinch map will annihilate all but the diagonal blocks of this matrix corresponding to the various eigensubspaces. Independently of the pinch map, each of these diagonal blocks can themselves be diagonalized by choosing an appropriate eigenbasis $|v_y\rangle$. Let us fix this basis to be the eigenbasis of σ. Then the action of the pinch map will leave a diagonal matrix with elements $\sum_x R(x)|\langle u_x|v_y\rangle|^2 = \sum_x R(x)W(x,y) = P_Y(y)$. The eigenvalues of σ are of course $Q_Y(y)$, and therefore

$$\beta_\alpha(\mathcal{P}_\sigma[\rho],\sigma) = \beta_\alpha(P_Y, Q_Y) \tag{12.48}$$

for any $\alpha \in [0,1]$ and this choice of eigenbasis of σ. By monotonicity it follows that, again for any $\alpha \in [0,1]$,

$$\beta_\alpha(P_{XY}, Q_{XY}) \le \beta_\alpha(P_Y, Q_Y)\,. \tag{12.49}$$

Now combine (12.47), (12.48), and (12.49) and use (9.13) to obtain

$$\beta_\alpha(\rho,\sigma) \ge \frac{1}{\gamma}\left(\frac{\alpha}{\nu(\sigma)} - \mathrm{Tr}[P_{XY}\{P_{XY} > \gamma Q_{XY}\}]\right). \tag{12.50}$$

Then move to the i. i. d. scenario by making the replacements $\rho \leftarrow \rho^{\otimes n}$, $\sigma \leftarrow \sigma^{\otimes n}$, $P_{XY} \leftarrow P_{X^nY^n}$, and $Q_{XY} \leftarrow Q_{X^nY^n}$. Set $\gamma = 2^{n(D(P_{XY},Q_{XY})+\delta)}$ for some $\delta > 0$ and define $T_n = \mathrm{Tr}[P_{X^nY^n}\{P_{X^nY^n} > \gamma Q_{X^nY^n}\}]$. Then the limit procedure as $n \to \infty$ and then as $\delta \to 0$ produces

$$\lim_{n\to\infty} \frac{-1}{n}\log\beta_\alpha(\rho^{\otimes n},\sigma^{\otimes n}) \le D(\rho,\sigma) - \lim_{\delta\to 0}\lim_{n\to\infty}\frac{1}{n}\log\left(\frac{\alpha}{\nu(\sigma^{\otimes n})} - T_n\right), \tag{12.51}$$

again using (12.8). To dispose of the last term, write

$$-\frac{1}{n}\log\left(\frac{\alpha}{\nu(\sigma^{\otimes n})} - T_n\right) = -\frac{1}{n}\log\alpha + \frac{1}{n}\log\nu(\sigma^{\otimes n}) - \frac{1}{n}\log\left(1 - \frac{1}{\alpha}T_n\nu(\sigma^{\otimes n})\right). \tag{12.52}$$

The first of these tends to zero as $n \to \infty$, since α does not vary with n.

Meanwhile, the second term deals with the number of distinct eigenvalues of $\sigma^{\otimes n}$. Each such eigenvalue is an n-fold product of the at most d eigenvalues of σ, so the distinct possible $\sigma^{\otimes n}$ eigenvalues are specified by the number of each of the possible σ eigenvalue factors. This is often called the *type* of the sequence of eigenvalues, and the set of all sequences of the same type is the *type class*. As each eigenvalue factor can occur zero up to n times, there are at most $(n+1)^d$ types. Hence the second term is upper-bounded by $\frac{d}{n}\log(n+1)$, which also vanishes in the limit as $n \to \infty$.

Finally, we come to the third term. By the Hoeffding bound (2.26), T_n goes to zero as e^{-cn} for some constant $c > 0$. Then the argument to the logarithm tends to 1 as $n \to \infty$, since $\nu(\sigma^{\otimes n})$ grows only polynomially in n. Therefore the third term also vanishes in this limit, and the proof of the converse is complete.

Exercise 12.8. Show additivity of the relative entropy by using Stein's lemma.

12.4 The data processing inequality

With Stein's lemma in hand, monotonicity of the relative entropy follows immediately from the monotonicity of β_α. For every pair of states $\rho, \sigma \in \mathrm{Stat}(\mathcal{H}_A)$ and every quantum channel $\mathcal{E}_{B|A}$, we have $\beta_\alpha(\rho^{\otimes n}, \sigma^{\otimes n}) \leq \beta_\alpha(\mathcal{E}_{B|A}[\rho_A]^{\otimes n}, \mathcal{E}_{B|A}[\sigma_A]^{\otimes n})$. Applying Stein's lemma yields

$$D(\mathcal{E}_{B|A}[\rho_A], \mathcal{E}_{B|A}[\sigma_A]) \leq D(\rho, \sigma)\,. \tag{12.53}$$

Monotonicity of $D(\rho, \sigma)$ can be used to find a bound on the relative entropy in terms of $\beta_\alpha(\rho, \sigma)$ by considering the optimal measurement for β_α. Consider the two-outcome measurement (QC) channel $\mathcal{M}_{X|A}$ specified by the POVM $\{\Lambda, \mathbb{1} - \Lambda\}$. For $\alpha = \mathrm{Tr}[\Lambda\rho]$ and $\beta = \mathrm{Tr}[\Lambda\sigma]$, monotonicity implies

$$\begin{aligned} D(\rho_A, \sigma_A) &\geq D(\mathcal{M}_{X|A}[\rho_A], \mathcal{M}_{X|A}[\sigma_A]) = D((\alpha, 1-\alpha), (\beta, 1-\beta)) \\ &= -h_2(\alpha) + \alpha \log \frac{1}{\beta} + (1-\alpha) \log \frac{1}{1-\beta} \geq \alpha \log \frac{1}{\beta} - h_2(\alpha)\,. \end{aligned} \tag{12.54}$$

Here $h_2(p) := -p \log p - (1-p) \log(1-p)$ is the *binary entropy*. Choosing Λ to be the optimal test in $\beta_\alpha(\rho, \sigma)$ gives $D(\rho, \sigma) \geq -\alpha \log \beta_\alpha(\rho, \sigma) - h_2(\alpha)$.

Exercise 12.9. Let ρ_{XB} be an arbitrary CQ state with classical X. Prove *Fano's*[5] *inequality*:

$$H(X|B)_\rho \leq h_2(P_{\mathrm{guess}}(X|B)_\rho) + (1 - P_{\mathrm{guess}}(X|B)_\rho) \log(|\mathcal{X}| - 1)\,. \tag{12.55}$$

Exercise 12.10. Using Taylor's[6] theorem, show that the binary relative entropy $d_2(p, q) := p \log(p/q) + (1-p) \log((1-p)/(1-q))$ satisfies $d_2(p, q) \geq \frac{2}{\ln 2}(p-q)^2$. Employing monotonicity for the optimal measurement in the distinguishability, prove *Pinsker's*[7] *inequality* for all quantum states ρ and σ:

$$D(\rho, \sigma) \geq \tfrac{2}{\ln 2} \delta(\rho, \sigma)^2\,. \tag{12.56}$$

5 Roberto Mario "Robert" Fano, 1917–2016.

6 Brook Taylor, 1685–1731.

7 Mark Semenovich Pinsker, 1925–2003.

Exercise 12.11. Following a similar argument for distinguishability, show that joint convexity and monotonicity of relative entropy are equivalent.

Proposition 12.8 (Monotonicity of mutual information and conditional entropy). *For any two quantum channels $\mathcal{E}_{A'|A}$ and $\mathcal{F}_{B'|B}$ and any quantum state ρ_{AB}, let $\rho'_{A'B'} = \mathcal{E}_{A'|A} \otimes \mathcal{F}_{B'|B}[\rho_{AB}]$. Then $I(A' : B')_{\rho'} \leq I(A : B)_\rho$. Furthermore, if $\mathcal{E}_{A'|A}$ is unital, then $H(A'|B')_{\rho'} \geq H(A|B)_\rho$.*

These inequalities are often called *data processing* inequalities, since processing data (random variables or quantum states) only increase entropy or decrease mutual information. They are also equivalent to the *strong subadditivity* of the entropy, which is subadditivity for conditional entropy:

$$H(AB|C)_\rho \leq H(A|C)_\rho + H(B|C)_\rho\,. \tag{12.57}$$

By (12.32) strong subadditivity is positivity of the conditional mutual information.

When ρ_{ABC} is a CQ state with classical C, strong subadditivity follows from usual subadditivity using Property 3 of Proposition 12.5. In the general case, it is easy to work out the statement of strong subadditivity is equivalent to $H(B|AC)_\rho \leq H(B|C)_\rho$ (or $H(A|BC)_\rho \leq H(A|C)_\rho$), which is just monotonicity under the partial trace map. By the Stinespring representation, monotonicity for the partial trace implies monotonicity for all quantum channels.

By (12.17) monotonicity of the conditional entropy immediately implies that the conditional entropy is a concave function of the quantum state: For $\rho_{AB} = \sum_x P_X(x)\sigma_{AB}(x)$,

$$\sum_x P_X(x)\, H(A|B)_{\sigma_{AB}(x)} \leq H(A|B)_\rho\,. \tag{12.58}$$

Therefore the conditional entropy of separable states is necessarily positive; this generalizes (12.28). However, the converse statement does not hold: There exist examples of nonseparable states with positive conditional entropy.

By data processing we can also infer continuity of the conditional entropy.

Proposition 12.9 (Continuity of the conditional entropy). *For any two bipartite density operators ρ_{AB} and σ_{AB}, let $\delta(\rho_{AB}, \sigma_{AB}) = \varepsilon$. Then*

$$\left|H(A|B)_\rho - H(A|B)_\sigma\right| \leq 2\varepsilon \log|A| + (1+\varepsilon)\, h_2\!\left(\frac{\varepsilon}{1+\varepsilon}\right). \tag{12.59}$$

Proof. Define $\varphi = \frac{1}{\varepsilon}\{\rho - \sigma\}_+$, which must be a normalized density operator, and then $\theta = \frac{1}{1+\varepsilon}(\sigma + \varepsilon\varphi)$. Observe that $\theta \geq 0$ and $\mathrm{Tr}[\theta] = 1$, as well as $\rho \leq (1+\varepsilon)\theta$. Therefore $\tau = \frac{1}{\varepsilon}((1+\varepsilon)\theta - \rho)$ is positive and normalized, thus giving another ensemble decomposition of θ:

$\theta = \frac{1}{1+\varepsilon}(\rho + \varepsilon\tau)$. Now consider the CQ extensions of the two ensemble decompositions,

$$\theta'_{ABX} = \frac{1}{1+\varepsilon}(\rho_{AB} \otimes |0\rangle\langle 0|_X + \tau_{AB} \otimes |1\rangle\langle 1|_X), \tag{12.60}$$

$$\theta_{ABX} = \frac{1}{1+\varepsilon}(\sigma_{AB} \otimes |0\rangle\langle 0|_X + \varphi_{AB} \otimes |1\rangle\langle 1|_X). \tag{12.61}$$

By construction, $\theta'_{AB} = \theta_{AB}$, and therefore $H(A|B)_\theta = H(A|B)_{\theta'}$. The conditional entropy can be bounded from above and below by entropy quantities involving X; specifically,

$$H(A|BX)_{\theta'} \le H(A|B)_{\theta'} = H(A|B)_\theta \le H(A|BX)_\theta + H(X|B)_\theta. \tag{12.62}$$

The lower bound is just data processing, and for the upper bound, we appeal to $H(X|AB)_\theta \ge 0$. By the chain rule this is equivalent to $H(AB)_\theta \le H(ABX)_\theta$, which implies $H(A|B)_\theta \le H(AX|B)_\theta$ and then $H(A|B)_\theta \le H(A|BX)_\theta + H(X|B)_\theta$, both by further applications of the chain rule. From positivity of the mutual information we can replace $H(X|B)_\theta$ with $H(X)_\theta$ to get $H(A|BX)_{\theta'} \le H(A|BX)_\theta + H(X)_\theta$. For the specific states in question, this is just

$$\frac{1}{1+\varepsilon}H(A|B)_\rho + \frac{\varepsilon}{1+\varepsilon}H(A|B)_\tau \le h_2\left(\frac{1}{1+\varepsilon}\right) + \frac{1}{1+\varepsilon}H(A|B)_\sigma + \frac{\varepsilon}{1+\varepsilon}H(A|B)_\varphi. \tag{12.63}$$

Rearranging and using the crude dimension upper and lower bounds on the conditional entropy gives

$$H(A|B)_\rho - H(A|B)_\sigma \le 2\varepsilon \log|A| + (1+\varepsilon)h_2\left(\frac{1}{1+\varepsilon}\right). \tag{12.64}$$

Interchanging ρ and σ gives the absolute value in the continuity statement. □

12.5 Additional exercises

Exercise 12.12. By the data processing inequality, the mutual information of two binary random variables X and Y decreases when Y is subjected to any channel. For the output Z of BSC(p) with input Y, show the following *strong data processing inequality*:

$$I(X:Y) \le (1-2p)I(X:Z). \tag{12.65}$$

Hint: Decompose BSC(p) into the identity channel and the channel with completely mixed output.

Exercise 12.13. Extend the previous result to the quantum depolarizing channel.

Exercise 12.14. Show that $\min_{\sigma_B} D(\rho_{AB}, \rho_A \otimes \sigma_B) = I(A : B)$, i. e., ρ_B is the optimal choice in the minimization. Similarly, show that $H(A|B)_\rho = -\min_{\sigma_B} D(\rho_{AB}, \mathbb{1}_A \otimes \sigma_B)$.

Exercise 12.15. Show that for any state ρ, $D(\rho, \sigma) \geq D(\rho, \tau)$ if $0 \leq \sigma \leq \tau$. Use this to show that log is operator monotone.

Exercise 12.16. Using operator monotonicity of log, show that $D(\tau, \sigma) \leq 0$ for $\tau \leq \sigma$. Use this to show that $H(X|B)_\rho \geq 0$ for classical X and an arbitrary CQ state ρ_{XB}, as in Proposition 12.5.

Exercise 12.17. Show the following complement to Fano's inequality: For any CQ state ρ_{XB},

$$H(X|B)_\rho \geq \log \frac{1}{P_{\text{guess}}(X|B)_\rho} \,. \tag{12.66}$$

Hint: use the optimal variables for $P_{\text{guess}}(X|B)_\rho$ *from* (11.6), *operator monotonicity of* log, *and the variational expression for* $H(A|B)_\rho$ *from Exercise* 12.14.

Exercise 12.18. Consider a CQ state $\rho_{XB^n} = \sum_x P_X(x)|x\rangle\langle x|_X \otimes \theta_{B_1}(x) \otimes \cdots \otimes \theta_{B_n}(x)$ for some arbitrary distribution P_X and states $\theta(x)$. Show that $\lim_{n\to\infty} \frac{1}{n} H(B^n)_\rho = H(B_1|X)_\rho$. Thus the correlations among the B systems induced by X exactly reduce the per-copy entropy of the collection B^n from $H(B_1)$ to $H(B_1|X)$.

12.6 Notes and further reading

Entropy as we have defined was first defined by Gibbs in the context of classical statistical mechanics [108] and later extended to the quantum setting by von Neumann [293]. Shannon introduced entropy as a quantification of uncertainty or information [258]. The first quote at the top of the chapter was recounted by Shannon to Tribus in [283]. The latter was relayed by the author's father. For more on classical information theory, see Cover and Thomas [64] and MacKay [198].

Kullback and Leibler [175] introduced the relative entropy, while the form of the quantum relative entropy is due to Umegaki [287]. Klein [162] had already established a more general result, which implies positivity of Umegaki's relative entropy much earlier. Relation (12.8) to the relative entropy of two probability distributions is due to Nussbaum and Szkoła [212]. Due to the noncommutation of ρ and σ, many other variants that capture the notion of a "quantum likelihood ratio $\frac{\rho}{\sigma}$" are possible, e. g., $D_{\text{BS}}(\rho, \sigma) = \text{Tr}[\rho \log(\rho^{1/2}\sigma^{-1}\rho^{1/2})]$ due to Belavkin and Staszewski [18]. Crucially, though, Stein's lemma and the chain rules hold for the Umegaki form.

The triangle inequality for the von Neumann entropy was proven (and named) by Araki and Lieb [7] and was one of the first uses of the purification of a quantum state (as in the present proof). Strong subadditivity was first shown by Lieb and Ruskai [190] using a result known as Lieb's concavity theorem [189]. Later Lindblad [191] used Lieb's concavity theorem to establish joint convexity of the relative entropy, which implies monotonicity. Fano's inequality was reported in his early book on information theory [97]. A bound relating relative entropy and distinguishability was originally found by Pinsker [223] and independently improved to the version stated here by Csiszár [65], Kullback [174], and Kemperman [159]. The continuity of conditional entropy is Winter's improvement [306] on the original statement from Alicki and Fannes [4].

In the field of statistics, "Stein's lemma" refers to something different, but our use is the common one in quantum information theory. In fact, Proposition 12.7 was first stated by Chernoff, who attributed the result to Stein [56]. Apparently, Stein denied ever having proved the statement [156]. The proof here follows Polyanskiy and Wu [225]. The quantum version was shown by Hiai and Petz [139], who showed achievability, and Ogawa and Nagaoka [214], who established the converse statement. That Stein's lemma implies monotonicity and the former has a simple proof was stressed by Bjelaković and Siegmund-Schultze [40]. Our approach to quantum Stein's lemma follows Li in the achievability part [187]. Following Hiai and Petz, Ogawa and Hayashi [213] make use of the pinching inequality $\rho \leq \nu(\sigma)\mathcal{P}_\sigma[\rho]$ as we do, though in the achievability argument.

13 Uncertainty relations

> It must have been one evening after midnight when I suddenly remembered my conversation with Einstein and particularly his statement, "It is the theory which decides what we can observe." I was immediately convinced that the key to the gate that had been closed for so long must be sought right here. ... The right question should therefore be: Can quantum mechanics represent the fact that an electron finds itself approximately in a given place and that it moves approximately with a given velocity, and can we make these approximations so close that they do not cause experimental difficulties?[1]
>
> Werner Heisenberg

One of the appealing aspects of the study of quantum information is that in trying to obtain tight bounds on various kinds of protocols, we are forced to sharpen existing results from quantum mechanics. In the process, we learn a bit more about quantum mechanics itself. An example is given by various uncertainty relations. Recall the uncertainty *principle*, which is a statement along the lines of

Complementary physical properties, like position and momentum, cannot be simultaneously known precisely.

The job of uncertainty *relations* is to make a precise statement in this direction. (There can be many such useful statements; there is not necessarily one uncertainty relation to rule them all.)

Uncertainty relations are mathematical statements, and in quantum mechanics, we have to be especially careful to make sure that mathematical quantities we use are meaningful in some way. One way to do this is to try to formulate a statement that has a direct operational meaning in that it says a particular process is constrained in some way. This distinction is a bit like the various forms of the second law. One formulation of the second law, due to Carathéodory, pertains more to the mathematical formulation of the theory: "In every neighborhood of any state ρ of an adiabatically enclosed system, there are states inaccessible from ρ." The notion of adiabatic accessibility is not so immediate. But the Kelvin[2]–Planck formulation is more direct: "It is impossible

---

1 Es mag an jenem Abend gegen Mitternacht gewesen sein, als ich mich plötzlich auf mein Gespräch mit Einstein besann und mich an seine Äußerung erinnerte: »Erst die Theorie entscheidet darüber, was man beobachten kann.« Es war mir sofort klar, daß der Schlüssel zu der so lange verschlossenen Pforte an dieser Stelle gesucht werden müsse. ... Die richtige Frage mußte also lauten: Kann man in der Quantenmechanik eine Situation darstellen, in der sich ein Elektron ungefähr—das heißt mit einer gewissen Ungenauigkeit—an einem gegebenen Ort befindet und dabei ungefähr—das heißt wieder mit einer gewissen Ungenauigkeit—eine vorgegebene Geschwindigkeit besitzt, und kann man diese Ungenauigkeiten so gering machen, daß man nicht in Schwierigkeiten mit dem Experiment gerät? [134]

2 William Thomson, 1st Baron Kelvin, 1824 –1907.

https://doi.org/10.1515/9783110570250-013

to devise a cyclically operating device, the sole effect of which is to absorb energy in the form of heat from a single thermal reservoir and to deliver an equivalent amount of work."

13.1 Guessing games

We can construct two concrete uncertainty relations that are more in the Kelvin–Planck vein and which will illuminate why quantum error correction and key distribution are possible. Instead of the task of delivering some amount of work, consider the following two guessing games played by Alice and Bob. The overarching goal of both games is for Bob to prepare a quantum system in a state such that he can predict the outcome of either of two complementary measurements made by Alice on that system, measurements in conjugate bases. We refer to these as measurements of "amplitude" and "phase", denote the corresponding basis vectors $|z\rangle$ and $|\tilde{x}\rangle$, and the random variables describing the result as Z and X, respectively.

There are two possible complementary measurements that are relevant for the purposes of this book. The first, applicable for any dimension d of Alice's system, is projective measurement in the eigenbasis of the shift and clock operators from (4.18) and (4.19), respectively. The second is applicable when Alice's system is a collection of n qubits. In this case the amplitude (phase) measurement is measurement of every qubit in the basis of σ_z (σ_x).

For either choice of measurement, the two games are:

<table>
<tr><th colspan="2">Alice–Bob guessing games</th></tr>
<tr><th>Version 1</th><th>Version 2</th></tr>
<tr><td>1. Bob prepares a qubit A in any manner of his choosing and delivers it to Alice.</td><td>1. Bob prepares a qubit A in any manner of his choosing and delivers it to Alice.</td></tr>
<tr><td>2. Alice randomly chooses X or Z and announces her choice.</td><td>2. Bob announces his guesses for the outcomes of an X measurement and a Z measurement.</td></tr>
<tr><td>3. Bob announces his guess for the outcome of Alice's announced measurement.</td><td>3. Alice randomly chooses the X or Z measurement.</td></tr>
<tr><td>4. Alice performs the corresponding measurement on A.</td><td>4. Alice performs the corresponding measurement on A.</td></tr>
<tr><td>5. They compare her outcome with his guess.</td><td>5. They compare her outcome with his guess.</td></tr>
</table>

Steps 1, 4, and 5 are identical, whereas Steps 2 and 3 are interchanged in the two versions. In Version 1, Bob only has to deliver one guess, and in Version 2, two guesses. Let us consider Version 2 first. According to the uncertainty principle, it should be impossible to always win the game. Mathematically, we anticipate this because there is no joint eigenvector of X and Z. We could hope to show that there is no strategy to always win Version 2 by appealing to the Heisenberg–Robertson[3] relation relating the variances of the observables in a state $|\psi\rangle$ to the expectation of their commutator:

$$(\Delta X)_\psi (\Delta Z)_\psi \geq \frac{1}{2}\left|\langle [X, Z] \rangle_\psi\right| . \tag{13.1}$$

In this case, however, the bound is trivial. For qubits, the operators X and Z anticommute ($XZ + ZX = 0$), and so the right-hand side reduces to $|\langle XZ \rangle_\psi|$. Choosing $|\psi\rangle = |0\rangle$ immediately yields zero. We will have to take a different approach.

Now consider Version 1. Despite the similarities with Version 2, it is possible to win this game with certainty! Continuing with the qubit case, in step 1, Bob should prepare $|\Phi\rangle_{AB}$ and keep B for himself. In step 3, he makes the same measurement on B as Alice has announced in step 2. Clearly, due to the form of the state, if Alice chooses Z, then Bob's Z measurement outcome from B will be the same as Alice's from A. But note that for the X eigenstates $|\pm\rangle = \frac{1}{\sqrt{2}}(|0\rangle \pm |1\rangle)$, we have that

$$|\Phi\rangle_{AB} = \frac{1}{\sqrt{2}}\left(|+\rangle_A \otimes |+\rangle_B + |-\rangle_A \otimes |-\rangle_B\right) . \tag{13.2}$$

Hence, if Alice chooses X, then his measurement result will *also* be identical to hers. In this way, he can circumvent the straightforward reading of the uncertainty principle.

The difference between Versions 1 and 2 hinges on what is meant by "simultaneously" in the uncertainty principle. Version 2 corresponds to the more straightforward reading, since we directly demand both pieces of information from Bob. In Version 1, though, he has to be ready to guess either, which makes it seem that he would just need to have concrete guesses for both. But not quite. Instead, he has a "quantum guess" in the form of qubit B, since it will tell him either result (but not both at the same time). It turns out (and we will prove shortly) that it is only possible to certainly win the game by using a "quantum guess": If both guessing probabilities are 1, then the initial state Bob starts with must be maximally entangled. We might wonder if Bob can win Version 2 by making use of a "quantum guess", but alas the answer is no. These two guessing games will play an important role in constructing protocols for quantum communication over noisy channels (Version 1) and the security of QKD (Version 2).

3 Howard Percy Robertson, 1903–1961.

13.2 Entropic uncertainty relations

The limitations and possibilities of each version of the game can be captured by entropic uncertainty relations. Before we state the particular relations, let us consider how to model the situation. Any strategy Bob might employ for Version 2 can be modeled by a device that outputs one quantum and two classical values. For instance, he may pick an eigenstate of X at random, report the corresponding state as his X guess, and simply pick a Z guess also at random. He could also prepare a tripartite quantum state ψ_{ABC} and measure system B (C) to obtain the Z (X) guess.

Indeed, by Stinespring, every device producing two classical values can be regarded as first preparing a tripartite state and then measuring systems B and C. Importantly, the measurements on B and C commute, because they pertain to different systems. In contrast, since Bob reports his guess in Version 1 only after Alice announces the choice of measurement, he need only prepare a bipartite quantum state ψ_{AB}. As in the entangled-state protocol just mentioned, his measurement on B depends on Alice's choice of measurement. In this case, Bob's two measurements do not commute.

To state the entropic uncertainty relations, it is convenient to denote the conditional entropy of Z measurement on system A given B for the bipartite state ρ_{AB} by $H(Z_A|B)_\rho$. That is, for the pinch $\mathcal{P}_A$ of A in the $|z\rangle$ basis, $H(Z_A|B)_\rho = H(A|B)_{\mathcal{P}_A[\rho_{AB}]}$ Similarly, by $H(X_A|B)_\rho$ we denote the conditional entropy of the X measurement on A, which is just $H(A|B)_{\tilde{\mathcal{P}}_A[\rho_{AB}]}$, where $\tilde{\mathcal{P}}$ denotes the pinch in the $|\tilde{x}\rangle$ basis. Then the uncertainty relations read as follows.

Proposition 13.1 (Entropic uncertainty relations). *For any tripartite state ρ_{ABC},*

$$H(X_A|B)_\rho + H(Z_A|B)_\rho \geq \log|A| + H(A|B)_\rho \quad \textit{and} \tag{13.3}$$

$$H(X_A|B)_\rho + H(Z_A|C)_\rho \geq \log|A| . \tag{13.4}$$

The first inequality pertains to Version 1 of the guessing game. Since the conditional entropy of a quantum system can be negative, it is possible to decrease the right-hand side by choosing more entangled states. Indeed, the canonical maximally entangled state has a conditional entropy of $-\log|A|$ as we have seen, so in this case the bound becomes trivial. This corresponds to the protocol for winning using entanglement that we described above.

The uncertainty relation implies that every strategy for perfect guessing in Version 1 will necessarily involve a maximally entangled state by the properties of purifications in Proposition 6.2. Stated differently, the extent to which Bob can win the game (and have low conditional entropy of X and Z) implies that the original state ρ_{AB} must have a conditional entropy near the minimum possible value $-\log|A|$.

The second inequality pertains to Version 2 of the guessing game and ensures that no strategy will allow Bob to perfectly guess each time. Perfect guessing would lead to a conditional entropy of zero for both terms, which is impossible. Indeed, the total amount of entropy must be the logarithm of the dimension, so for qubits, there will always be one bit of entropy in Bob's guesses.

Before proceeding to the proof, it is worth noticing the tension between uncertainty relations and Bell's theorem. The uncertainty relations allow us to place constraints on incompatible observables, but reasoning about their properties is precisely what we learned to be wary of from the Bell scenario. The resolution is that the various entropies are computed from a fixed quantum state, which mediates the relationship between the incompatible observables. As remarked in Section 7.3.1, Bell's theorem is not an absolute prohibition on reasoning about incompatible observables, only that local hidden variables cannot be used for this purpose.

Now we turn to the proof. Its crux is that $\tilde{\mathcal{P}}_A \circ \mathcal{P}_A = \pi_A \operatorname{Tr}_A$, which is to say that sequentially measuring in the two bases produces a mixed state. This was established for amplitude and phase bases defined from the shift and clock operators in (5.6). That result for $d = 2$ implies the n-qubit version, simply by separately pinching each qubit in both bases. It is also therefore the case that the entropic uncertainty relations hold for any choice of complementary bases such that sequential measurement yields the mixed state, but we will not pursue this further.

Proof. First note that we only need to prove the first relation for pure ρ_{ABC}. For if ρ_{ABC} has a purification ρ_{ABCR} with nontrivial R and $H(X_A|B)_\rho + H(Z_A|CR)_\rho \geq \log|A|$, then the desired statement follows by monotonicity of the conditional entropy under partial trace of R.

So let us assume that ρ_{ABC} is pure. In this case the two statements imply each other, because it turns out that $H(Z_A|B)_\rho - H(Z_A|C)_\rho = H(A|B)_\rho$. To see this, first note that for $\rho_{ABC} = |\psi\rangle\langle\psi|_{ABC}$, we can always express $|\psi\rangle_{ABC}$ as

$$|\psi\rangle_{ABC} = \sum_z \sqrt{P_Z(z)}|z\rangle_A|\varphi_z\rangle_{BC} \tag{13.5}$$

for some normalized vectors $|\varphi_z\rangle_{BC}$. Thus, for every outcome z of the Z measurement on A, the conditional state of BC is a pure state. Therefore $H(B|Z_A)_\rho = H(C|Z_A)_\rho$. Using chain rules and $H(AB)_\rho = H(C)_\rho$, we have

$$\begin{aligned} H(Z_A|B)_\rho - H(Z_A|C)_\rho &= H(Z_AB)_\rho - H(B)_\rho - H(Z_AC)_\rho + H(C)_\rho \\ &= H(Z_AB)_\rho - H(Z_AC)_\rho + H(C)_\rho - H(B)_\rho \\ &= H(B|Z_A)_\rho - H(C|Z_A)_\rho + H(AB)_\rho - H(B)_\rho \\ &= H(A|B)_\rho . \end{aligned} \tag{13.6}$$

It remains to prove one of the relations. Writing the first in terms of relative entropy and rearranging terms reveals that it is equivalent to

$$D(\rho_{AB}, \pi_A \otimes \rho_B) \geq D(\tilde{\mathcal{P}}_A[\rho_{AB}], \pi_A \otimes \rho_B) + D(\mathcal{P}_A[\rho_{AB}], \pi_A \otimes \rho_B), \tag{13.7}$$

where $\tilde{\mathcal{P}}_A$ is the pinch of A in the $|\tilde{x}\rangle$ basis. Recall (12.23), which states

$$D(\rho_{AB}, \pi_A \otimes \rho_B) = D(\rho_{AB}, \mathcal{P}_A[\rho_{AB}]) + D(\mathcal{P}_A[\rho_{AB}], \pi_A \otimes \rho_B). \tag{13.8}$$

By monotonicity under $\tilde{\mathcal{P}}_A$ the first term on the right-hand side satisfies

$$D(\rho_{AB}, \mathcal{P}_A[\rho_{AB}]) \geq D(\tilde{\mathcal{P}}_A[\rho_{AB}], \tilde{\mathcal{P}}_A \circ \mathcal{P}_A[\rho_{AB}]) = D(\tilde{\mathcal{P}}_A[\rho_{AB}], \pi_A \otimes \rho_B). \tag{13.9}$$

Using $\tilde{\mathcal{P}}_A \circ \mathcal{P}_A = \pi_A \operatorname{Tr}_A$ completes the proof. □

The only inequality step in the proof is the pinch in the X eigenbasis and therefore states ρ_{AB} for which $\rho_{AB} = \tilde{\mathcal{P}}_A[\rho_{AB}]$ satisfy the uncertainty relations with equality. By interchanging X and Z in the proof the same is true when $\rho_{AB} = \mathcal{P}_A[\rho_{AB}]$. These are CQ states.

For the bipartite relation, these equality conditions are not terribly meaningful: When $\rho_{AB} = \mathcal{P}_A[\rho_{AB}]$, we have $H(Z_A|B)_\rho = H(A|B)_\rho$, and (13.3) reduces to $H(X_A|B)_\rho \geq \log|A|$. Equality must hold in this case, as $\log|A|$ is an upper bound on the conditional entropy. Moreover, when ρ_{AB} is classical on A in the Z basis, then naturally the X basis outcome is completely random, and therefore the entropy is maximal.

Translated to the tripartite relation, however, the equality conditions are more meaningful. If ρ_{ABC} is pure and such that $\rho_{AB} = \tilde{\mathcal{P}}_A[\rho_{AB}]$, then equality in (13.7) together with (13.6) gives equality in (13.4). Interchanging (X_A, B) and (Z_A, C), equality in (13.4) also holds if $\rho_{AC} = \mathcal{P}[\rho_{AC}]$. Equality also holds if $\mathcal{P}_A[\rho_{AB}] = \rho_{AB}$ or $\tilde{\mathcal{P}}_A[\rho_{AC}] = \rho_{AC}$, since in this case, one of the entropies is maximal, and by (13.6) the other is zero. Altogether, we have the following:

Proposition 13.2 (Equality conditions). *For pure ρ_{ABC}, if either ρ_{AB} or ρ_{AC} is a CQ state with classical A in either the X or Z basis, then equality holds in* (13.4).

Exercise 13.1. Consider a classical state $\rho_{AB} = \frac{1}{2}\sum_{u,z\in\mathbb{Z}_2} P_U(u)|z\rangle\langle z|_A \otimes |z+u\rangle\langle z+u|_B$, i. e., the classical variable B is the image of A under BSC(p) for $p = P_U(1)$. Take the purification to be $|\psi\rangle_{ABC_1C_2} = \frac{1}{\sqrt{2}}\sum_{u,z}\sqrt{P_U(u)}|z\rangle_A|z+u\rangle_B|u\rangle_{C_1}|z\rangle_{C_2}$ and show that after applying a CNOT gate from C_1 to C_2, the state becomes $|\psi'\rangle_{ABC_1C_2} = \frac{1}{\sqrt{2}}\sum_x |\tilde{x}\rangle_A Z_B^x|\Phi\rangle_{BC_1} Z_{C_2}^x|\theta\rangle_{C_1}$ for $|\theta\rangle = \sum_{u\in\mathbb{Z}_2}\sqrt{P_U(u)}|u\rangle$. Confirm that equality holds in (13.4). Hence the state in C_2 given X_A is the output of PSC(f) for $f = |1-2p|$, while the output in C_1 is the mixed state π. Thus the BSC and PSC are related by the uncertainty principle!

13.3 Guessing probability and fidelity uncertainty relations

13.3.1 Statement

Besides entropic formulations, we can find uncertainty relations for the two versions of the game in terms of guessing probabilities, recoverable entanglement, as well as how close a bipartite state is to being completely uncorrelated. The first bound is relevant to Version 1.

Proposition 13.3 (Bipartite guessing and entanglement bound). *For any bipartite state ρ_{AB} and measurements Λ_B and Γ_B, there exists a channel $\mathcal{E}_{A'|B}$ such that*

$$\begin{aligned} &\arccos F\big(\Phi_{AA'}, \mathcal{E}_{A'|B}[\rho_{AB}]\big) \\ &\quad \leq \arccos P_{\text{guess}}(Z_A|B)_{\rho,\Lambda} + \arccos P_{\text{guess}}(X_A|B)_{\rho,\Gamma} . \end{aligned} \tag{13.10}$$

Furthermore, $\mathcal{E}_{A'|B}$ has an explicit constructed in terms of Λ_B and Γ_B.

Note that on the left-hand side we have the square root of $R_{\text{ent}}(A|B)_{\rho,\mathcal{E}}$. The bound says something very similar to (13.3): High guessing probability for both measurements implies that the joint state ρ_{AB} is close to being entangled. But the entanglement recovery operation in the bound can be constructed from the underlying POVMs, which will prove useful in constructing quantum communication protocols using quantum error-correcting codes in Chapter 19.

Exercise 13.2. Choosing suitable measurements Λ and Γ, show that the bound is satisfied but not tight for bipartite states diagonal in the Bell basis.

For Version 2 of the guessing game, we can show the following:

Proposition 13.4 (Tripartite guessing bound). *For any tripartite state ρ_{ABC} and any measurement Λ_B,*

$$F\big(\tilde{\mathcal{P}}_A[\rho_{AC}], \pi_A \otimes \rho_C\big) \geq P_{\text{guess}}(Z_A|B)_{\rho,\Lambda} , \tag{13.11}$$

$$F\big(\mathcal{P}_A[\rho_{AC}], \pi_A \otimes \rho_C\big) \geq P_{\text{guess}}(X_A|B)_{\rho,\Lambda} . \tag{13.12}$$

If the guess for the Z_A measurement using Λ_B is good, then Bob's guess for X_A using C will be poor no matter what measurement he uses, since the fidelity of the CQ state relevant for the X_A measurement is correspondingly close to being completely uncorrelated. The statement here is the same as (13.4), just phrased using different quantities, and as there, a poor guessing probability of Z_A does not imply much about guessing X_A. It is possible to transform these relations into a statement involving only guessing probabilities, as in the following exercise, but the formulation involving closeness to an uncorrelated state already shows that simultaneous perfect guessing is impossible. This formulation will prove useful in establishing the security of QKD in Chapter 20.

Exercise 13.3. Show that for $d = |A|$, (13.11) implies

$$\sqrt{P_{\text{guess}}(X_A|R)_\rho} + \sqrt{d-1}\sqrt{1 - P_{\text{guess}}(X_A|R)_\rho} \geq \sqrt{d}\, P_{\text{guess}}(Z_A|B)_\rho \,. \tag{13.13}$$

13.3.2 Proof of the tripartite bound

The setup for the proofs of both propositions is quite similar, so we prove both in the remainder of this section. For a bipartite state ρ_{AB}, let $|\psi\rangle_{ABR}$ be any purification. Using the eigenstates $|z\rangle$ of Z_A, we can write

$$|\psi\rangle_{ABR} = \sum_z |z\rangle_A |\varphi(z)\rangle_{BR} \tag{13.14}$$

for $|\varphi(z)\rangle_{BR} = {}_A\langle z|\psi\rangle_{ABR}$. As in (6.17), the isometry $U_{A'B|B} = \sum_z \sqrt{\Lambda_B(z)} \otimes |z\rangle_{A'}$ implements the Λ_B measurement coherently, storing the result in system $A' \simeq A$. Ideally, the state $U_{A'B|B}|\psi\rangle_{ABR}$ would be equal to

$$|\psi'\rangle_{AA'BR} = \sum_z |z\rangle_A |z\rangle_{A'} |\varphi(z)\rangle_{BR} \,. \tag{13.15}$$

The overlap of the two states is simply $\sum_z \langle\varphi(z)|\sqrt{\Lambda_B(z)}|\varphi(z)\rangle_{BR}$, and using the fact that $\sqrt{\Lambda} \geq 0$ for $\Lambda \geq 0$ gives

$$F(|\psi'\rangle_{AA'BR}, U_{A'B|B}|\psi\rangle_{ABR}) \geq P_{\text{guess}}(Z_A|B)_{\rho,\Lambda} \,. \tag{13.16}$$

Observe that $|\psi'\rangle$ is such that tracing out A' leaves a CQ state on ABR with diagonal A in the Z_A basis. Then from monotonicity under partial trace of $A'B$ we have the following, which is important enough to highlight as a proposition.

Proposition 13.5. *For any tripartite pure state ρ_{ABR} and measurement Λ_B,*

$$F(\rho_{AR}, \mathcal{P}_A[\rho_{AR}]) \geq P_{\text{guess}}(Z_A|B)_{\rho,\Lambda} \,. \tag{13.17}$$

This bound states that a high guessing probability of Z_A using B implies the joint state of A and the purification R is correspondingly close to a CQ state with diagonal A in the Z_A basis.

Proposition 13.4 now follows easily. For such CQ states, it is clear that measurement in the X_A basis will yield a random outcome, that is, by monotonicity of the fidelity under $\tilde{P}_A$, we further obtain

$$F(\tilde{\mathcal{P}}_A[\rho_{AR}], \pi_A \otimes \rho_R) \geq P_{\text{guess}}(Z_A|B)_{\rho,\Lambda} \,. \tag{13.18}$$

Here we again use $\tilde{\mathcal{P}}_A \circ \mathcal{P}_A = \pi_A \operatorname{Tr}_A$. To adapt the statement to an arbitrary tripartite state ρ_{ABC}, observe that system C must be contained in the purification R. By Proposition 6.2, ρ_{ABR} also purifies ρ_{ABC}. Tracing out the irrelevant parts of R and again using monotonicity gives (13.11). Interchanging X_A and Z_A gives (13.12).

The above argument for the tripartite bound works for both choices of amplitude and phase bases considered at the outset. For Proposition 13.3, slightly different arguments are required for the two choices. We will first consider the clock and shift case, and then detail which alterations are required for the n-qubit version after the entire argument is complete.

13.3.3 Proof of the bipartite bound

To establish Proposition 13.3, we first take an apparent shortcut, which is simpler, gets very close to the desired statement, and will turn out to be useful in Chapter 19. Then we modify the argument to give the desired bound. The first step is to apply (13.16) to $|\psi'\rangle_{AA'BR}$ from (13.15) itself, but for the X_A measurement. First, express the state as $|\psi'\rangle_{AA'BR} = \sum_x |\tilde{x}\rangle_A |\theta(x)\rangle_{A'BR}$ for $|\theta(x)\rangle_{A'BR} = {}_A\langle\tilde{x}|\psi'\rangle_{AA'BR}$. Abusing notation somewhat, let us overload Z_A, and let it also denote the clock operator, so that $Z_A^j = \sum_{k\in\mathbb{Z}_d} \omega^{jk}|k\rangle\langle k|$. That is, Z_A denotes the amplitude observable as well as the random variable resulting from its measurement, depending on the context. Notice that

$$\begin{aligned}
|\theta(x)\rangle_{A'BR} &= \sum_z \langle\tilde{x}|z\rangle |z\rangle_{A'} |\varphi(z)\rangle_{BR} = \frac{1}{\sqrt{d}} \sum_z \omega^{-xz} |z\rangle_{A'} |\varphi(z)\rangle_{BR} \\
&= \frac{1}{\sqrt{d}} \sum_z Z_{A'}^{-x} |z\rangle_{A'} |\varphi(z)\rangle_{BR} = \frac{1}{\sqrt{d}} Z_{A'}^{-x} |\psi\rangle_{A'BR},
\end{aligned} \tag{13.19}$$

where $|\psi\rangle_{A'BR}$ is just the same as $|\psi\rangle_{ABR}$ with system A' replacing A. Hence

$$|\psi'\rangle_{AA'BR} = \frac{1}{\sqrt{d}} \sum_x |\tilde{x}\rangle_A \otimes Z_{A'}^{-x} |\psi\rangle_{A'BR}. \tag{13.20}$$

Now consider any measurement $\Gamma'_{A'B}(x)$ on $A'B$ and a somewhat nonstandard coherent implementation

$$V_{A''A'B|A'B} = \sum_x \sqrt{\Gamma'_{A'B}(x)} \otimes |-\tilde{x}\rangle, \tag{13.21}$$

where arithmetic inside the ket is modulo d. Here the measurement result is stored as $-x$ in the X eigenbasis, which will prove convenient. The ideal output would be

$$|\psi''\rangle_{AA'A''BR} = \sum_x |\tilde{x}\rangle_A |-\tilde{x}\rangle_{A''} |\theta(x)\rangle_{A'BR} = \frac{1}{\sqrt{d}} \sum_x |\tilde{x}\rangle_A |-\tilde{x}\rangle_{A''} Z_{A'}^{-x} |\psi\rangle_{A'BR}. \tag{13.22}$$

By (13.16) we therefore have

$$F(|\psi''\rangle_{AA'A''BR}, V_{A'A''B|A'B}|\psi'\rangle_{AA'BR}) \geq P_{\text{guess}}(X_A|A'B)_{\psi',\Gamma'} . \tag{13.23}$$

In light of (13.22), applying the unitary $W_{A'A''} = \sum_x |x\rangle\langle x|_{A''} \otimes Z^x_{A'}$ to $|\psi''\rangle$ yields $W_{A'A''}|\psi''\rangle_{AA'A''BR} = \frac{1}{\sqrt{d}} \sum_x |\tilde{x}\rangle_A |-\tilde{x}\rangle_{A''}|\psi\rangle_{A'BR}$. A straightforward calculation shows that this state is simply $|\Phi\rangle_{AA''} \otimes |\psi'\rangle_{A'BR}$. Thus by isometric invariance of the fidelity under W, (13.23) implies

$$F(|\Phi\rangle_{AA''} \otimes |\psi\rangle_{A'BR}, W_{A'A''}V_{A'A''B|A'B}|\psi'\rangle_{AA'BR}) \geq P_{\text{guess}}(X_A|A'B)_{\psi',\Gamma'} . \tag{13.24}$$

Again using isometric invariance, we can combine this fidelity with $F(|\psi'\rangle, U|\psi\rangle)$ in the triangle inequality (10.22) to obtain

$$\begin{aligned} &\arccos F(|\Phi\rangle_{AA''}|\psi\rangle_{A'BR}, WVU|\psi\rangle) \\ &\quad \leq \arccos F(|\Phi\rangle_{AA'}|\psi\rangle_{A'BR}, WV|\psi'\rangle) + \arccos F(|\psi'\rangle, U|\psi\rangle) . \end{aligned} \tag{13.25}$$

Then the bounds (13.16) and (13.24) give

$$\begin{aligned} &\arccos\, F(|\Phi\rangle_{AA''}|\psi\rangle_{A'BR}, WVU|\psi\rangle) \\ &\quad \leq \arccos P_{\text{guess}}(X_A|A'B)_{\psi',\Gamma'} + \arccos P_{\text{guess}}(Z_A|B)_{\rho,\Lambda} . \end{aligned} \tag{13.26}$$

This is nearly the bound we wish to prove, modulo interchanging the names A'' and A'. Tracing out $A'BR$ will only increase the fidelity on the left-hand side, and the entanglement recovery map $\mathcal{E}_{A'|B}$ is given by (changing the name to fit the current convention)

$$\mathcal{E}_{A''|B}[\rho_{AB}] = \text{Tr}_{A'}[WVU\rho_{AB}(WVU)^*] . \tag{13.27}$$

The measurement $\Gamma'_{A'B}$ potentially makes use of the fact that the Λ_B measurement has already taken place. This is reflected in the fact that the $\Gamma'_{A'B}$ POVM elements are operators on $A'B$, not just B. Nevertheless, we can construct an appropriate $\Gamma'_{A'B}$ from a measurement Γ_B acting only on B for which

$$P_{\text{guess}}(X_A|A'B)_{\psi',\Gamma'} \geq P_{\text{guess}}(X_A|B)_{\rho,\Gamma} . \tag{13.28}$$

In particular, for $\tilde{\Pi}(x) = |\tilde{x}\rangle\langle\tilde{x}|$, define

$$\Gamma'_{A'B}(x) = \sum_y \tilde{\Pi}_{A'}(y - x) \otimes \Gamma_B(y) . \tag{13.29}$$

This measurement yields the difference between the Γ measurement on B and an X measurement on A'. Notice that since $Z^x\tilde{\Pi}(y)Z^{-x} = \tilde{\Pi}(x+y)$, $Z^x_{A'}\Gamma'_{A'B}(x)Z^{-x}_{A'} = \Gamma'_{A'B}(0)$, and this is the operator determining $P_{\text{guess}}(X_A|B)_{\rho,\Gamma}$. Using (13.20) therefore gives

$$\begin{aligned} P_{\text{guess}}(X_A|A'B)_{\psi',\Gamma'} &= \text{Tr}\Big[\Big(\sum_x \tilde{\Pi}_A(x)\otimes\Gamma'_{A'B}(x)\Big)\psi'_{AA'BR}\Big] \\ &= \langle\psi|\Gamma'_{A'B}(0)|\psi\rangle_{A'BR} = P_{\text{guess}}(X_A|B)_{\rho,\Gamma}\,. \end{aligned} \tag{13.30}$$

Using this Γ' in (13.26) completes the proof of (13.10).

Now we examine the case of the amplitude and phase bases defined by tensor products of σ_x and σ_z. Here the two bases are related not by the Fourier transform, but by the Hadamard transform $|\tilde{x}\rangle = \frac{1}{\sqrt{2^n}}\sum_{z\in\mathbb{Z}_2^n}(-1)^{x\cdot z}|z\rangle$ for $x\in\mathbb{Z}_2^n$. Instead of using the clock operator to define the amplitude observable Z_A, now let $Z_A^x = \sigma_z^{x_1}\otimes\cdots\otimes\sigma_z^{x_n}$ for $x\in\mathbb{Z}_2^n$. The entire collection of operators $\{Z_A^x\}_{x\in\mathbb{Z}_2^n}$ is required to specify a specific eigenstate $|z\rangle$ for $z\in\mathbb{Z}_2^n$; that is, the eigenvalues of the amplitude "observable" Z_A are effectively elements of $\mathbb{Z}_2^n$, not $\mathbb{C}$ as for observables properly defined. Nevertheless, we still refer to Z_A as the amplitude observable. The calculation in (13.19) goes through as before, though of course $Z_A^{-x} = Z_A^x$. Now there is no need to store the result of measuring x as $|\tilde{x}\rangle$, or actually there is no difference between $|-\tilde{x}\rangle$ and $|\tilde{x}\rangle$. The state $\frac{1}{\sqrt{2^n}}\sum_{x\in\mathbb{Z}_2^n}|\tilde{x}\rangle|\tilde{x}\rangle$ is just $|\Phi\rangle^{\otimes n}$ for the two-qubit entangled state $|\Phi\rangle$, which is also a 2^n-dimensional maximally entangled state. With these modifications, the rest of the proof goes through as before.

13.4 Notes and further reading

The uncertainty principle was introduced by Heisenberg [132] in 1927. He elaborated on the argument in a letter to Pauli [133] and later in [134]. See also Werner and Farrelly [300] for a recent discussion on Heisenberg's thinking and the context of his original paper on the uncertainty principle. Robertson established the more well-known form [243]. Everett [95] and Hirschman [140] conjectured an entropic uncertainty relation, which was later proven by Białynicki-Birula and Mycielski [39] using just established results in Fourier analysis by Beckner [14, 15]. Białynicki-Birula and Mycielski also considered entropic relations for angular momentum and phase. Deutsch [76] investigated for entropic relations for arbitrary discrete observables. Kraus [172] improved Deutsch's bound for complementary observables as we have considered here and conjectured that the improvement holds generally. Maassen and Uffink [196] proved the statement. The extension of the result to classical side information was considered by Hall [116]. Renes and Boileau [239] showed the result for complementary observables with quantum side information, drawing on techniques from Christandl and Winter [60]. Berta et al. [34] proved the relation for general projective observables. A good overview is the review of Coles et al. [63] The nonentropic statements are taken from [237] by the author.

Part III: **Information processing protocols**

14 Data compression

> The essence of a quote is the compression of a mass of thought and observation into a single saying.
>
> John Morley, 1st Viscount Morley of Blackburn

Having completed our study of the tools needed to analyze information processing protocols, we can finally proceed to do so. We begin in this chapter with the tasks of compression, either of classical or of quantum data. The goal, of course, is to reduce the amount of space needed to store some particular information. For instance, consider compression of text, this text even. In the entire text, there is a vast amount of redundancy in several senses. For one, it is not usually necessary to write every letter in a word for the reader to determine which word was meant. Secondly, after reading a decent amount of the text, it becomes easier to predict what the next word will be, given the immediate context.

There are a multitude of ways of removing redundancy from data. Here we are interested in the possibilities and limitations of *fixed-length, approximately lossless protocols*. Starting with the first adjective, "fixed-length" refers to the fact that the compression operation on a sequence of inputs has a fixed output size (number of bits) for a fixed input size. All inputs are mapped to shorter sequences. Because the decompressor makes use of the probability distribution of the input, recovery is possible even though some data is missing. In variable length coding, by contrast, some inputs are mapped to shorter outputs, while some are made longer, with the idea that the likely inputs are the ones that become shorter. The familiar *zip* and *gzip* software applications are examples of variable-length compressors.

"Approximately lossless" means that the goal is to recover the exact input sequence, at least with high probability. This is the standard approach for digital data we consider. For continuous signals, such as audio, recovering the exact waveform is neither necessary nor even really possible with finite-precision devices. Given some kind of metric on the space of possible signals, we can specify that the signal only needs to be recovered up to a certain precision. Examples of such *lossy compression* are the jpeg image, mp3 audio, and mp4 video formats.

Finally, the specific protocols we will consider here are *block protocols* in that they read all of the input data before outputting the compressed version. The converses also apply to *streaming protocols*, which read in the data sequentially and begin outputting the compressed version after only a short lag. Streaming protocols are especially important for large data, e. g., audio and video.

https://doi.org/10.1515/9783110570250-014

14.1 Compression of classical data

14.1.1 Setup and basic properties

Given a random variable X with probability distribution P_X, the goal of data compression is to transform X to Y so that $|Y| \leq |X|$ and yet X can be approximately recovered from Y in the sense of Figure 14.1. This task is also known as *source coding*. It could be that $X = X_1, \dots, X_n$, but we will first consider the "monolithic" case of a structureless X as it includes this specialization. As remarked in the introduction, the case of a structureless X is often referred to as the "one-shot" scenario.

More formally, a (k, ε) compression scheme for X consists of a compression map $C_{Y|X}$ with $|Y| = k$ and a decompression map $D_{X|Y}$, such that $\delta(P_{XX'}, D_{X|Y} \circ C_{Y|X}[P_{XX'}]) \leq \varepsilon$, where $P_{XX'}(x, x') = P_X(x)\delta_{x,x'}$. By (11.2) this is equivalent to saying that the probability of successfully guessing the value of X from the compressed value Y is at least $1 - \epsilon$. Clearly, there is a tradeoff between k and ε; they cannot both be small. Define $L^\star_\varepsilon(X)_P$ to be the smallest k such that there exists a (k, ε) compression protocol for X.

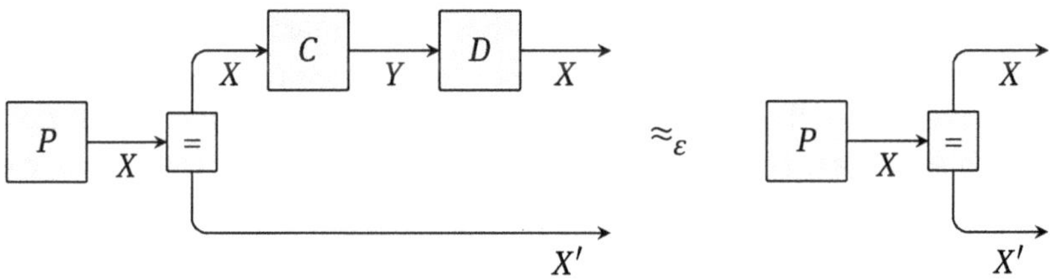

Figure 14.1: A (k, ε) compression protocol of classical data X with $k = |Y|$. The random variable Y is the compressed version of X, since the decompressor D can recreate the particular output of the source (stored in the copy of X) from it. The equality symbol = means that $X = X'$.

We should be careful not to confuse the compression scenario with approximating a source. That is, the approximation condition is not that $\delta(P_X, D_{X|Y} \circ C_{Y|X}P_X) \leq \varepsilon$, but rather $\delta(P_{XX'}, D_{X|Y} \circ C_{Y|X}[P_{XX'}]) \leq \varepsilon$. The variable X' is used as a reference value to compare with the output of the decompressor. Without it, the task of simulating P_X by $D_{X|Y} \circ C_{Y|X}P_X$ is trivial absent any further restrictions. The compressor can just discard the input, and the decompressor can just output a new $X \sim P_X$. (Requiring that the decompressor be deterministic would change this conclusion.)

For the task of compression, though, deterministic compression and decompression are sufficient to achieve any (k, ε) protocol that is achievable by stochastic means. If the compressor uses randomness, then by Proposition 3.3 it is really $C_{Y|X} = \sum_r P_R(r) C_{Y|XR=r}$ for a deterministic channel $C_{Y|XR}$. Then the error will just be the average over R, since the joint output distribution over XX' is

$$Q_{XX'} = D_{X|Y} \circ C_{Y|X}[P_{XX'}] = \sum_r P_R(r) D_{X|Y} \circ C_{Y|XR=r}[P_{XX'}]\,. \tag{14.1}$$

So we may pick the value of R such that $Q_{XX'R=r} = D_{X|Y} \circ C_{Y|XR=r}[P_{XX'}]$ has the smallest error and just use that one. An entirely similar argument applies to the decompressor.

14.1.2 One-shot bounds

To characterize the tradeoff between k and ε for a given source P_X, we can establish the following bounds on $L^\star_\varepsilon(X)_P$.

Proposition 14.1 (Compression bounds). *For any random variable X distributed according to P_X,*

$$\beta_{1-\varepsilon}(P_X, \mathbb{1}_X) \leq L^\star_\varepsilon(X)_P \leq \min_{\eta\in(0,\varepsilon]} \frac{1}{\eta}\beta_{1-\varepsilon+\eta}(P_X, \mathbb{1}_X) + 1\,. \tag{14.2}$$

The lower bound here is the *converse bound*, since it limits the possibilities of any protocol. The upper bound is the *achievability bound*, as it is established by constructing a particular protocol. In this case the bounds very nearly match, as we will see shortly.

Proof. Let us start with the converse bound. Assume that $C_{Y|X}$ and $D_{X|Y}$ constitute a (k, ε) protocol. Therefore $Q_{XX'}$ as defined in (14.1) passes the equality test $T_{XX'}(x, x') = \delta_{x,x'}$ with probability $1-\varepsilon$: $T_{XX'} \cdot Q_{XX'} \geq 1-\varepsilon$. Defining $P_{XY} = C_{Y|X'}[P_{XX'}]$ and $\hat{T}_{XY} = D^T_{X'|Y}[T_{XX'}]$ (recall that the transpose is the adjoint here), we have $\hat{T}_{XY} \cdot P_{XY} \geq 1-\varepsilon$. Hence $\hat{T}_{XY}$ is feasible for $\beta_{1-\varepsilon}(P_{XY}, Q_{XY})$ for all positive Q_{XY}, meaning that

$$\beta_{1-\varepsilon}(P_{XY}, Q_{XY}) \leq \hat{T}_{XY} \cdot Q_{XY}\,. \tag{14.3}$$

Now let $E_{X'X|X}$ be the map that just copies X to X' and notice that $P_{XY} = C_{Y|X'} \circ E_{X'X|X}[P_X]$. Choosing $Q_{XY} = C_{Y|X'} \circ E_{X'X|X}[\mathbb{1}_X]$, it follows from monotonicity of β_α that

$$\beta_{1-\varepsilon}(P_X, \mathbb{1}_X) \leq \beta_{1-\varepsilon}(P_{XY}, Q_{XY})\,. \tag{14.4}$$

Moreover, by design $Q_{XY}(x, y) = C_{Y|X=x}(y)$, so that $Q_{XY} \leq \mathbb{1}_{XY}$. Therefore we have

$$\hat{T}_{XY} \cdot Q_{XY} \leq \hat{T}_{XY} \cdot \mathbb{1}_{XY} = T_{XX'} \cdot D_{X'|Y}[\mathbb{1}_{XY}] = \sum_{xy} D_{X'|Y=y}(x) = |Y|. \tag{14.5}$$

Combining (14.3), (14.4), and (14.5) gives $\beta_{1-\varepsilon}(P_X, \mathbb{1}_X) \leq |Y|$ and therefore the lower bound in (14.2).

The achievability statement is somewhat simpler. Intuitively, the compressor should simply keep the values of X with the highest probability and discard the rest, mapping them to any one of the high probability x values. The total probability of the inputs that are not discarded should be at least $1-\varepsilon$. We therefore take the compressor to all those x with $P(x) \geq r$ to themselves for an appropriate value of r, and all other inputs to be mapped to any fixed element of high-probability inputs. The total size of the alphabet is now reduced, as some x values are not used. The decompressor, meanwhile, simply outputs the input value.

The compressor is characterized by the projector $\{P \geq r\mathbb{1}\}$ with r chosen small enough so that $\mathrm{Tr}[\{P \geq r\mathbb{1}\}P] \geq 1-\varepsilon$. Let us see how we can ensure this is the case by suitable choice of the compressor output size. The size of the compressed output is simply $k = \mathrm{Tr}[\{P \geq r\mathbb{1}\}]$, and normalization of P requires that $kr \leq 1$. From (9.13) we get a lower bound on $\mathrm{Tr}[\{P \geq r\mathbb{1}\}P]$: $\mathrm{Tr}[\{P \geq r\mathbb{1}\}P] \geq \alpha - r\beta_\alpha(P, \mathbb{1})$. Combining with the previous statement gives $\mathrm{Tr}[\{P \geq r\mathbb{1}\}P] \geq \alpha - \frac{1}{k}\beta_\alpha(P, \mathbb{1})$. Choosing $\alpha = 1-\varepsilon+\eta$ for some $\eta \in (0, \varepsilon]$ and $k = \lceil \frac{1}{\eta}\beta_{1-\varepsilon+\eta}(P, \mathbb{1})\rceil$ ensures that the right-hand side is at least $1-\varepsilon$ and k is an integer, as intended. The optimal choice for k is $\min_{\eta\in(0,\varepsilon]}\lceil \frac{1}{\eta}\beta_{1-\varepsilon+\eta}(P, \mathbb{1})\rceil$. Using $\lceil y \rceil \leq y+1$ for $y \geq 0$ gives the achievability bound in (14.2). □

14.1.3 Optimal asymptotic i. i. d. rate

One case of interest is that the random variable to be compressed X^n is i. i. d. of length n, i. e., $P_{X^n} = P_X^{\times n}$ as in (2.20). The length n is usually referred to as the *blocklength*. In this setting, it is more convenient to work with the logarithm of the compression size. Even better is the *rate* R of the compression scheme, $R = \frac{1}{n}\log k$. For fixed P_X, there are tradeoffs between R, n, and ε, e. g., small ε certainly constrains how large R can be. We are most interested in what rates are possible in the case of large n and small ϵ. The optimal rate for fixed ε and n is simply

$$R(P_X, \varepsilon, n) := \frac{1}{n}\log L_\varepsilon^\star(X^n)_{P^{\otimes n}}. \tag{14.6}$$

Appealing to Stein's lemma, we can infer that the optimal compression rate in the limit as $n \to \infty$ is in fact independent of ε and equal to the entropy $H(X)_P$. This is Shannon's source coding theorem.

Proposition 14.2 (Shannon's source coding theorem). *For arbitrary i. i. d. $X^n \sim P^{\times n}$ and all $\varepsilon \in (0,1)$, $\lim_{n\to\infty}\frac{1}{n}L_\varepsilon^\star(X^n)_{P^{\otimes n}} = H(X)_P$.*

Proof. From the converse bound we obtain $R(P_X, \varepsilon, n) \geq \frac{1}{n}\log\beta_{1-\varepsilon}(P_{X^n}, \mathbb{1}_{X^n})$. By (12.33) this is $-D(P_X, \mathbb{1}_X) = H(X)_P$. To obtain a matching upper bound from the achievability statement, choose $\eta = \varepsilon/2$. The prefactor $2/\varepsilon$ and additional term $+1$ vanish in the limit as $n \to \infty$. □

Note that the zero-error rate, with ε fixed to zero from the outset, is quite different. Indeed, unless some values of X have zero probability, nontrivial compression is impossible. The stronger version of Stein's lemma in (12.39) ensures that the entropy is still an achievable rate even when taking ε to be exponentially decreasing in n (as long as the base of the exponent is not too large). For all practical purposes, this is equivalent to zero-error performance.

On the other hand, we may have thought that allowing a somewhat large error could lead to compression at a lower asymptotic rate. This is evidently not the case, and compressing a source below the entropy rate will result in very large error. The error will increase at least exponentially in the blocklength n, since the optimal rate even for compression with error exponentially close to 1 is still the entropy.

14.2 Compression of quantum data

14.2.1 Setup and basic properties

Now we turn to the task of compressing quantum data. Here, though, there is less intuition about what this means, or what this task should accomplish. As just discussed in the previous section, in classical compression, we want to obtain the actual value of X, not just replicate the machine that generates X distributed according to P_X. Put differently, if there exists some random variable Y that is correlated with X in some way, then this correlation should survive the compression and decompression operations. Formulating the definition of a compression protocol using $X' = X$ ensures that this is the case, as any other Y can be generated from X'.

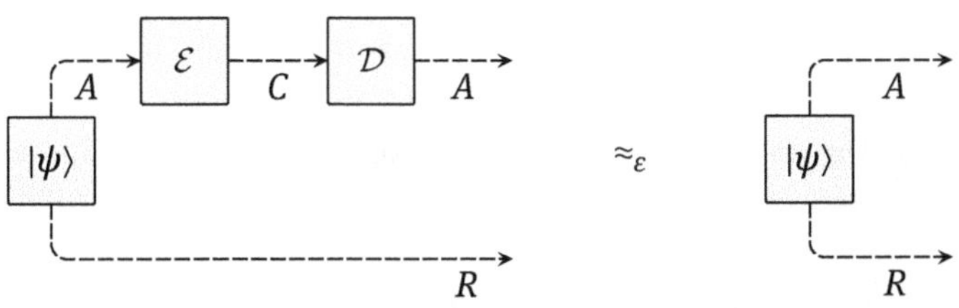

Figure 14.2: A (k, ε) protocol for compression of quantum data in A with $k = |C|$. In contrast to the classical case, here we also want to retain correlations or entanglement of the system A with its purification R. Dashed lines indicate quantum systems.

The quantum analog is now more clear. Consider a source described by a density operator ρ_A on system A. Again, there might be other systems correlated or entangled with A. To ensure these correlations are preserved, we should take a purification $|\psi\rangle_{AR}$ to be the ideal output and measure the quality of the actual output by the fidelity. Then any actual correlations or entanglement with A will be preserved, since all extensions of ρ_A can be generated by a purification. Specifically, we will use the entanglement fidelity from (10.35). A (k, ε) protocol for compression of quantum data A with density operator ρ_A then consists of a compressor $\mathcal{E}_{C|A}$ and decompressor $\mathcal{D}_{A|C}$ such that $F_{\text{ent}}(\rho, \mathcal{D} \circ \mathcal{E}) \geq 1 - \varepsilon$. This is depicted in Figure 14.2. Again, we define $L_\varepsilon^\star(A)_\rho$ to be the smallest k such that a (k, ε) protocol exists.

14.2.2 Achievability from classical compression

We could prove an achievability bound along the same lines as in Proposition 14.1, namely by projecting onto the subspace of eigenstates with large enough eigenvalues. However, it is important to appreciate that every classical compression protocol can be transformed into a quantum compression protocol, not merely that particular one. The strategy is essentially the same: Apply the classical compression map designed for the eigenvalue distribution to the eigenbasis of the quantum state. Doing so, we can show that $L_\varepsilon^\star(A)_\rho \leq L_\varepsilon^\star(X)_P$, where P_X is the distribution of eigenvalues of ρ. This is the first instance of a protocol *reduction* mentioned in the introduction. We will make extensive use of reductions in the coming chapters.

There are many ways to realize a classical channel as a quantum channel, and we must be careful to pick the correct one for the reduction to work. Examining the possibilities carefully will reveal that quantum compression is not so much about reducing the size of the quantum data as it is about maximizing the amount of quantum information that can be deleted.

To illustrate, let us first make the wrong realization of the classical compressor as a quantum compressor. Suppose that the eigenvalues of ρ_A have the distribution P_X and we would like to recycle the compressor for X. As we just saw in Section 14.1.1, any (k, ε) compression protocol can already be made to have a deterministic compressor and decompressor. In particular, we would like to construct a unitary that implements the compression function f. However, f is not necessarily one-to-one, so we cannot just map $|x\rangle_A$ to $|f(x)\rangle_C$. Instead, we can append the input to the output and define $U_{A'C|A} = \sum_x |f(x)\rangle_C |x\rangle_{A'} \langle x|_A$.

To see why this is the wrong choice, imagine for the sake of argument that the decompressor manages to perfectly reconstruct the basis states from C back to A. Then, applied to the purified input, the compression protocol results in a state of the form

$$|\theta\rangle_{AA'R} = \sum_x \sqrt{P_X(x)} |x\rangle_A |x\rangle_{A'} |x\rangle_R \,. \tag{14.7}$$

Now the trouble is clear: This state does not have high fidelity with the ideal purification ψ_{AR}, since now there is a copy of the x value in the system A'. More precisely, the fidelity of the actual and ideal outputs is just $\sum_x P_X(x)^2$, which is only large if the state to be compressed is nearly a pure state to begin with (and for which $k = 1$ is possible).

In retrospect, leaving a copy of the input X value lying around at the compressor was a too crude approach, since this information interferes with the entanglement we are trying to preserve.

We can extend f to a reversible function more frugally as follows. First, for a given ordering of the values of x, we can define the sequence $\mathcal{X}_y = \{x : f(x) = y\}$. Then we can define $g(x)$ to be the position of x in $\mathcal{X}_f(x)$, starting our count from zero, say. With both $f(x)$ and $g(x)$, we can reconstruct x, so the map $x \to (f(x), g(x))$ is reversible and can be implemented by a unitary operation. One small detail is that the sequences $\mathcal{X}_y$ do not all have the same size for different values of y, so this map is not so well-defined, but we can just define g to map $\mathcal{X}$ to $\mathcal{Z} = 0, \dots, \max_y |\mathcal{X}_y| - 1$, and the reversible version $\bar{f}$ of f becomes $\bar{f} : \mathcal{X} \to \mathcal{Y} \times \mathcal{Z}$. The unitary, or more precisely isometric, implementation can be defined as $V_{CB|A} = \sum_x |f(x)\rangle_B |g(x)\rangle_C \langle x|_A$.

It is convenient to slightly alter the definition of $\mathcal{X}_y$ so that the x appear in order of their probability, the largest first. The job of the decompressor is to map y to $\bar{f}^{-1}(y, 0)$. Then the output of the protocol is

$$
\begin{aligned}
|\psi'\rangle_{ACR} &= \sum_x \sqrt{P_X(x)} |\bar{f}^{-1}(f(x), 0)\rangle_A |g(x)\rangle_C |b_x\rangle_R \\
&= \sum_{x:g(x)=0} \sqrt{P_X(x)} |x\rangle_A |x\rangle_R |0\rangle_C + |\text{fail}\rangle\,,
\end{aligned}
\tag{14.8}
$$

where $|\text{fail}\rangle$ denotes the terms in the superposition with $g(x) \neq 0$. The overlap with $|\psi\rangle_{AR}|0\rangle_C$ is precisely $\sum_{x:g(x)=0} P_X(x) \geq 1-\varepsilon$. Therefore we have established a reduction of quantum compression to classical compression.

Proposition 14.3. *For any quantum state ρ_A, let P_X be the distribution of its eigenvalues. Then for every (k, ε) compression protocol for P_X, there exists a (k, ε) protocol for ρ_A that can be constructed from the classical protocol. Hence $L^\star_\varepsilon(A)_\rho \leq L^\star_\varepsilon(X)_P$.*

Observe that the state $|\psi'\rangle_{ACR}$ produced by the protocol is such that C is essentially a pure state. Thus we may regard the task of the compressor as not so much to squeeze all of the information into a system B with entropy as high as possible, but rather to maximize the size of system C, the output with zero entropy. Because the compression operation is reversible, the part that is not pure must contain all the correlation and entanglement with the purification R.

The failure of the initial approach can now be appreciated from a different angle via the action of the compression maps. Consider the output B, the only part the decompressor will see, under $U_{A'B|A}$ versus $V_{BC|A}$. Their action is identical on inputs of the form $|x\rangle\langle x|_A$; each produces $|f(x)\rangle\langle f(x)|_B$, but they differ on "off-diagonal" inputs $|x\rangle\langle x'|$ for $x' \neq x$. Our original choice $U_{A'B|A}$ simply annihilates them, whereas $V_{BC|A}$ produces $|f(x)\rangle\langle f(x')|\delta(g(x), g(x'))$. The ability to keep the off-diagonal elements is crucial for maintaining entanglement.

14.2.3 Converse

Given the relation $L^\star_\varepsilon(A)_\rho \le L^\star_\varepsilon(X)_P$ from Proposition 14.3, we may wonder if quantum compression is possible at much lower rate than classical compression. The following converse is nearly identical to the classical converse, and hence this is not the case.

Proposition 14.4. *Every (k,ε) compression scheme for a state ρ_A satisfies $k \ge \beta_{(1-\varepsilon)^2}(\rho_A, \mathbb{1}_A)$. Equivalently, $k \ge \beta_{(1-\varepsilon)^2}(P_X, \mathbb{1}_X)$ for the distribution P_X of eigenvalues of ρ_A.*

Note that $(1-\varepsilon)^2 \ge 1-2\varepsilon$, so $k \ge \beta_{1-2\varepsilon}(P_X, \mathbb{1}_X)$ also holds; this form is more immediately comparable to the achievability statement in Proposition 14.1.

Proof. The proof is based on constructing a feasible test for $\beta_{1-\varepsilon}$ using the fact that $F_{\mathrm{ent}}(\rho, \mathcal{D}\circ\mathcal{E})^2 \ge (1-\varepsilon)^2$. Suppose D_m and E_j are the Kraus operators of the decompressor and compressor, respectively; he we will just use subscripts as the system names can be omitted. Further, let Π_m be the projector onto the image of D_m. Note that this space cannot be larger than $|B|$ by construction, and therefore $\mathrm{Tr}[\Pi_m] \le k$ for all m. Turning to the entanglement fidelity, by the Cauchy–Schwarz inequality we have

$$\begin{aligned} F_{\mathrm{ent}}(\rho, \mathcal{D}\circ\mathcal{E})^2 &= \sum_{jm} |\mathrm{Tr}[\rho \Pi_m D_m E_j]|^2 \le \sum_{jm} \mathrm{Tr}[\Pi_m \rho \Pi_m]\, \mathrm{Tr}[D_m E_j \rho E_j^* D_m^*] \\ &\le \mathrm{Tr}[\Pi_{m^\star}\rho] \sum_{jm} \mathrm{Tr}[D_m E_j \rho E_j^* D_m^*] = \mathrm{Tr}[\Pi_{m^\star}\rho]\,, \end{aligned} \tag{14.9}$$

where $\Pi_{m^\star}$ is such that $\mathrm{Tr}[\Pi_{m^\star}\rho] \ge \mathrm{Tr}[\Pi_m\rho]$ for all m. Therefore $\Pi_{m^\star}$ is feasible for $\beta_{(1-\varepsilon)^2}(\rho, \mathbb{1})$, and we have the desired bound. The equivalent form follows by Exercise 9.13. □

It is interesting to note that the converse bound also holds if we consider protocols in which the compressor can also transmit some amount of classical information to the decompressor as well. In this case the compressor has Kraus operators of the form $E(j,y) \otimes |y\rangle\langle y|$, and the operators in the entanglement fidelity will have an additional y dependence and the expression an additional summation over y. However, the range bound on the Kraus operators of the decompressor still holds, and so the proof goes through. Therefore classical communication does not enable any savings in the amount of quantum information required for compression.

14.2.4 Optimal asymptotic i. i. d. rate

The optimal rate $R(\rho, \varepsilon, n)$ of compressing $\rho^{\otimes n}$ with entanglement fidelity at least $1-\varepsilon$ is defined in precisely the same manner as in the classical case. Again, the entropy is the optimal asymptotic rate, which is Schumacher's[1] source coding theorem. Here it follows from Stein's lemma.

Proposition 14.5 (Schumacher's source coding theorem). *For arbitrary quantum states ρ_A and all $\varepsilon \in (0,1)$, $\lim_{n\to\infty} \frac{1}{n} L_\varepsilon^\star(A^n)_{\rho^{\otimes n}} = H(A)_\rho$.*

14.3 Notes and further reading

The fact that the entropy sets the compressibility of an i. i. d. source was discovered by Shannon [258]. Compression of general sources was considered by Han [120] using the "information spectrum" approach pioneered by Han and Verdú [119]. The spectrum in question is the different possible values of the likelihood ratio; see the overview by Han for much more detail [121]. Hayashi [126] treats the fixed-length problem of classical compression in a similar setting as here. Schumacher [251] considered the i. i. d. problem in the quantum context, coining the word "qubit." The converse in Proposition 14.4 is an adaptation of an argument given by Datta and Leditzky [69], who considered the one-shot problem.

1 Benjamin Schumacher.

15 Classical communication over noisy channels

> The fundamental problem of communication is that of reproducing at one point, either exactly or approximately, a message selected at another point.
>
> Claude Shannon

Let us now (re)turn to the fundamental problem of information theory, noisy channel coding. Consider communication of a block of binary data sequentially over BSC(p). Assuming that the noise encountered in each use of the channel is independent, each bit will be corrupted with probability p, making direct communication very unreliable. Instead, we attempt to *encode* the data we wish to transmit into a longer sequence, building in some redundancy such that the original input can be reliably *decoded* from the noisy output. For instance, the simplest coding scheme is the *repetition code*, in which the input is simply transmitted multiple times and the decoder takes the majority vote of the channel output. When a single bit is transmitted three times, for instance, two errors have to occur for decoding to fail. This occurs with probability $O(p^2)$, an improvement on the original error rate of p.

We can also think of coding as making reliable communication possible because not all channel inputs are used to transmit information, instead of by adding redundancy. In the three-bit repetition code, to continue the example, only 000 or 111 are input to the three uses of the BSC. For each input, the noisy channel generates some output distribution or quantum state, and the fewer of these there are, the easier they are to distinguish.

15.1 Setup and basic properties

We now formulate the problem more precisely, picking up from Section 8.1. We begin with reliable communication over a CQ channel. Suppose we would like to use a noisy CQ channel $\mathcal{N}_{B|X}$ to send messages from an alphabet $\mathcal{M}$. Often, $\mathcal{N}_{B|X}$ is n i. i. d. instances of some simple channel, for instance, $\mathcal{N} = \mathrm{BSC}(p)^{\otimes n}$ as above, but for now we formulate the problem for a single arbitrary channel, the "one-shot" setting.

To ensure that the messages are received reliably, we first encode them with an encoder $\mathcal{E}_{X|M}$. The output X of the encoder is sent over the channel to the receiver, which applies a decoding operation $\mathcal{D}_{M'|B}$. The encoder–decoder pair specifies the protocol, and we denote by k the size of $\mathcal{M}$. The goal of the protocol is to simulate the identity channel, as depicted in Figure 15.1. A protocol with message alphabet $\mathcal{M}$ that achieves a distinguishability $\delta(\mathcal{D} \circ \mathcal{N} \circ \mathcal{E}, \mathbb{1}) \leq \varepsilon$ is called a $(k, \varepsilon)_{\mathrm{wc}}$ protocol. By Exercise 9.26 the distinguishability is related to the worst-case probability of error for the channel $\mathcal{D} \circ \mathcal{N} \circ \mathcal{E}$.

Clearly, there is some kind of limit on possible combinations of k and ε for a given channel $\mathcal{N}_{B|X}$. For instance, if the channel is very noisy but there are a lot of messages,

https://doi.org/10.1515/9783110570250-015

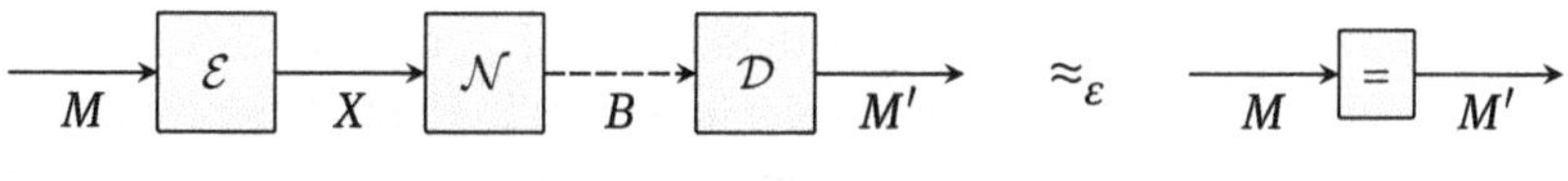

Figure 15.1: A (k, ε) protocol for classical communication over the CQ channel $\mathcal{N}_{B|X}$ with $k = |M|$.

then we expect that it will be difficult to determine which message was transmitted, that is, the error rate ε will be large if k is and vice versa.

It turns out to be much easier to analyze protocols defined in terms of average error with the assumption that the messages to be transmitted are uniformly distributed, i. e., $1 - P_{\text{agree}}(\mathcal{D} \circ \mathcal{N} \circ \mathcal{E})$. A protocol with k messages that is ε-good in this average sense is called a (k, ε) protocol. Denote by $M_\varepsilon^\star(B|X)_\mathcal{N}$ the largest k such that there exists a (k, ε) protocol for $\mathcal{N}_{B|X}$.

Using the average error is actually no real loss of generality for the following reasons. First, any encoder that is ε-good even for the worst-case input is certainly ε-good in the average case, if not better. Therefore a converse that applies to the average case also applies to the worst case.

Second, for the average case scenario, we might as well assume that the encoder is deterministic. More specifically, any stochastic encoding scheme with error ε can always be converted to a deterministic encoder with an error no larger than ε. This follows by regarding the encoder as a convex combination of deterministic maps, which we can do by Proposition 3.3. Then $P_{\text{agree}}(\mathcal{D}\circ\mathcal{N}\circ\mathcal{E})$ contains an average over the choice of deterministic encoder, and therefore if $P_{\text{agree}}(\mathcal{D} \circ \mathcal{N} \circ \mathcal{E}) \geq 1 - \varepsilon$, then at least one of the deterministic choices satisfies this bound. So we might as well use that encoder.

The outputs of a deterministic encoder define a *code* or *codebook*, and the particular outputs x_m for input m are *codewords* of the error-correcting code. The above argument does not rule out the possibility that the encoder maps two different messages to the same codeword, but this is of course unwise. We will henceforth assume that there is a one-to-one correspondence between messages and codewords when referring to a (k, ε) *code* (as opposed to the more general (k, ε) *protocol*).

Exercise 15.1. Show that the three-bit repetition code over BSC(p) is a $(2, p^2(3 - 2p))$ code. How does this generalize to n channel uses? What is the error rate for n uses of BEC(q)?

Finally, if we find a (k, ε) code in the sense of average error, then we can in principle convert it to $(\frac{1}{2}k, 2\varepsilon)_{\text{wc}}$ code for the worst-case error simply by throwing away the worst half of the codewords. This is sometimes called *expurgation*. To see this, split the set of codewords into two equal-sized subsets such that the error probability of each codeword in the first set is smaller than any error probability in the second (that is, split the codewords ordered by error probability at the median). Call the average error probabilities for the two sets ε_- and ε_+, respectively, and notice that $\varepsilon_- + \varepsilon_+ = 2\varepsilon$. This equation cannot hold if the largest error probability in the better set is larger than 2ε, for then ε_+

would be larger than 2ε. Thus we have constructed a set of $\frac{1}{2}k$ codewords with worst-case error smaller than 2ε. This may seem like a large loss in the code size, but since code size is usually specified logarithmically and measured in bits, i. e., $\log_2 k$ bits for size k, this only represents the loss of one bit at a price of doubling the error.

15.2 Converse

A simple but very useful converse bound relating ε, k, and $\mathcal{N}_{B|X}$ follows by setting up an appropriate hypothesis test involving a good code. The idea is that if a code has a small error probability, then the input and output are highly correlated, so it should be easy to distinguish the joint input–output distribution from a distribution with no correlation whatsoever.

More formally, suppose we have a (k,ε) protocol for average error. The above argument notwithstanding, let us first consider the possibility of stochastic encoding. Let $\varphi_B(x) = \mathcal{N}_{B|X=x}$ be the output of the channel for input x, and let $\vartheta_B(m) = \sum_{x\in\mathcal{X}} \varphi_B(x)\mathcal{E}_{X|M=m}(x)$ be the output of the encoder and channel for input m. Now consider the case of transmitting a message chosen uniformly at random and let $\rho_{MB} = \frac{1}{k}\sum_m |m\rangle\langle m|_M \otimes \vartheta_B(m)$.

Since the protocol is ε-good, the output of the decoder matches the actual message with probability no smaller than $1-\varepsilon$. With $\Pi_{MM'} = \sum_m |m,m\rangle\langle m,m|_{MM'}$, this is just the statement $\mathrm{Tr}[\Pi_{MM'}\mathcal{D}_{M'|B}[\rho_{MB}]] \geq 1-\varepsilon$. Therefore $\Lambda_{MB} = \mathcal{D}^{\star}_{M'|B}[\Pi_{MM'}]$ is feasible in $\beta_{1-\varepsilon}(\rho_{MB},\tau_{MB})$ for all $\tau_{MB} \geq 0$, so that $\beta_{1-\varepsilon}(\rho_{MB},\tau_{MB}) \leq \mathrm{Tr}[\Lambda_{MB}\tau_{MB}]$. Choose an uncorrelated state $\tau_{MB} = \pi_M \otimes \sigma_B$ for any σ_B. Since the decoder acts only on B, we have

$$\begin{aligned}\mathrm{Tr}[\Lambda_{MB}\rho_M \otimes \sigma_B] &= \mathrm{Tr}\Big[\Pi_{MM'}\sum_{m'}\frac{1}{k}|m'\rangle\langle m'|_M \otimes \mathcal{D}_{M'|B}[\sigma_B]\Big] \\ &= \frac{1}{k}\sum_m \langle m|\mathcal{D}_{M'|B}[\sigma_B]|m\rangle_{M'} = \frac{1}{k}\mathrm{Tr}[\sigma_B] = \frac{1}{k}.\end{aligned} \tag{15.1}$$

The argument works for any σ_B, and therefore we obtain the inequality

$$\max_{\sigma_B} \beta_{1-\varepsilon}(\rho_{MB},\rho_M \otimes \sigma_B) \leq \frac{1}{k}. \tag{15.2}$$

This inequality makes our intuition about good protocols precise, showing that ε-good protocols have a discrimination error less than the inverse of the message size.

However, we do not yet have the desired converse, a bound involving k and ε that depends only on $\mathcal{N}_{B|X}$. The left-hand side depends on the encoder $\mathcal{E}_{X|M}$ through ρ_{MB}. To obtain the converse, restrict attention to deterministic injective encoders, so that we are dealing with a (k,ε) code. Defining $\rho_{XB} = \frac{1}{k}\sum_m |x_m\rangle\langle x_m|_X \otimes \varphi_B(x_m)$, applying the inverse of the encoder to X results in ρ_{MB}. (To make this operation well-defined on all x, suppose that x not in the code are randomly mapped to M.) Similarly, applied to

$\rho_X \otimes \sigma_B$, the inverse of the encoding operation results in $\rho_M \otimes \sigma_B$. Therefore by the data processing inequality we obtain

$$\max_{\sigma_B} \beta_{1-\varepsilon}(\rho_{XB}, \rho_X \otimes \sigma_B) \leq \frac{1}{k}. \tag{15.3}$$

The state ρ_{XB} is defined with a uniform distribution P_X over the codewords x_m, and therefore this expression still depends on the details of the code. But optimizing over P_X removes the dependence, leaving the desired converse bound.

Proposition 15.1 (Noisy channel coding converse). *For any CQ channel $\mathcal{N}_{B|X}$, every (k, ε) code satisfies*

$$\min_{P_X} \max_{\sigma_B} \beta_{1-\varepsilon}(\rho_{XB}, \rho_X \otimes \sigma_B) \leq \frac{1}{k}, \tag{15.4}$$

where $\rho_{XB} = \sum_x P_X(x)|x\rangle\langle x|_X \otimes \varphi_B(x)$ for $\varphi_B(x) = \mathcal{N}_{B|X=x}$.

The left-hand side depends on ε and the channel $\mathcal{N}_{B|X}$, but not on the details of the encoder or decoder, while the right-hand side evidently only depends on the number of messages k. The bound has the behavior we mentioned, intuitively expected of a converse bound, that increasing ε loosens the bound on k. This follows because larger ε lead to a larger set of feasible tests. Hence the left-hand side of the bound decreases, meaning that larger k are in principle possible.

The difficulty in working with (15.4) is the minimization over P_X. Whereas every choice of σ_B will produce a converse bound, the bound requires the optimal P_X. In the i. i. d. analysis to come in Section 15.4, we will deal with this for general channels by making use of entropic chain rules.

When the channel is symmetric in a suitable sense, it turns out that $\beta_\alpha(\rho_{XB}, \rho_X \otimes \sigma_B)$ is independent of P_X, at least for symmetric σ_B. In particular, a channel $\mathcal{N}_{B|X}$ with outputs $\varphi_B(x)$ is symmetric if for all $x, x' \in \mathcal{X}$, there exists a unitary U_B such that $\varphi_B(x') = U_B \varphi_B(x) U_B^*$. For instance, the BSC, PSC, and BEC are all symmetric in this sense, though the Z channel is not.

Proposition 15.2. *For any channel $\mathcal{N}_{B|X}$ symmetric in the sense just described and any state σ_B invariant under the symmetry operations U_B, we have the following for all $\alpha \in [0, 1]$ and $x \in \mathcal{X}$:*

$$\beta_\alpha(\rho_{XB}, \rho_X \otimes \sigma_B) = \beta_\alpha\big(\varphi_B(x), \sigma_B\big). \tag{15.5}$$

Proof. Consider $\beta_\alpha(\varphi_B(x), \sigma_B)$. By the form of the primal optimization in (9.4) it follows that this quantity is the same for all x, simply because the optimal POVM element $\Lambda^\star(x)$ can be made feasible for x' by using $U_B \Lambda_B^\star(x) U_B^*$. Then the upper bound in (15.5) follows from (9.17). For the lower bound, suppose $\mu^\star(x)$ and $\theta_B^\star(x)$ are optimal in the dual formulation (9.12), so that

$$\mu^\star(x)\varphi_B(x) - \sigma_B \le \theta^\star_B(x) \quad \text{and} \tag{15.6}$$
$$\mu^\star(x)\alpha - \mathrm{Tr}[\theta^\star_B(x)] = \beta_\alpha(\varphi_B(x), \sigma_B)\,. \tag{15.7}$$

Applying U_B for some x' gives $\mu^\star(x)\varphi_B(x') - \sigma_B \le U_B\theta^\star_B(x)U^*_B$, meaning that $\mu^\star(x)$ and $U_B\theta^\star_B(x)U^*_B$ are feasible for $\beta_\alpha(\varphi_B(x'),\sigma)$. As we have already established that $\beta_\alpha(\varphi_B(x),\sigma_B)$ is independent of x, it follows that $\mu^\star(x)$ and $U_B\theta^\star_B(x)U^*_B$ are an optimal choice of variables. In particular, $\mu^\star = \mu^\star(x')$ for arbitrary $x' \in \mathcal{X}$ is optimal for all x. Therefore $\mu^\star P_X(x)\varphi_B(x) - P_X(x)\sigma_B \le P_X(x)\theta^\star_B(x)$ for all $x \in \mathcal{X}$, which is the statement

$$\mu^\star\rho_{XB} - \rho_X \otimes \rho_B \le \sum_{x\in\mathcal{X}} P_X(x)|x\rangle\langle x|_X \otimes \theta^\star_B(x)\,. \tag{15.8}$$

Thus $\beta_\alpha(\rho_{XB}, \rho_X \otimes \sigma_B) \ge \mu^\star\alpha - \sum_{x\in\mathcal{X}} P_X(x)\,\mathrm{Tr}[\theta^\star(x)] = \beta_\alpha(\varphi_B(x),\sigma)$ for all $x \in \mathcal{X}$. □

Choosing a symmetric σ_B (i. e., $\sigma_B = U_B\sigma_B U^*_B$ for all symmetry operations U_B) and applying (15.5) to (15.4) gives the bound

$$\max_{\sigma_B \text{ symm.}} \beta_{1-\varepsilon}(\varphi_B(x),\sigma_B) \le \frac{1}{k} \quad \forall x \in \mathcal{X}\,. \tag{15.9}$$

A similar converse bound holds when considering the worst-case error probability.

Exercise 15.2. Using the fact that $\min_{z\in\mathcal{Z}} P_Z(z) \le \frac{1}{|Z|}$ for arbitrary P_Z, show that for every $(k,\varepsilon)_{\mathrm{wc}}$ code,

$$\max_{\sigma_B} \min_{x\in\mathcal{X}} \beta_{1-\varepsilon}(\varphi_B(x),\sigma_B) \le \frac{1}{k}\,. \tag{15.10}$$

The following example shows that the converse bound is sometimes tight.

Exercise 15.3. For $\mathrm{BSC}(p)^{\otimes 3}$, show that $k \le 2$ for a code with error probability $p^2(3-2p)$ when $p \in (0, 1/2]$. What does the converse imply for $p = 0$ and $p > 1/2$?

15.3 Achievability

Now let us show the existence of a (k,ε) code for average error for which the relationship between k and ε is very nearly the same as in (15.4). In comparison to the converse, to establish achievability requires thinking more carefully about the code and the operation of the decoder. We first motivate the choices we will make in the subsequent formal argument.

Following Shannon, the approach is to show the existence of an ε-good code C by showing that the error rate suitably averaged over all codes is less than ε. Then there must be at least one good ε-good code. Moreover, we are free to use any distribution over codes in this argument. Note, though, that the argument is nonconstructive, and we will not learn anything about the structure of the code or codes that satisfy the achievability bound we derive. This is unfortunate, but our main goal here is actually to show that the converse bound is tight.

This is often called the *random coding argument*. It is a method of proving achievability, not a design of the encoder. The encoder is still a deterministic function from messages to codewords. Moreover, the probability distribution used in the averaging over codes has nothing to do with the uniform prior probability of messages to be transmitted.

For a given code C, the choice of decoder is ostensibly clear: Use the optimal measurement to achieve $P_{\text{guess}}(M|B)_\rho$. However, the behavior of this *maximum a posteriori* (MAP) decoder is difficult to analyze in the random coding approach. We will make do with the pretty good measurement.

Now we turn to the formal argument, where we show the following:

Proposition 15.3 (Noisy channel coding achievability). *For any CQ channel $\mathcal{N}_{B|X}$ and error $\varepsilon \in [0,1]$, there exists a (k,ε) code with*

$$\frac{1}{k} \leq \min_{\eta\in[0,\varepsilon]} \min_{P_X} \frac{1}{\eta}\beta_{1-\varepsilon+\eta}(\omega_{XB}, \omega_X \otimes \omega_B) \tag{15.11}$$

for $\omega_{XB} = \sum_x P_X(x)|x\rangle\langle x|_X \otimes \varphi_B(x)$.

Proof. For a given code C with codewords $\{x_m\}_{m\in C}$, the probability of successfully decoding under the pretty good measurement is just $\Pr[M' = M] = \frac{1}{k}Q(\rho_{XB}, \rho_X \otimes \rho_B)$ for $\rho_{XB} = \sum_m |x_m\rangle\langle x_m|_X \otimes \varphi_B(x_m)$. Consider the average of the successful decoding over all possible codes, which we denote by $\langle \Pr[M' = M]\rangle_C$. The probability distribution of codes is generated by picking each codeword according to some P_X, independently and identically to all other codewords. That is, the probability that the code contains the codewords $x_1, x_2, \ldots, x_k$, denoted $P_{X_1\cdots X_k}(x_1,\ldots,x_k)$, satisfies $P_{X_1\cdots X_k}(x_1,\ldots,x_k) = P_X(x_1)P_X(x_2)\cdots P_X(x_k)$. Joint convexity of Q (Proposition 11.4) implies

$$\langle \Pr[M' = M]\rangle_C \geq \frac{1}{k}Q(\langle\rho_{XB}\rangle_C, \langle\rho_X \otimes \rho_B\rangle_C)\,. \tag{15.12}$$

The average in the first argument gives

$$\begin{aligned}\langle\rho_{XB}\rangle_C &= \sum_{x_1,\ldots,x_n} P_{X_1X_2\cdots X_k}(x_1,\ldots,x_k)\frac{1}{k}\sum_{m=1}^{k}|x_m\rangle\langle x_m|_X \otimes \varphi_B(x_m)\\ &= \frac{1}{k}\sum_{m=1}^{k}\sum_{x_m} P_X(x_m)|x_m\rangle\langle x_m|_X \otimes \varphi_B(x_m)\\ &= \sum_x P(x)|x\rangle\langle x|_X \otimes \varphi_B(x) = \omega_{XB}\,.\end{aligned} \tag{15.13}$$

The average in the second is slightly more involved:

$$\begin{aligned}\langle\rho_X\otimes\rho_B\rangle_C &= \sum_{x_1,\dots,x_n} P_{X_1X_2\cdots X_k}(x_1,\dots,x_k)\frac{1}{k^2}\sum_{m,m'}|x_m\rangle\langle x_m|\otimes\varphi_B(x_{m'})\\ &= \frac{1}{k}\omega_{XB} + \sum_{x_1,\dots,x_n} P_{X_1X_2\cdots X_k}(x_1,\dots,x_k)\frac{1}{k^2}\sum_{m'\neq m}|x_m\rangle\langle x_m|\otimes\varphi_B(x_{m'})\\ &= \frac{1}{k}\omega_{XB} + \frac{1}{k^2}\sum_{m'\neq m}\sum_{x_m,x_{m'}} P_X(x_m)P_X(x_{m'})|x_m\rangle\langle x_m|_X\otimes\varphi_B(x_{m'})\\ &= \frac{1}{k}(\omega_{XB} + (k-1)\omega_X\otimes\omega_B)\,.\end{aligned} \tag{15.14}$$

Here the first term in the second equality comes from the cases $m' = m$, for which the calculation is the same as that for the first argument. Using the fact that $Q(\rho, c\sigma) = \frac{1}{c}Q(\rho,\sigma)$, it follows that

$$\langle\Pr[M' = M]\rangle_C \geq Q(\omega_{XB}, \omega_{XB} + (k-1)\omega_X\otimes\omega_B)\,. \tag{15.15}$$

Finally, we relate the bound to something involving $\beta_\alpha(\omega_{XB}, \omega_X\otimes\omega_B)$. Consider the two-outcome POVM $\{\Lambda_{XB}, \mathbb{1}_{XB} - \Lambda_{XB}\}$ for Λ_{XB} optimal in $\beta_\alpha(\omega_{XB}, \omega_X\otimes\omega_B)$ for arbitrary $\alpha\in[0,1]$. For notational convenience, set $\theta_{XB} = \omega_{XB} + (k-1)\omega_X\otimes\omega_B$. Monotonicity of Q under this measurement implies $Q(\omega_{XB}, \theta_{XB}) \geq Q(P,R)$ for $P = (\mathrm{Tr}[\Lambda_{XB}\omega_{XB}], \mathrm{Tr}[(\mathbb{1}_{XB} - \Lambda_{XB})\omega_{XB}])$, and similarly for R, using θ_{XB}. From the form of Q it follows that $Q(P,R)\geq p_1^2/r_1$ for the first entry p_1 of P and the first entry r_1 of R. In this case, $p_1 = \alpha$ and $r_1 = \beta_\alpha(\omega_{XB},\theta_{XB})$. Hence the bound on the probability of successful transmission becomes

$$\langle\Pr[M' = M]\rangle_C \geq \alpha^2(\alpha + (k-1)\beta_\alpha(\omega_{XB}, \omega_X\otimes\omega_B))^{-1}\,. \tag{15.16}$$

The average error probability is therefore guaranteed to be less than ε if α and k are chosen so that the right-hand side is at least $1-\varepsilon$. After some algebra, it can be verified that this condition holds for all $\alpha\in[1-\varepsilon,1]$ and all k such that $k\leq 1+\alpha\frac{\alpha+\varepsilon-1}{1-\varepsilon}\beta_\alpha(\omega_{XB},\omega_X\otimes\omega_B)^{-1}$. For notational convenience, call the upper bound $1+t$; the best choice of k is thus $k = \lfloor t+1\rfloor$, since k should be an integer. Then, to simplify the bound, write $k = \lfloor t+1\rfloor \geq t$, which is

$$\frac{1}{k}\leq\frac{1-\varepsilon}{\alpha(\alpha+\varepsilon-1)}\beta_\alpha(\omega_{XB},\omega_X\otimes\omega_B)\,. \tag{15.17}$$

This bound holds for arbitrary $\alpha\in[1-\varepsilon,1]$, so we may pick α leading to the minimal value. We are also free to pick any P_X in the construction of ω_{XB}. Letting $\alpha = 1-\varepsilon+\eta$ for $\eta\in[0,\varepsilon]$ and loosening the bound slightly by using $1-\varepsilon\leq 1-\varepsilon+\eta$ completes the proof. □

In the above argument, specifically (15.14), we do not actually need the distribution on codebooks to be such that the codewords are i. i. d. Instead, it is enough if the codewords are identically distributed but only *pairwise independent*, since (15.14) only involves pairs of codewords. An example is given by *linear codes*. Taking the input alphabet $\mathcal{X}$ to be an Abelian[1] group so that addition is defined, e. g., $\mathcal{X} = \mathbb{Z}_2^n$, we can define linear codes to be those that satisfy $x_j + x_k \in C$ for all $x_j, x_k \in C$. The repetition code is a (nearly trivial) example. It can be shown that a uniformly random choice of a linear code results in pairwise independence of the codewords. Therefore linear codes achieve (15.11), at least for uniform P_X.

15.4 Coding for i. i. d. channels

15.4.1 Capacity

In the i. i. d. setting of $\mathcal{N}_{B|X}^{\otimes n}$ the optimal rate for given ε is

$$R(\mathcal{N}_{B|X}, \varepsilon, n) := \frac{1}{n} \log M^\star(\mathcal{N}_{B|X}, \varepsilon, n) \,. \tag{15.18}$$

In the limit as $n \to \infty$, the optimal rate becomes the *ε-capacity* $C(\mathcal{N}_{B|X}, \varepsilon)$. Usually, we are interested in the limit as $\varepsilon \to 0$, which defines the capacity $C(\mathcal{N}_{B|X}) := \lim_{\varepsilon\to 0} C(\mathcal{N}_{B|X}, \varepsilon)$.

As with data compression, the zero-error behavior is quite different from the limit as $\varepsilon \to 0$. It is easy to see that for a channel like the BSC, it is impossible to transmit any input with zero error, and hence its *zero-error capacity* is zero: $C(\mathrm{BSC}(p), 0) = 0$ for $p > 0$. Allowing instead a vanishingly small error, we can show the following:

Proposition 15.4 (Classical capacity of CQ channels). *For any CQ channel $\mathcal{N}_{B|X}$ and associated state $\omega_{XB} = \sum_{x\in\mathcal{X}} P_X(x)|x\rangle\langle x|_X \otimes \varphi_B(x)$,*

$$C(\mathcal{N}_{B|X}) = \max_{P_X} I(X:B)_\omega \,. \tag{15.19}$$

Proof. Start with achievability and choose $P_{X^n} = P_X^{\times n}$ for any P_X, $\varepsilon \in (0,1)$, and $\eta \in (0,\varepsilon)$ in (15.11). Using Stein's lemma, we have

$$\begin{aligned} \lim_{n\to\infty} \frac{-1}{n}(\log\eta - \log k) &\geq \max_{P_X} \lim_{n\to\infty} \frac{-1}{n} \log \beta_{1-\varepsilon+\eta}(\omega_{XB}^{\otimes n}, (\omega_X \otimes \omega_B)^{\otimes n}) \\ &= \max_{P_X} I(X:B)_\omega \,. \end{aligned} \tag{15.20}$$

The restriction to $\varepsilon \in (0,1)$ ensures that the first term on the left-hand side vanishes in the limit, whereas the second gives $C(\mathcal{N}_{B|X}, \varepsilon)$. Therefore we have $C(W, \varepsilon) \geq \max_{P_X} I(X{:}B)_\omega$ for all $\varepsilon \in (0,1)$.

1 Niels Henrik Abel, 1802–1829.

The upper bound, from the converse, is more difficult, since we now have to ensure that an i. i. d. choice for P_{X^n} is optimal. Instead of using Stein's lemma, here we employ Fano's inequality (12.54) to convert the bound to one involving the relative entropy and then use chain rules.

Choosing $\sigma_{B^n} = \omega_B^{\otimes n}$ and using (12.54) in (15.4) gives

$$\log k \le \max_{P_{X^n}} \frac{I(X^n{:}B^n)_\omega + h_2(\varepsilon)}{1-\varepsilon}. \tag{15.21}$$

Dividing by n and taking the limit as $n \to \infty$ removes the second term. A further limit as $\varepsilon \to 0$ then yields

$$C(\mathcal{N}_{B|X}) \le \lim_{n\to\infty} \frac{1}{n} \max_{P_{X^n}} I(X^n{:}B^n)_\omega. \tag{15.22}$$

Now we reduce this to the *single-letter* expression (involving only $n = 1$) given in (15.19). By the chain rule we have

$$I(X^n : B^n) = H(B^n) - H(B^n|X^n) = H(B^n) - \sum_{j=1}^{n} H(B_j|X^n B_1^{j-1}). \tag{15.23}$$

Since each channel use is independent of all the others, the output B_j depends only on X_j. Hence $H(B_j|X^n B_1^{j-1}) = H(B_j|X_j)$. Using subadditivity for the first term gives

$$I(X^n : B^n) \le \sum_{j=1}^{n} H(B_j) - H(B_j|X_j) = \sum_{j=1}^{n} I(X_j : B_j). \tag{15.24}$$

Thus we have

$$\begin{aligned} C(W) &\le \lim_{n\to\infty} \max_{P_{X^n}} \frac{1}{n} \sum_{j=1}^{n} I(X_j : B_j)_{\omega_{X_j B_j}} \\ &= \lim_{n\to\infty} \frac{1}{n} \sum_{j=1}^{n} \max_{P_{X_j}} I(X_j : B_j)_{\omega_{X_j B_j}} = \max_{P_X} I(X : B)_\omega. \end{aligned} \tag{15.25}$$

The first equality comes from the fact that the mutual information in the jth term only depends on the marginal distribution P_{X_j}, so we can replace any distribution P_{X^n} by the product of its marginals in the maximization. □

The upper bound only matches the lower bound in the case $\varepsilon \to 0$, which is the statement of the *weak converse*: Codes with rates above the capacity cannot achieve vanishing error. We might expect that allowing a finite error increases the possible rate that can be achieved, i. e., that $C(\mathcal{N}_{B|X}, \varepsilon)$ increases with ε. However, the maximum achievable rate is in fact independent of ε, a statement known as the *strong converse*:

$$C(\mathcal{N}_{B|X}, \varepsilon) = \max_{P_X} I(X : B)_\omega \quad \forall \varepsilon \in (0,1). \tag{15.26}$$

Thus the capacity signals a kind of phase transition in the behavior of error rate of optimal codes. Codes with rates below the capacity can achieve essentially zero error, but as soon as the rate surpasses the capacity, the error rate is forced to one. The strong converse can in fact be obtained from the converse bound in (15.4), but the derivation is much lengthier than that of the weak converse, and we will not cover it here.

Looking back, we can now appreciate *why* entropy and mutual information play such a pivotal role in this problem. The hypothesis testing converse already shows that the likelihood ratio of the joint distribution to the product of its marginals is relevant. This is a random variable, so any finite-size bound has to extract some property or other from it. However, by the law of large numbers, in the asymptotic i. i. d. limit the random variable tends to its average value, the relative entropy.

Importantly, the optimization over P_X in the capacity expression is a convex optimization, that is, the function $f : P_X \mapsto I(X : B)_\omega$ for a fixed channel $\mathcal{N}_{B|X}$ is concave. To see this, suppose the random variable Y determines the distribution P_X, define $\omega_{XB}(y) = \sum_{y\in\mathcal{Y}} P_{X|Y=y}(x)|y\rangle\langle y|_Y \otimes \varphi_B(x)$, and consider the joint state $\omega_{XYB} = \sum_{y\in\mathcal{Y}} P_Y(y)\,\omega_{XB}(y)$. Observe that $H(B|XY)_\omega = H(B|X)_\omega$, since the channel output $\varphi_B(x)$ is only a function of the channel input x. Then we have

$$\begin{aligned} I(X : B)_\omega &= H(B)_\omega - H(B|XY)_\omega \\ &\geq H(B|Y)_\omega - H(B|XY)_\omega = \sum_{y\in\mathcal{Y}} P_Y(y)\, I(X : B)_{\omega(y)}. \end{aligned} \tag{15.27}$$

This is useful when the channel is symmetric. When two different distributions $P_{X|Y=y}$ lead to the same value of the mutual information, it is advantageous in the optimization to consider their uniform mixture.

Exercise 15.4. Determine the capacities of BSC(p), BEC(q), and PSC(f).

Exercise 15.5. Consider the channel $W : \mathbb{Z}_n \to \mathbb{Z}_n$ that sends any input $x \in \mathbb{Z}_n$ to itself with probability $1-p$ and to $x \pm 1$ with probability $p/2$ each, where arithmetic is modulo n. Observe that W is covariant with respect to shifts of the input in the sense that for the channel V_y that takes x to $x+y$, $W \circ V_y = V_y \circ W$. Using this symmetry, show that the optimal input distribution in the capacity is the uniform distribution and determine the capacity. What is the capacity of the related channel that sends x to $x+1$ with probability p and x to itself with probability $1-p$?

15.4.2 Finite-blocklength bounds

One of the nice features of having one shot bounds in terms of β_α is that it is possible to fairly accurately determine the optimal code lengths at fixed error ε for certain channels at modest blocklengths, well away from the asymptotic limit. An example is depicted for the BSC, BEC, and PSC in Figure 15.2. We see that the bounds are quite close together but not in a particular hurry to arrive at the asymptotic limit.

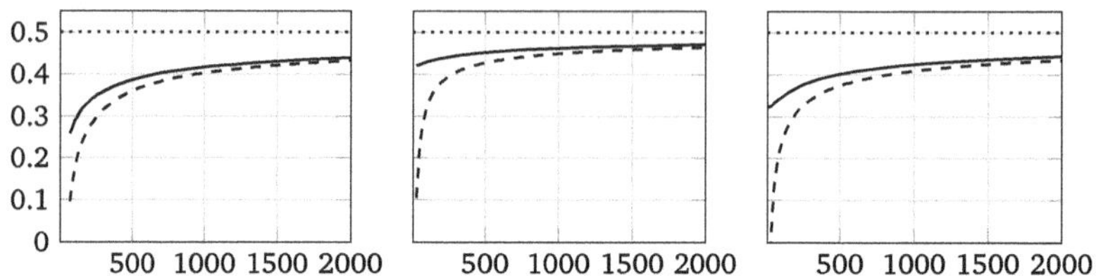

(a) BSC(0.11) (b) BEC(0.5) (c) PSC(0.78)

Figure 15.2: Upper (solid) and lower (dashed) bounds on the optimal code rate versus blocklength for the three indicated channels. In each case the target error rate of the code is $1/10^3$, and the parameters are chosen such that the capacity of each channel is 1/2, indicated by the dotted line.

In each case the calculation is simplified by choosing $\sigma_B = \omega_B$ in (15.4), $\eta = \varepsilon/2$ in (15.11), and then appealing to (15.5) to remove the optimization over the prior distribution. Hence it remains to compute $\beta_\alpha(\varphi(0)^{\otimes n}, \bar{\varphi}^{\otimes n})$ for $\alpha = 1-\varepsilon$ and $\alpha = 1-\varepsilon/2$, where $\varphi(j)$ are the outputs of the individual CQ channel in question, and $\bar{\varphi} = \frac{1}{2}(\varphi(0) + \varphi(1))$. For the BSC, $\bar{\varphi} = \pi$, and the necessary calculation is already done in Exercise 9.16. For the BEC(q), $\varphi(0)$ is the distribution $(1-q, 0, q)$, and $\bar{\varphi}$ is the distribution $(\frac{1-q}{2}, \frac{1-q}{2}, q)$. By Exercise 9.15 the problem reduces to $\beta_\alpha((1-q,q), (\frac{1-q}{2}, q))$. This can be computed in the same manner as suggested in Exercise 9.16.

The noncommuting outputs of the pure state channel make the calculation more involved, though still tractable because $\varphi(0)$ is a pure state. Let us investigate the general problem of computing $\beta_\alpha(|\psi\rangle\langle\psi|, \sigma)$ for a d-dimensional state $|\psi\rangle$ and arbitrary positive operator σ, as it provides another example of working with SDPs. We will then return to the specifics of the pure state channel afterward.

In the dual formulation of (9.12), we have $\beta_\alpha(|\psi\rangle\langle\psi|, \sigma) = \max\{\mu\alpha - \theta\}$ for $\theta = \{\mu|\psi\rangle\langle\psi| - \sigma\}_+$. It happens that θ will be rank one for the following reason. Consider the $(d-1)$-dimensional subspace of vectors orthogonal to $|\psi\rangle$ and the two-dimensional subspace corresponding to the two largest eigenvalues λ_1 and λ_2 of $\mu|\psi\rangle\langle\psi| - \sigma$. Since the sum of their dimensions is larger than d, there must be a vector common to both subspaces, call it $|\xi\rangle$. Then we have $\lambda_2 \le \langle\xi|(\mu|\psi\rangle\langle\psi| - \sigma)|\xi\rangle = -\langle\xi|\sigma|\xi\rangle \le 0$. Hence only $\lambda_1 \ge 0$, and θ has rank one, as claimed.

Exercise 15.6. Extend the argument and prove the *Weyl inequalities* that the eigenvalues of two Hermitian operators A and B on $\mathbb{C}^d$ satisfy $\lambda_{i+j-1}(A+B) \le \lambda_i(A) + \lambda_j(B)$ for $i+j-1 \le d$.

Therefore the optimization becomes $\beta_\alpha(|\psi\rangle\langle\psi|, \sigma) = \max\{\mu\alpha - \lambda : \mu|\psi\rangle\langle\psi| \le \sigma + \lambda\mathbb{1}, \lambda \ge 0, \mu \ge 0\}$. Conjugating the constraint inequality by $(\sigma+\lambda\mathbb{1})^{-1/2}$ (a valid expression since $\sigma + \lambda\mathbb{1} \ge 0$ for $\lambda \ge 0$) gives $\mu(\sigma+\lambda\mathbb{1})^{-1/2}|\psi\rangle\langle\psi|(\sigma+\lambda\mathbb{1})^{-1/2} \le \mathbb{1}$, which implies $\mu\langle\psi|(\sigma+\lambda\mathbb{1})^{-1}|\psi\rangle \le 1$. This inequality implies the original constraint just by reversing the steps, and hence $\beta_\alpha(|\psi\rangle\langle\psi|, \sigma) = \max\{\mu\alpha - \lambda : \mu\langle\psi|(\sigma+\lambda\mathbb{1})^{-1}|\psi\rangle \le 1\}$. The best choice of μ is clear from the constraint. Writing $\sigma = \sum_x s_x|x\rangle\langle x|$ with the eigenvalues s_x of σ and $|\psi\rangle = \sum_x \psi_x|x\rangle$, we end up with

$$\beta_\alpha(|\psi\rangle\langle\psi|, \sigma) = \max\left\{\alpha\left(\sum_x \frac{\psi_x^2}{s_x+\lambda}\right)^{-1} - \lambda : \lambda \ge 0\right\}. \tag{15.28}$$

This is a convex optimization in one real variable, a considerable simplification. It can be easily solved by finding the λ for which the derivative is zero; by convexity we can be sure that there is only one such value. The condition on the optimal λ^* is

$$\alpha \sum_x \frac{\psi_x^2}{(s_x+\lambda^*)^2} = \left(\sum_x \frac{\psi_x^2}{s_x+\lambda^*}\right)^2. \tag{15.29}$$

Now we can apply this general result to the problem at hand. The outputs of PSC(f) can be expressed as $|\varphi(j)\rangle = \sqrt{p}|0\rangle + (-1)^j\sqrt{1-p}|1\rangle$ with $p = \frac{1+f}{2}$, which implies that $\bar\varphi$ is the diagonal operator with eigenvalues p and $1-p$. The state $|\varphi(0)\rangle^{\otimes n}$ is invariant under all permutations and is therefore an element of the *symmetric subspace*, the span of such vectors. Fortunately, this subspace has dimension $n+1$, and again by Exercise 9.15 we only need to concern ourselves with the projection of $\bar\varphi^{\otimes n}$ on this subspace. Hence the number of terms in (15.29) is vastly reduced from 2^n to $n+1$. Were this not the case, the calculation would be intractable.

A convenient basis for the symmetric subspace is simply the normalized vectors $|b_t\rangle$ of all superpositions of strings of type t, that is, $|b_t\rangle = \binom{n}{t}^{-1/2}\sum_s U_s|1\dots10\dots0\rangle$, where the state has precisely t 1s, and U_s are unitary operators, which permute the n systems. Therefore we have

$$|\varphi(0)\rangle^{\otimes n} = \left(\sqrt{p}|0\rangle + \sqrt{1-p}|1\rangle\right)^{\otimes n} = \sum_{j=0}^{n}\sqrt{p^j(1-p)^{n-j}\binom{n}{j}}|b_j\rangle. \tag{15.30}$$

To determine the projection of $\bar\varphi^{\otimes n}$, first observe that $\bar\varphi^{\otimes n}$ can be expressed as $\sum_{j=0}^n p^j(1-p)^{n-j}\Pi(j)$, where $\Pi(j)$ is the projector onto the subspace of type j. The projector onto the symmetric subspace is $\Pi_{\text{sym}} = \sum_t |b_t\rangle\langle b_t|$. As $\langle b_t|\Pi(j)|b_t\rangle = \delta_{j,t}\frac{1}{\binom{n}{t}}$, it follows that $\Pi_{\text{sym}}\bar\varphi^{\otimes n}\Pi_{\text{sym}} = \sum_{j=0}^n p^j(1-p)^{n-j}|b_j\rangle\langle b_j|$. Thus, to compute β_α, we need only take $\psi_j^2 = \binom{n}{j}p^j(1-p)^{n-j}$ and $s_j = p^j(1-p)^{n-j}$ in (15.29), which can then be solved numerically.

15.5 Classical coding over quantum channels

15.5.1 Recycling the CQ result

The setup for communication over CQ channels can be immediately applied to the task of classical communication over quantum channels. As discussed in Section 8.1, the only change we need to make to the definition of a (k, ε) protocol is that the encoder $\mathcal{E}_{A|M}$ now has a quantum output.

In this context a general encoding map can be decomposed into a convex combination of maps that output pure states. Suppose the outputs are $\rho_A(m)$ and choose a decomposition for each into pure states: $\rho_A(m) = \sum_x P_{X|M=m}(x)|\psi_{x,m}\rangle\langle\psi_{x,m}|$. The encoder can then be regarded as being composed of two parts: first, a classical channel, which sends m to the pair (x_m, m), and subsequently the CQ channel taking a pair (x, m) to $|\psi_{x,m}\rangle$. To the first channel we can apply Proposition 3.3, and, in particular, the probability for the map with the specific choice $\{x_m\}_m$ will be $\Pi_{m\in\mathcal{M}} P_{X|M=m}(x_m)$. Therefore, as for CQ channels, for any (k, ε) protocol, there exists a pure-state output encoder with the same or better error probability.

Observe that concatenating any state preparation map $\mathcal{S}_{A|X}$ from some alphabet $\mathcal{X}$ with the channel $\mathcal{N}_{B|A}$ results in a CQ channel $\mathcal{N}'_{B|X} = \mathcal{N}_{B|A} \circ \mathcal{S}_{A|X}$. Therefore we may apply the one-shot CQ achievability result of Proposition 15.3, combining the optimization over P_X and $\mathcal{S}_{A|X}$ into a single optimization over CQ states on XA. However, we cannot proceed in this fashion in the one-shot converse (Proposition 15.1). The latter half of the proof does not go through: We cannot ensure that $\mathcal{S}_{A|X}$ is invertible to apply the data processing inequality to obtain (15.3).

15.5.2 Capacity expression

Unlike the case of classical communication over CQ channels, we have no satisfactory treatment of the relation between M and ε for communication over quantum channels. As we will see, the difficulty is the possibility of entangled inputs to the channel. We will settle for results on the capacity of the channel in the asymptotic limit, even though here again the results are not satisfactory in that we have no single-letter formula.

Nevertheless, the capacity expression is at least similar looking. In analogy with the case of CQ channels, we might expect that the capacity is given by the *Holevo information*

$$\chi(\mathcal{N}_{B|A}) = \max_{\rho_{XA}} I(X : B)_{\mathcal{N}_{B|A}[\rho_{XA}]}, \tag{15.31}$$

where the maximization is over all CQ states ρ_{XA} with classical X. Indeed, this rate is achievable by the argument above. In the i. i. d. scenario, simply prepend a fixed state preparation map $\mathcal{S}_{A|X}$ to each channel $\mathcal{N}_{B|A}$ and make use of the achievability argument in the proof of Proposition 15.4 for $\mathcal{N}'_{B|X} = \mathcal{N}_{B|A} \circ \mathcal{S}_{A|X}$. The optimizations over P_X and $\mathcal{S}_{A|X}$ can be combined into an optimization over CQ states ρ_{XA}.

By the same argument the rate $\frac{1}{2}\chi(\mathcal{N}_{B|A}^{\otimes 2})$ is achievable simply by fixing a state preparation map $\mathcal{S}_{A_1A_2|X}$ for inputs to pairs of channels. The factor of 1/2 accounts for the two uses of the channel. This argument works for any n, and therefore the *regularized* Holevo information $\chi_{\text{reg}}(\mathcal{N}_{B|A}) := \lim_{n\to\infty} \frac{1}{n}\chi(\mathcal{N}_{B|A}^{\otimes n})$ is also an achievable rate. Fortunately, we can stop here, as this quantity is in fact the capacity. Since the choice of preparation map $\mathcal{S}_{A^n|X^n}$ is included in the optimization, (15.22) from the first step of the converse proof in Proposition 15.4 applies to $\mathcal{N}'_{B^n|X^n} = \mathcal{N}_{B|A}^{\otimes n} \circ \mathcal{S}_{A^n|X^n}$. The remainder of the converse proof, the single-letterization, fails due to the presence of arbitrary $\mathcal{S}_{A^n|X^n}$. In this sense, we are unable to completely remove the dependence of the capacity bound on the encoding operation.

Proposition 15.5 (Classical capacity of quantum channels). *For an arbitrary quantum channel $\mathcal{N}_{B|A}$, the capacity $C_C(\mathcal{N}_{B|A})$ to transmit classical information is given by*

$$C_C(\mathcal{N}) = \lim_{n\to\infty} \frac{1}{n}\chi\big(\mathcal{N}_{B|A}^{\otimes n}\big). \tag{15.32}$$

We can read the single-letterization in the proof of Proposition 15.4 as showing that the Holevo information for CQ channels is *additive*, $\frac{1}{n}\chi(\mathcal{N}_{B|X}^{\otimes n}) = \chi(\mathcal{N}_{B|X})$ for any n. The same proof technique fails to establish additivity in the case of quantum channels for good reason: There exist explicit examples for which the Holevo information is indeed superadditive. The examples make use of entangled inputs to pairs of channels, though we will not go into details here. Thus regularization is in general necessary in a capacity expression of this form. Whether or not some other single-letter formula exists for the classical capacity of quantum channels remains an open question.

15.5.3 Properties of the Holevo information

Entanglement is responsible for nonadditivity because the Holevo information is additive for entanglement-breaking channels. To show this, we need only consider $\chi(\mathcal{N}_{B_1|A_1} \otimes \mathcal{N}'_{B_2|A_2})$ for an entanglement-breaking channel $\mathcal{N}_{B_1|A_1}$ and an arbitrary channel $\mathcal{N}'_{B_2|A_2}$. Suppose that the optimal input state is $\rho_{XA_1A_2} = \sum_{x\in\mathcal{X}} P_X(x)|x\rangle\langle x|_X \otimes \varphi_{A_1A_2}(x)$, and let $\omega_{XB_1B_2} = \mathcal{N}_{B_1|A_1} \otimes \mathcal{N}'_{B_2|A_2}[\rho_{XA_1A_2}]$. Since the output of $\mathcal{N}_{B_1|A_1}$ is always separable, for some states $\sigma_{B_1}(x,y)$ and $\theta_{A_2}(x,y)$ and conditional distributions $P_{Y|X}$, we have

$$\omega_{XB_1B_2} = \sum_{x\in\mathcal{X}} \sum_{y\in\mathcal{Y}} P_X(x)P_{Y|X=x}(y)|x\rangle\langle x|_X \otimes \sigma_{B_1}(x,y) \otimes \mathcal{N}_{B_2|A_2}[\theta_{A_2}(x,y)]. \tag{15.33}$$

Let $\omega_{XYB_1B_2}$ be the extension that includes a classical system storing the value Y. Observe that $\omega_{XYB_1B_2}$ can be generated from the marginal ω_{XYB_2} by the quantum channel $\mathcal{E}_{B_2|XY}$, which conditionally generates $\sigma_{B_2}(x,y)$ given the values of x and y stored in X and Y, respectively. Now consider the Holevo information, which is $I(X:B_1B_2)_\omega$ by assumption. By subadditivity of entropy of B_1B_2 it follows that $I(X:B_1B_2)_\omega \le I(X:B_1)_\omega + I(XB_1:B_2)_\omega$. From data processing we can upper bound the second term by $I(XYB_1:B_2)_\omega = I(XY:B_2)_\omega$. Therefore

$$\chi(\mathcal{N}_{B_1|A_1} \otimes \mathcal{N}'_{B_2|A_2}) \le I(X:B_1)_\omega + I(XY:B_2)_\omega \le \chi(\mathcal{N}_{B_1|A_1}) + \chi(\mathcal{N}'_{B_2|A_2}), \tag{15.34}$$

and additivity follows. In particular, QC channels, i. e., measurements, like CQ channels have a single-letter expression for the classical capacity.

However, despite the apparent similarities of the Holevo information and the expression for the capacity of a CQ channel, the Holevo information is not a convex optimization. As shown above in (15.27), the function $f : \rho_{XA} \mapsto \chi(\mathcal{N}_{B|A})$ is concave in the X distribution of ρ_{XA} for fixed conditional states in A. However, it is a convex function of the conditional states for fixed distributions on X. Therefore computing the capacity for general entanglement-breaking channels and even measurements is considerably more difficult than for CQ channels.

Exercise 15.7. Show that $\rho_{XA} \mapsto \chi(\mathcal{N}_{B|A})$ is convex for fixed ρ_X.

15.6 Notes and further reading

The coding theorem for i. i. d. classical channels is of course due to Shannon [258]; the quote is from the opening of this paper. One-shot bounds were developed by Hayashi [127] and Polyanskiy, Poor, and Verdú [224], though the idea of hypothesis testing for the converse goes back to Nagaoka [208]. Hayashi [128] mentions that around 2000 Nagaoka had the idea of organizing all topics in information theory around the binary hypothesis testing quantity. As will become evident in subsequent chapters, we are very much working in this tradition. See also the recent lecture notes of Polyanskiy and Wu [225].

The classical capacity of quantum channels was slower to arrive. Holevo [144] gave an upper bound on the capacity in 1973, but a matching achievability statement was not found until the late 1990s by Schumacher and Westmoreland [254] and Holevo [142]. One-shot bounds for CQ channel coding were found by Hayashi and Nagaoka [129]. The one-shot converse for CQ channel coding presented here is from Wang and Renner [296], which is based on the aforementioned [224]. The achievability proof using the pretty good measurement is a modification of the construction by Beigi and Gohari [16]. Hastings [122] showed that the Holevo information is superadditive.

16 Information reconciliation

> The astonishment of life is the absence of any appearances of reconciliation between the theory and the practice of life.
>
> Ralph Waldo Emerson

Recall the bipartite scenario in Section 11.1 of a classical random variable X correlated with a quantum system B. There the goal was to determine X as well as possible by just making a measurement on B. We can imagine that X is held by Alice and B by Bob. Suppose now that Alice helps Bob determine X by computing a random variable Y as a (possibly stochastic) function of X and transmitting it to him. The goal is to make Bob's guessing probability close to unity. This task is known as information reconciliation because in the case of classical B, we can view the task as reconciling the classical value B with that of X (i. e., making the former equal to the latter). It can also be regarded as classical compression or source coding with *side information*, because Bob can make use of the side information B at the decoder. The latter interpretation is especially clear from the depiction in Figure 16.1.

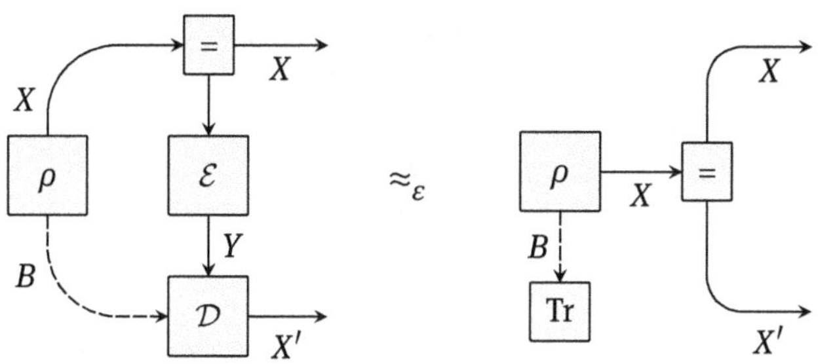

Figure 16.1: A (k, ε) protocol for compression of classical data X relative to quantum side information B with $k = |Y|$. The random variable Y is the compressed version of X, since the decompressor $\mathcal{D}$ can recreate the particular output of the source (stored in X) from it. The partial trace operation Tr can be regarded as trashing the B system.

Clearly, Alice could just send Bob the entire X, but the goal is to complete the task with Y as small as possible. It is perhaps surprising that such a scheme is possible at all, since Alice does not know exactly what Bob needs to know. For example, suppose that X and B are two classical bit strings of length n; Alice's string is random, and Bob's differs in at most t positions. If Alice knew in *which* t positions Bob's string differed from hers, then the protocol would be simple. However, even by sending a sufficient amount of essentially random information about her string (in the form of the output of a randomly chosen function), Bob can combine this information with his string to determine Alice's string.

https://doi.org/10.1515/9783110570250-016

As in the case of channel coding, we can gain a better understanding by considering the protocol from Bob's point of view. In general, his system B is in one of a set of states $\{\varphi_B(x)\}$, but he is unsure which. The set itself is known, but not the particular state. Furthermore, the states are generally not distinguishable, so he cannot just measure the system to determine x with high reliability. The information he receives from Alice narrows the set of possible states, making the distinguishing task simpler. Since both parties know the joint state produced by the source, for each x, Alice also knows how likely Bob is to correctly determine that x. She just needs to sufficiently narrow the set possible states at his end to make the guessing probability close to unity.

16.1 Setup and basic properties

The formal setup is very similar to that of compression of classical information in Section 14.1.1. The decompressor just needs to be modified to use the side information. Consider a CQ state

$$\rho_{XB} = \sum_{x\in\mathcal{X}} P_X(x)|x\rangle\langle x|_X \otimes \varphi_B(x) \tag{16.1}$$

for some probability distribution P_X and set of states $\varphi_B(x)$. Let $P_{XX'}(x,x') = P_X(x)\delta(x,x')$, and denote by $\rho_{XX'B}$ the CCQ state with X' containing a copy of X. Then a (k,ε) information reconciliation protocol for ρ_{XB} consists of an encoder or compressor $\mathcal{E}_{Y|X}$ and a decoder or decompressor $\mathcal{D}_{X'|YB}$ such that $\delta(P_{XX'}, \mathcal{D}_{X'|YB} \circ \mathcal{E}_{Y|X}[\rho_{XX'B}]) \le \varepsilon$. By (11.2) the distinguishability requirement is equivalent to the guessing probability of X given Y and B being at least $1-\varepsilon$. Let $L_\varepsilon^\star(X|B)_\rho$ be the smallest k such that a (k,ε) protocol exists for ρ_{XB}.

The compressor might as well be deterministic, just as in the case of compression with no side information. The same holds for the decompressor when the side information is classical, i. e., when all the $\varphi_B(x)$ commute. Given $B = b$ and $Y = y$, the decompressor simply picks $\operatorname{argmax} P_{X|B=b,Y=y}(x)$. When B is a quantum system, the channel $\mathcal{D}_{X'|BY}$ is a measurement, or rather a sequence of measurements, one for each value of $Y = y$. That is, the POVM elements $\Gamma_{BY}(x)$ have the form

$$\Gamma_{BY}(x) = \sum_y \Lambda_B(x,y) \otimes |y\rangle\langle y|_Y \tag{16.2}$$

for some operators $\Lambda_B(x,y)$, which are positive and satisfy $\sum_x \Lambda_B(x,y) = \mathbb{1}_B$ for all y. The measurement consists of two steps: The value of Y is first determined, and then the measurement with POVM elements $\{\Lambda_B(x,y)\}_{x\in\mathcal{X}}$ is performed on B.

16.2 Converse

The main question of the converse is as follows: For a given ρ_{XB}, how large does Y have to be such that Bob's guessing probability is at least $1-\varepsilon$? We can find a constraint based on using any possible compression scheme to construct a hypothesis testing measurement for distinguishing between the source state ρ_{XB} and a particular uncorrelated operator.

Proposition 16.1 (Converse to compression of classical data). *Given any CQ state ρ_{XB}, every (k,ε) compression protocol satisfies*

$$k \geq \max_{\sigma\in\mathrm{Stat}(\mathcal{H}_B)} \beta_{1-\varepsilon}(\rho_{XB}, \mathbb{1}_X \otimes \sigma_B)\,. \tag{16.3}$$

Proof. Let $\mathcal{E}_{XY|X}$ be an extension of the compression operation that retains a copy of the input X. For the conditional distributions $P_{Y|X=x}$ of the compressor output for input $X=x$, define

$$\omega_{XYB} = \mathcal{E}_{YX|X}[\rho_{XB}] = \sum_{x,y} P_X(x)P_{Y|X=x}(y)|x\rangle\langle x|_X \otimes |y\rangle\langle y|_Y \otimes \varphi_B(x)\,. \tag{16.4}$$

By monotonicity of β_α it follows that for any state σ_B,

$$\beta_{1-\varepsilon}(\rho_{XB}, \mathbb{1}_X \otimes \sigma_B) \leq \beta_{1-\varepsilon}(\omega_{XYB}, \mathcal{E}_{XY|X}[\mathbb{1}_X] \otimes \sigma_B)\,. \tag{16.5}$$

Since $\mathcal{E}_{YX|X}[\mathbb{1}_X] = \sum_{x,y} P_{Y|X=x}(y)|x\rangle\langle x|_X \otimes |y\rangle\langle y|_Y$, it satisfies $\mathcal{E}_{YX|X}[\mathbb{1}_X] \leq \mathbb{1}_{XY}$, and therefore

$$\beta_{1-\varepsilon}(\omega_{XYB}, \mathcal{E}_{XY|X}[\mathbb{1}_X] \otimes \sigma_B) \leq \beta_{1-\varepsilon}(\omega_{XYB}, \mathbb{1}_{XY} \otimes \sigma_B)\,. \tag{16.6}$$

Using the decompressor, define the operator $\Gamma_{XYB} = \sum_{x,y} |x\rangle\langle x|_X \otimes |y\rangle\langle y|_Y \otimes \Lambda_B(x,y)$. It satisfies $\mathrm{Tr}[\Gamma_{XYB}\omega_{XYB}] \geq 1-\varepsilon$ by assumption and is therefore feasible for the optimization in $\beta_{1-\varepsilon}(\omega_{XYB}, \mathbb{1}_{XY} \otimes \sigma_B)$. Thus

$$\begin{aligned}\beta_{1-\varepsilon}(\omega_{XYB}, \mathbb{1}_{XY} \otimes \sigma_B) &\leq \mathrm{Tr}[\Gamma_{XYB}(\mathbb{1}_{XY} \otimes \sigma_B)] \\ &= \sum_{x,y} \mathrm{Tr}[\Lambda_B(x,y)\,\sigma_B] = \sum_y \mathrm{Tr}[\sigma_B] = k\,.\end{aligned} \tag{16.7}$$

Since the sequence of inequalities holds for arbitrary σ_B, the converse follows. □

16.3 Achievability

16.3.1 Statement

To show a nearly matching achievability bound, we again make use of Shannon's random coding argument and average over the choice of compression function. We again use the pretty good measurement for the decompressor.

Proposition 16.2. *For every CQ state ρ_{XB} and $\varepsilon \in [0,1]$, there exists a (k,ε) information reconciliation protocol such that*

$$k \leq \min_{\eta\in[0,\varepsilon]} \frac{1}{\eta}\beta_{1-\varepsilon+\eta}(\rho_{XB}, \mathbb{1}_X \otimes \rho_B) + 1. \tag{16.8}$$

Proof. For convenience, let $\rho_{XB} = \sum_x |x\rangle\langle x|_X \otimes \varphi_B(x)$, absorbing the probability of x into the normalization of $\varphi_B(x)$. Then the compressor creates the state $\omega_{XYB} = \sum_x |x\rangle\langle x|_X \otimes |f(x)\rangle\langle f(x)|_Y \otimes \varphi_B(x)$. The decompressor applies the pretty good measurement appropriate for the observed value $y = f(x)$. For a fixed y, the POVM elements are $\Lambda_B(x,y) = \theta_B(y)^{-1/2}\varphi_B(x)\theta_B(y)^{-1/2}$ for $\theta_B(y) = \sum_{x:f(x)=y}\varphi_B(x)$, plus an additional element $\mathbb{1}_B - \{\theta_B(y) < \mathbb{1}_B\}$ if necessary. Moreover, we take $\Lambda_B(x,y) = 0$ when $y \neq f(x)$. Then we may regard this collection as a POVM on YB with elements $\Gamma_{YB}(x) = \sum_y |y\rangle\langle y|_Y \otimes \Lambda_B(x,y)$.

For any $X = x$, $\Pr[X' = x]$ satisfies $P_X(x)\Pr[X' = x] = Q(\varphi_B(x), \theta_B(f(x)))$ with Q from (11.30). Averaging over the choice of f and using joint convexity of Q gives $P_X(x)\Pr[X' = x] \geq Q(\varphi_B(x), \langle\theta_B(f(x))\rangle_f)$. Since the choice of f is random, we have

$$\begin{aligned}\langle\theta_B(f(x))\rangle_f &= \Big\langle \sum_{x'} \delta_{f(x),f(x')}\varphi_B(x')\Big\rangle_f = \varphi_B(x) + \Big\langle \sum_{x'\neq x} \delta_{f(x),f(x')}\varphi_B(x')\Big\rangle_f \\ &= \varphi_B(x) + \frac{1}{k}(\rho_B - \varphi_B(x)) = \frac{k-1}{k}\varphi_B(x) + \frac{1}{k}\rho_B\,.\end{aligned} \tag{16.9}$$

The value of $f(x)$ is completely random, as is the value of $f(x')$. Therefore the chance that they are equal is simply the inverse of the size of the output alphabet. Averaging over $X = x$ gives the following bound on the overall probability of success:

$$\begin{aligned}\Pr[X' = X] &\geq k\sum_{x\in\mathcal{X}} Q(\varphi_B(x), (k-1)\varphi_B(x) + \rho_B) \\ &= kQ(\rho_{XB}, (k-1)\rho_{XB} + \mathbb{1}_X \otimes \rho_B)\,.\end{aligned} \tag{16.10}$$

Now we proceed as in the proof of the achievability of noisy channel coding. Let Λ_{XB} be the optimal POVM element in $\beta_\alpha(\rho_{XB}, \mathbb{1}_X \otimes \rho_B)$ for any $\alpha \in [0,1]$. Monotonicity of Q under the binary POVM $\{\Lambda_{XB}, \mathbb{1}_{XB} - \Lambda_{XB}\}$ gives $\langle\Pr[X' = X]\rangle_F \geq Q(P,R)$ for $P = (\alpha, 1-\alpha)$ and $R = ((k-1)\alpha + \beta_\alpha, (k-1)(1-\alpha) + |X| - \beta_\alpha)$. Since $Q(P,R) \geq p_1^2/r_1$, we obtain

$$\langle\Pr[X' = X]\rangle_f \geq k\alpha^2((k-1)\alpha + \beta_\alpha)^{-1}\,. \tag{16.11}$$

The bound still holds when replacing the $k-1$ factor in the inverse with k. Then we want to choose k such that $\alpha(1+\frac{1}{k}\frac{1}{\alpha}\beta_\alpha) \geq 1-\varepsilon$, which translates to

$$k \geq \beta_\alpha \frac{1-\varepsilon}{\alpha(\alpha-(1-\varepsilon))}. \tag{16.12}$$

Hence α must be no smaller than $1-\varepsilon$ for the denominator to be positive, so set $\alpha = 1-\varepsilon+\eta$ for $\eta \in [0,\varepsilon]$. Choosing

$$k = \left\lceil \frac{1-\varepsilon}{\eta(1-\varepsilon+\eta)} \beta_{1-\varepsilon+\eta}(\rho_{XB}, \mathbb{1}_X \otimes \rho_B) \right\rceil \tag{16.13}$$

therefore ensures that $\langle \Pr[X' = X] \rangle_f \geq 1-\varepsilon$. Using $\lceil x \rceil \leq x+1$ for $x \geq 0$, the ceiling function can be removed in the bound. We are still free to optimize η, and again we loosen the bound slightly for simplicity. □

Observe that (16.8) reduces to (14.2) for trivial B, that is, when the $\varphi_B(x)$ are all identical. However, the protocols in this case are somewhat different. Here the compressed output is generated by a function f, and because the decoder uses the pretty good measurement, upon learning the value of $Y = y$, it samples an x from the distribution $P_{X|Y=y}$. Nevertheless, for randomly chosen functions, the size of Y ensures that there is essentially only one possible high-probability x compatible with any given y. In the end, though two protocols are designed differently, they function very similarly.

16.3.2 Universal hashing

As in channel coding, it is not necessary to resort to completely random functions for the compressor. Instead, in (16.9), it is only necessary that $\langle \delta_{f(x),f(x')} \rangle_F = \frac{1}{k}$ for all $x' \neq x$. Since we consider a uniformly random choice, this is equivalent to saying that the number of functions for which $f(x) = f(x')$ for any $x' \neq x$ is n_f/k, where n_f is the number of functions. A collection of functions for which the collision probability is at most $1/k$ is called a *universal family of hash functions*.

A particularly convenient choice for our purposes in later chapters will be to take $\mathcal{X}$ and $\mathcal{Y}$ to be the linear spaces $\mathbb{Z}_2^n$ and $\mathbb{Z}_2^m$, respectively, for which the set of all *surjective linear functions* forms a universal family. (We could also choose $\mathbb{Z}_p^n$ for prime p.) A direct way to see this is to consider the kernels associated with the functions. Each one will be of dimension precisely $\dim(\mathcal{X}) - \dim(\mathcal{Y}) = n-m$ by the rank-nullity theorem. By symmetry the probability of $f(x) = 0$ for a random f and $x \neq 0$ is equivalent to the probability of the same event for fixed f and random $x \neq 0$. This is just the probability for x to be in the fixed kernel of f, which is just $\frac{2^{n-m}-1}{2^n-1} = 2^{-m}\frac{1-2^{m-n}}{1-2^{-n}} \leq 2^{-m}$. The -1 reflects the fact that $x \neq 0$. For the above achievability argument, we need only make a small change to use this set of functions. In (16.9), k should be replaced by the precise value $2^m \frac{1-2^{-n}}{1-2^{m-n}}$, carried through to (16.11), and then lower bounded by 2^m.

Exercise 16.1. Show that the set of all linear functions $f : \mathbb{Z}_2^n \to \mathbb{Z}_2^m$ with $m \leq n$ forms a universal family of hash functions.

16.3.3 Syndrome decoding

Linear hash functions fit very well for reconciliation of X using classical side information X' related to X via additive noise. Consider the case of $X, X' \in \mathbb{Z}_2^n$ with $P_{XX'}(x, x') = \frac{1}{|X|} P_Y(x + x')$ for the i. i. d. Bernoulli distribution P_Y with parameter p, that is, X' is the output of BSC(p) for input X. To determine X from X', it is sufficient (and also necessary) to determine Y, meaning that we simply need a hash of Y as just described above. The hash of Y can be easily obtained using a linear hash function $\hat{f}$, since $\hat{f}(x) + \hat{f}(x') = \hat{f}(y)$.

This is the basis of *syndrome decoding*. The decoder computes $\hat{f}(x')$ and adds the message from the compressor to obtain the *syndrome* $\hat{f}(y)$. This information enables the decoder to diagnose the particular error y using the error model P_Y. Hence for this case, we have the achievability bound $\frac{1}{\eta}\beta_{1-\varepsilon+\eta}(P_Y, \mathbb{1}_Y)+1$ on the size of the hash. This is essentially the same as $\beta_{1-\varepsilon}(P_{XX'}, \mathbb{1}_X \otimes P_{X'})$ by monotonicity of $\beta_{1-\varepsilon}$ under the reversible operation $(x, x') \to (x, x + x')$.

16.4 Reconciliation of i. i. d. sources

The i. i. d. analysis is by now routine. For any CQ state ρ_{XB} and $\varepsilon \in (0, 1)$, we have

$$\lim_{n\to\infty} \log \frac{L_\varepsilon^\star(X^n|B^n)_{\rho^{\otimes n}}}{n} = H(X|B)_\rho . \tag{16.14}$$

Exercise 16.2. Prove the statement.

Exercise 16.3. Consider a tripartite CCQ state ρ_{XZB} in which Alice holds X, Charlie holds Z, and Bob holds B. Show that in the i. i. d. scenario, Alice and Charlie can separately compress X and Z relative to B so that Bob can reconstruct both X and Z by receiving information at rate $H(XY|B)_\rho$. *Hint: Apply the chain rule for the entropy expression and take inspiration from the result.*

16.5 Notes and further reading

The quote is from Emerson's 1850 essay "Montaigne; or, The Skeptic", reprinted in [93]. The name "information reconciliation" comes from the cryptographic setting of reconciling secret keys, which we will encounter in Chapter 20. In the context of classical information theory, it is considered compression with side information and was first implicitly studied by Slepian and Wolf [268] in their analysis of distributed compression (Exercise 16.3). The optimal compression rate for i. i. d. sources with quantum side information was found by Devetak and Winter [79]. The one-shot case was considered by Renes and Renner [240]. The converse bound here is from Tomamichel and Hayashi [281], and the achievability statement is a modification of an argument by Beigi and Gohari [16].

17 Entanglement distillation

> A good story cannot be devised; it has to be distilled.
>
> Raymond Chandler

Suppose two separated parties Alice and Bob would like to implement the teleportation protocol to transfer an arbitrary quantum system from one to the other, but they only share an imperfectly entangled state ρ_{AB}. A natural approach to construct a protocol capable of performing the task from the resources at hand is to first try to convert the state ρ_{AB} into something approximating a maximally entangled state, and then just execute the usual teleportation protocol. This simplified task is known as entanglement distillation, which we first encountered in Chapter 8. Since their original aim is to transmit quantum information, this approach makes sense only if Alice and Bob employ LOCC operations. In this chapter, we will study entanglement distillation using one-way communication for a particular class of states ρ_{AB}. The case of arbitrary states is notably more complicated and will be taken up in Chapter 19.

17.1 Setup and basic properties

Before specializing to the states of interest, let us recall the setup of any entanglement distillation procedure using local operations and one-way classical communication from Section 8.6. It consists of a map $\mathcal{E}_{QY|A}$ on Alice's side that outputs a quantum system Q and a classical value Y along with a map $\mathcal{D}_{Q'|YB}$ on Bob's side, which takes Y and B as inputs and outputs a quantum system Q'. Ideally, the output would be the maximally entangled state, as depicted in Figure 17.1. For simplicity, we again quantify the approximation by the squared fidelity

$$F(\Phi_{QQ'}, \mathcal{D}_{Q'|YB} \circ \mathcal{E}_{QY|A}[\rho_{AB}])^2 = \mathrm{Tr}[\Phi_{QQ'} \mathcal{D}_{Q'|YB} \circ \mathcal{E}_{QY|A}[\rho_{AB}]] \,. \tag{17.1}$$

A pair $(\mathcal{E}, \mathcal{D})$ is a (k, ε) entanglement distillation protocol when the squared fidelity is greater than $1 - \varepsilon$ and $|Q| = k$. The largest amount k of *distillable entanglement* from a given state ρ_{AB} for a specified ε is denoted by $E^{\star}_{\varepsilon}(A|B)_{\rho}$.

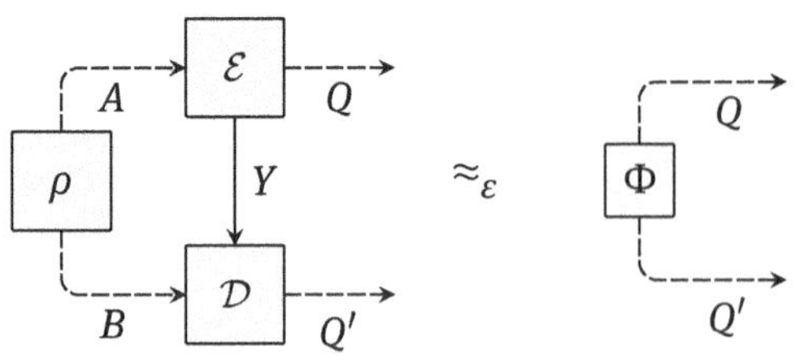

Figure 17.1: A (k, ε) protocol for entanglement distillation of ρ_{AB} by one-way classical communication from A to B with $k = |Y|$.

https://doi.org/10.1515/9783110570250-017

17.2 Converse

A simple converse bound for general entanglement distillation protocols is given by the following:

Proposition 17.1 (PPT converse on entanglement distillation). *For any bipartite state ρ_{AB}, every (k,ε) entanglement distillation protocol satisfies*

$$\max_{\sigma\in PPT} \beta_{1-\varepsilon}(\rho_{AB},\sigma_{AB}) \le \frac{1}{k}. \tag{17.2}$$

Proof. Suppose $\mathcal{E}$ and $\mathcal{D}$ constitute a (k,ε) protocol for ρ_{AB}, and consider $\beta_{1-\varepsilon}(\rho_{AB},\sigma_{AB})$ for some PPT state σ_{AB}. The operator $\Lambda_{AB} = \mathcal{E}^* \circ \mathcal{D}^*[\Phi_{QQ'}]$ is feasible for the optimization, since the output is nearly entangled. Hence

$$\beta_{1-\varepsilon}(\rho_{AB},\sigma_{AB}) \le \mathrm{Tr}[\Lambda_{AB}\sigma_{AB}] = \mathrm{Tr}[\Phi_{QQ'}\mathcal{D}\circ\mathcal{E}[\sigma_{AB}]]. \tag{17.3}$$

The state $\theta_{QQ'} = \mathcal{D}_{Q'|YB}\circ\mathcal{E}_{QY|A}[\sigma_{AB}]$ is PPT if σ_{AB} is. To see this, use the Choi representation to compute the partial transpose of M':

$$\begin{aligned}\mathcal{T}_{Q'}[\theta_{QQ'}] &= \mathrm{Tr}[D_{Q'YB}^{T_{Q'}} E_{QYA}\sigma_{AB}^{T}] \\ &= \mathrm{Tr}[D_{Q'YB}^{T_{Q'B}} E_{QYA}\sigma_{AB}^{T_A}] = \mathrm{Tr}[D_{Q'YB}^{T} E_{QYA}\hat{\sigma}_{AB}^{T}],\end{aligned} \tag{17.4}$$

where $\hat{\sigma}_{AB} = \sigma_{AB}^{T_B}$. In the last equality, we also use the fact that the transpose does not affect the subsystem Y of $D_{Q'YB}$ or E_{QYA}, since it is classical. Since $\hat{\sigma}_{AB}$ is positive and $D_{Q'YB}^{T}$ represents a completely positive map, the Choi representation theorem implies $\mathcal{T}_{Q'}[\theta_{QQ'}] \ge 0$. Then it follows by Proposition 8.1 that $\beta_{1-\varepsilon}(\rho_{AB},\sigma_{AB}) \le 1/k$. □

Although we did not precisely define distillation protocols involving two-way communication, it is easy to see that the bound applies to them as well. Feasibility is clearly not affected by having more general LOCC operations. Moreover, all LOCC operations preserve the PPT property of the input state. This follows since in the Choi operator argument above the map $\mathcal{E}$ can be replaced by all operations but the very last, and the argument goes through as before.

17.3 Achievability for a special case

Now we turn to the particular case that ρ_{AB} is such that $P_{\text{guess}}(Z_A|B)_\rho = 1$ for Z_A (either of) the amplitude observables considered in Chapter 13. From (13.10), the uncertainty relation relevant for Version 1 of the uncertainty game, it follows that to create entanglement, it is enough to create a state such that Bob can predict the conjugate observables X_A and Z_A on Alice's system. Hence the special case here is that the job is half done already.

This case includes distillation from pure bipartite states, which is known as *entanglement concentration* in the literature. Using the Schmidt basis to define Z_A, it is immediately apparent that $P_{\text{guess}}(Z_A|B)_\rho = 1$. This case also includes Bell-diagonal states plagued by phase noise, e. g., $\rho_{AB} = \sum_x P(x) Z_B^x \Phi_{AB} Z_B^{-x}$ for Z_B from (4.19).

17.3.1 Linear hashing

In principle, we only need to perform information reconciliation on the X_A information relative to side information B to ensure a high guessing probability for both amplitude and phase. This is another example of a reduction. However, there is an immediate hurdle to overcome if we follow this strategy: the backaction of the measurement Alice performs. Surely, she cannot just measure X_A and then compute the compressed output, for it would leave no quantum systems on her end to be entangled with Bob's systems. Put differently, measuring X_A would completely destroy Bob's Z_A guessing probability, since Alice's system would now be an X_A eigenstate. What is needed is a means to generate the output needed for X_A information reconciliation without too badly damaging Bob's Z_A information.

One solution is to base the compressor in information reconciliation on a linear function. As remarked at the end of Section 16.3, random surjective linear functions are sufficient for the achievability statement. Let us first see how this allows Alice to perform her part of the reconciliation protocol and leave some part of her system to be entangled with Bob.

Suppose that Alice's system A consists of n qubits (the case of n d-dimensional systems is similar for prime d). We can choose the n-qubit amplitude and phase observables from Chapter 13. Then the surjective linear hash function $\breve{g} : \mathbb{Z}_2^n \to \mathbb{Z}_2^m$ can be represented by an $m \times n$ matrix $\breve{G}$ with entries in $\mathbb{Z}_2$ such that $\breve{g}(x) = \breve{G}x$, where in the latter we interpret $x \in \mathbb{Z}_2^n$ as a column vector. We choose this notation to be consistent with the discussion of quantum error correcting codes to come in Section 19.2. The rows of $\breve{G}$ must be linearly independent since $\breve{g}$ is surjective; otherwise, row reduction will yield one or more zero rows, implying that the image of $\breve{G}$ is not the entirety of $\mathbb{Z}_2^m$. Therefore we can find $n-m$ additional linearly independent row vectors, yielding an $n \times n$ matrix G. As all of its rows are linearly independent, G is invertible. The last $n-m$ rows of G define a function $\bar{g} : \mathbb{Z}_2^n \to \mathbb{Z}_2^{n-m}$, and the function $x \mapsto \breve{g}(x) \oplus \bar{g}(x)$ is invertible. Here $\oplus$ denotes the concatenation of the vectors, i. e., an element of the direct sum of the underlying vector spaces.

More abstractly, the kernel of $\breve{g}$ defines a subspace W of $V = \mathbb{Z}_2^n$, and its outputs label the cosets V/W. We have assumed that $\breve{g}$ has rank m, so the kernel is of dimension $n-m$ by the rank-nullity theorem. The function $\bar{g}$ maps $x \in V$ to the kernel of $\breve{g}$. Any surjective map suffices to make $x \mapsto \breve{g}(x) \oplus \bar{g}(x)$ invertible, since V can be decomposed into the direct sum of W and the subspace of its cosets V/W. Note that the matrix representation of $\breve{g}$ consists of rows that span $W^\perp$, the orthogonal complement of the kernel of $\breve{g}$, while that of $\bar{g}$ consists of rows spanning $(V/W)^\perp$, the orthogonal complement of the subspace of cosets V/W.

Now consider the unitary operator $U_A = \sum_{x\in\mathbb{Z}_2^n} |\widetilde{Gx}\rangle\langle\tilde{x}|_A$. The first m rows of G correspond to $\breve{g}$, so applying U and measuring the first m qubits in the $|\tilde{x}\rangle$ basis produces $\breve{g}(x)$ for input state $|\tilde{x}\rangle$, just would measuring each qubit in the $|\tilde{x}\rangle$ basis and computing $\breve{g}(x)$ directly. Therefore, given the output of the measurement of the first m qubits, Bob can implement the decoder of the information reconciliation protocol for X_A relative to his side information B.

The crux of the approach here is that U also acts as a linear function in the amplitude basis as well. Applied to an amplitude basis input $|z\rangle$, U gives

$$\begin{aligned} U|z\rangle &= \sum_{x\in\mathbb{Z}_2^n} |\widetilde{Gx}\rangle\langle\tilde{x}|z\rangle = \frac{1}{\sqrt{2^n}}\sum_{x\in\mathbb{Z}_2^n}(-1)^{x\cdot z}|\widetilde{Gx}\rangle \\ &= \frac{1}{2^n}\sum_{x,z'\in\mathbb{Z}_2^n}(-1)^{x\cdot z}(-1)^{Gx\cdot z'}|z'\rangle \\ &= \frac{1}{2^n}\sum_{x,z'\in\mathbb{Z}_2^n}(-1)^{x\cdot(z+G^Tz')}|z'\rangle = |(G^T)^{-1}z\rangle\,. \end{aligned} \tag{17.5}$$

The function implemented by U in the amplitude basis is simply the inverse of the transpose of G. Measuring the last $n-m$ qubits of the output in the amplitude basis defines a linear function $\bar{f}:\mathbb{Z}_2^n\to\mathbb{Z}_2^{n-m}$. *This information is not affected by the measurement of the first m qubits in the phase basis.* The matrix representation of the function consists of rows that span V, the kernel of $\breve{g}$, and so $\bar{f}$ maps to $(V/W)^\perp$. In the context of a given $\breve{g}$, we call $\bar{f}$ the *dual function* (not to be confused with the dual optimization).

17.3.2 One-shot bound

Suppose then that $P_{\text{guess}}(Z_A|B)_\rho = 1$ and we have an (m,ε) information reconciliation protocol for X_A relative to B with a linear compression function $\breve{g}$. Then Alice's part of the entanglement distillation protocol for ρ_{AB} is as follows. She first implements $\breve{g}$ via U, measures the first m qubits in the phase basis, and transmits the result Y to Bob. Denote the remaining $n-m$ unmeasured qubits system by Q and the postmeasurement state by ρ'_{YQB}. Since $P_{\text{guess}}(Z_A|B)_\rho = 1$ and $Z_Q = \bar{f}(Z_A)$, it follows that $P_{\text{guess}}(Z_Q|B)_{\rho'} = 1$. Because the reconciliation protocol is ε-good, $P_{\text{guess}}(X_Q|BY)_{\rho'} = 1-\varepsilon$. Therefore by (13.26) Bob can use the decoding measurement from the reconciliation protocol to construct an entanglement recovery map $\mathcal{D}_{Q'|BY}$ such that $F(\Phi_{QQ'},\mathcal{D}_{Q'|BY}[\rho_{QBY}]) \geq 1-\varepsilon$.

Thus we have shown the following for $|A| = 2^n$, $n\in\mathbb{N}$. The case of $|A| = d^n$ for prime d is similar, and any A can be embedded into n qubits or n qudits without changing $P_{\text{guess}}(Z_A|B)$.

Proposition 17.2. *Given a state ρ_{AB} such that $P_{\text{guess}}(Z_A|B)_\rho = 1$, let $(\breve{g}, \Lambda_B)$ be the compression function and decompression measurement of a (k, ε) information reconciliation protocol for X_A relative to B, with a surjective linear function $\breve{g}$. Then there exists an $(|A|/k, \varepsilon(2-\varepsilon))$ entanglement distillation protocol $(\mathcal{E}_{QY|A}, \mathcal{D}_{Q'|BY})$ such that $\mathcal{E}$ can be constructed from the compression function $\breve{g}$, and $\mathcal{D}$ can be constructed from the reconciliation measurement Λ_B.*

The following one-shot bounds for the special case follow from Proposition 17.1 and by combining Proposition 17.2 with Proposition 16.2.

Proposition 17.3 (One-shot entanglement distillation bounds, special case). *For any bipartite state ρ_{AB} with $P_{\text{guess}}(Z_A|B)_\rho = 1$, we have the following bounds for all $\varepsilon \in (0,1)$:*

$$\beta_{1-\varepsilon}(\rho_{AB}, \mathcal{P}_A[\rho_{AB}]) \le \frac{1}{E_\varepsilon^\star(A|B)_\rho} \le \min_{\eta\in[0,\varepsilon]} \frac{4}{\varepsilon}\beta_{1-\frac{\varepsilon}{4}}(\rho_{AB}, \mathcal{P}_A[\rho_{AB}]) + 1. \tag{17.6}$$

Proof. The lower bound here is simply the choice $\sigma_{AB} = \mathcal{P}_A[\rho_{AB}]$ in (17.2). For the upper bound, observe that by (13.16) any ρ_{AB} for which $P_{\text{guess}}(Z_A|B)_\rho = 1$ can be transformed into $\psi'_{AA'B}$ from (13.15) by action solely on B. On B Bob coherently implements the guessing measurement for Z_A, storing the result in A'. Starting from a purification $|\psi\rangle_{ABR}$ of ρ_{AB}, this produces $|\psi'\rangle_{AA'BR} = \sum_{z\in\mathbb{Z}_2^n} |z\rangle_A|z\rangle_{A'}|\varphi(z)\rangle_{BR}$ for some states $|\varphi(z)\rangle_{BR}$. The probability of $Z = z$ is encoded in the normalization of $|\varphi(z)\rangle$. Bob holds the systems $A'B$.

Now consider the achievability bound (16.8) for $\varepsilon/2$-good information reconciliation of X_A relative to $A'B$ to ensure ε-good entanglement distillation. Due to the form of $\psi'_{AA'B}$ evident in (13.20), i. e., $|\psi'\rangle = \frac{1}{\sqrt{2^n}}\sum_{x\in\mathbb{Z}_2^n} |\tilde{x}\rangle_A Z_A^x|\psi\rangle_{ABR}$, the state $\bar{\psi}_{XA'B}$ after measuring X_A is simply

$$\bar{\psi}_{XA'B} = \frac{1}{d}\sum_x |x\rangle\langle x|_X \otimes Z_{A'}^x \psi_{A'B} Z_{A'}^x . \tag{17.7}$$

Recall from the end of Section 13.3.3 that $Z_A = (\sigma_z)_{A_1} \otimes \cdots \otimes (\sigma_z)_{A_n}$. From (5.4), which also holds for the n-qubit Z_A operator, it follows that $\bar{\psi}_{A'B} = \mathcal{P}_{A'}[\psi_{A'B}]$. Therefore, applying the unitary $U_{XA'} = \sum_x |\tilde{x}\rangle\langle\tilde{x}|_A \otimes Z_{A'}^x$ to $\bar{\psi}_{XA'B}$ and $\mathbb{1}_X \otimes \bar{\psi}_{A'B}$ results in $\pi_X \otimes \psi_{A'B}$ and $\mathbb{1}_X \otimes \mathcal{P}_{A'}[\psi_{A'B}]$, respectively. Thus we have

$$\begin{aligned}\frac{1}{|X|}\beta_\alpha(\bar{\psi}'_{XA'B}, \mathbb{1}_X \otimes \psi'_{A'B}) &= \frac{1}{|X|}\beta_\alpha(\pi_X \otimes \psi_{A'B}, \mathbb{1}_X \otimes \mathcal{P}_{A'}[\psi_{A'B}]) \\ &= \beta_\alpha(\pi_X \otimes \psi_{A'B}, \pi_X \otimes \mathcal{P}_{A'}[\psi_{A'B}]) \\ &= \beta_\alpha(\psi_{A'B}, \mathcal{P}_{A'}[\psi_{A'B}]) .\end{aligned} \tag{17.8}$$

The second equality is just $c\beta_\alpha(\rho,\sigma) = \beta_\alpha(\rho, c\sigma)$, while the third is $\beta_\alpha(\rho\otimes\tau, \sigma\otimes\tau) = \beta_\alpha(\rho,\sigma)$ from Exercise 9.16. Choosing $\eta = \varepsilon/4$ gives the upper bound. □

17.3.3 Distillation from i. i. d. states

Once again, we can show matching upper and lower bounds on the optimal rate of entanglement distillation from the special class of state ρ_{AB} by appealing to Stein's lemma.

Proposition 17.4 (Optimal i. i. d. entanglement distillation rate, special case). *For all ρ_{AB} such that $P_{\text{guess}}(Z_A|B)_\rho = 1$ and every $\varepsilon \in (0,1)$,*

$$\lim_{n\to\infty} \frac{1}{n} \log E_\varepsilon^\star(A^n|B^n)_{\rho^{\otimes n}} = D(\rho_{AB}, \mathcal{P}_A[\rho_{AB}]) = -H(A|B)_\rho . \tag{17.9}$$

In view of Proposition 17.3, the only thing to show is the equality $D(\rho_{AB}, \mathcal{P}_A[\rho_{AB}]) = -H(A|B)_\rho$. This follows immediately from the chain rule of relative entropy in (12.23) since $H(Z_A|B)_\rho = 0$.

Exercise 17.1. Determine the optimal distillation rates for the states $\rho_{AB} = |\psi\rangle\langle\psi|_{AB}$ with $|\psi\rangle = \sqrt{1-p}|00\rangle + \sqrt{p}|11\rangle$ and $\sigma_{AB} = (1-p)\Phi_{AB} + pZ_B\Phi_{AB}Z_B$.

Exercise 17.2. What is the optimal distillation rate for a general pure state $|\psi\rangle_{AB}$?

17.4 Notes and further reading

The quotation from Chandler can be found in [55]. Entanglement concentration was introduced by Bennett et al. [25], who also determined the optimal rate in the asymptotic limit. One-shot concentration was studied by Datta and Leditzky [69]. General entanglement distillation was introduced by Bennett et al. [31], and the one-shot setting was investigated by Buscemi and Datta [49] as well as by Brandão and Datta [45]. The converse here is adapted from Fang et al. [96], which follows techniques of Rains [233].

18 Randomness extraction

> Anyone who considers arithmetical methods of producing random digits is, of course, in a state of sin.
>
> John von Neumann

Until now we have been interested in protocols which aim to create correlation or entanglement. However, extracting entropy or randomness is also useful, as randomness is utilized in many applications. Although the most important thing here is its usefulness in cryptography, which we will investigate in Chapter 20, randomness also plays a role in certain algorithms, e. g., in primality testing and simulation of physical systems via Monte Carlo methods.

As alluded to by the quotation above, the most we can expect from a deterministic procedure is to extract the randomness present in the input to the procedure. Randomness cannot be created by deterministic means. A simple example of a *randomness extractor* comes from von Neumann himself. Consider a sequence of bits with i. i. d. distribution such that each is biased toward 0 with probability p. The probability that two particular bits take the values $(0, 1)$ is the same as the probability that they take the values $(1, 0)$. Therefore, with probability $2p(1-p)$, one uniformly random bit can be generated from each pair of inputs. For a long sequence of bits, the total rate at which random bits are produced is thus $p(1 - p)$.

In this chapter, we will see that the ultimate limit is again given by the entropy of the source, which is $h_2(p)$ in this example. With an eye toward cryptographic application in Chapter 20, we consider the case of randomness extraction from sources correlated to information held by an adversary or eavesdropper. This task is also known as privacy amplification. In this case, randomness can be extracted at a rate given by the conditional entropy, even when the adversary holds quantum side information.

18.1 Setup and basic properties

Given a CQ state ρ_{ZE} with classical Z, the goal of randomness extraction is to create a state ρ_{YE} very close to $\pi_Y \otimes \rho_E$ by applying a suitable extractor function $f : Z \to Y$. This is depicted in Figure 18.1. The quality of the approximation is measured by the distinguishability $\delta(\rho_{YE}, \pi_Y \otimes \rho_E)$, and f is a (k, ε) extractor for ρ_{ZE} when $|Y| = k$ and $\delta(\rho_{YE}, \pi_Y \otimes \rho_E) \leq \varepsilon$. Let $K_\varepsilon^\star(Z|E)_\rho$ be the largest k such that a (k, ε) extractor exists for ρ_{ZE}. We require f to be a deterministic function. Otherwise, the problem is trivial: Just forget (trace out) Z and output a random Y; this is precisely the ideal output.

https://doi.org/10.1515/9783110570250-018

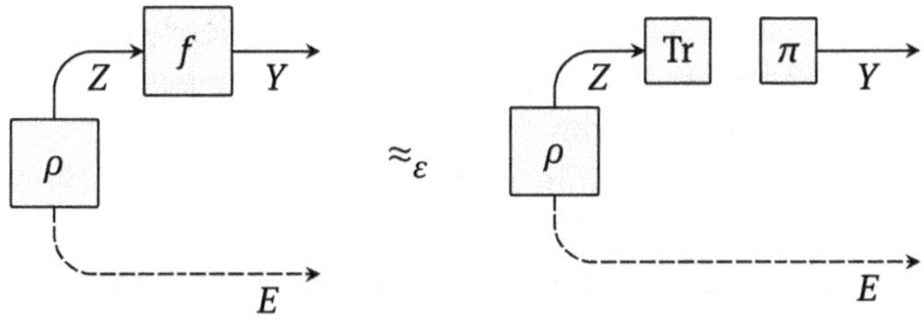

Figure 18.1: A (k, ε) protocol for randomness extraction from Z relative to side information E with $k = |Y|$.

18.2 Converse: from extraction to distillation

By making use of the quantum eraser from Section 6.6.4 any protocol for randomness extraction can be transformed into a protocol for entanglement distillation from a suitable state. This gives a reduction from the latter to the former. Thus the converse bound for distillation applies to randomness extraction, and there is no need to prove a separate converse specifically for extraction.

The particular setup is as follows. Suppose we are given a (k, ε) extractor f for Z relative to E in the CQ state ρ_{ZE}. To make it easier to connect with the discussion of entanglement distillation in Section 17.3, regard Z as a quantum system A such that ρ_{AE} is a CQ state, write $\rho_{AE} = \sum_z |z\rangle\langle z|_A \otimes \varphi_E(z)$ for some unnormalized states $\varphi_E(z)$, and let $|\psi\rangle_{AA'BE} = \sum_z |z\rangle_A |z\rangle_{A'} |\varphi(z)\rangle_{BE}$ be a purification of ρ_{AE}. Then we can show the following:

Proposition 18.1. *For every CQ state ρ_{AE}, any (k, ε) extractor for Z_A relative to E can be transformed into a (k, ε) entanglement distillation protocol using one-way communication from A to $A'B$ in the state $\psi_{AA'B}$.*

Proof. To $|\psi\rangle$ apply a unitary implementation of the map $z \mapsto (f(z), z)$, where $f(z)$ is stored in system Q, held by Alice. This produces the state

$$|\psi'\rangle_{QAA'BE} = \sum_z |f(z)\rangle_Q |z\rangle_A |z\rangle_{A'} |\varphi(z)\rangle_{BE}, \tag{18.1}$$

for which $\delta(\psi'_{QE}, \pi_Q \otimes \psi_E) \leq \varepsilon$. Therefore $F(\psi'_{QE}, \pi_Q \otimes \psi_E) \geq 1 - \varepsilon$. Clearly, $|\psi'\rangle$ is a purification of ψ'_{QE}, whereas $|\Phi\rangle_{QQ'}|\psi\rangle_{ABE}$ purifies $\pi_Q \otimes \psi_E$. Then by the properties of fidelity there exists an isometry V mapping $AA'B$ to $QAA'B$ such that

$$\langle\Phi|_{QQ'}\langle\psi|_{AA'BE} V_{Q'AA'B|AA'B}|\psi'\rangle_{QAA'BE} \geq 1 - \varepsilon. \tag{18.2}$$

Using this isometry, Alice and Bob can create a high-fidelity entangled state.

Thus knowing that the Q system is maximally mixed allows us to infer the existence of an operation that creates the desired state $|\Phi\rangle_{QQ'}$. This trick of inferring that entangled states can be created by showing that the marginal is completely mixed is quite widespread in quantum information theory. It goes under the name *decoupling* since Q is decoupled (independent) from everything else.

However, this does not yet yield an LOCC protocol, because V might require joint operations on $AA'B$. To show that there is an LOCC version of V, we return to $|\psi'\rangle_{QAA'BE}$ and apply the quantum eraser discussed in Section 6.6.4 to remove A. Then the same decoupling argument will apply, with the resulting isometry acting only on $A'B$.

Suppose Alice measures system A in the conjugate $|\tilde{x}\rangle$ basis from (4.20). Then the conditional state given outcome x is

$$_A\langle\tilde{x}|\psi'\rangle_{QAA'BE} = \frac{1}{\sqrt{d}}\sum_{xz}|f(z)\rangle_M Z_{A'}^{-x}|z\rangle_{A'}|\varphi(z)\rangle_{BE}\,. \tag{18.3}$$

As in Section 13.3.3, here we overload notation and let $Z_A^k = \sum_z \omega^{kz}|z\rangle\langle z|$ for $\omega = e^{2\pi i/d}$ and $d = |A|$. Alice can then inform Bob of the outcome x, at which point he can apply $Z_{A'}^x$ to remove the x dependence from the state. Thus, for every outcome x, the quantum state after these operations is

$$|\theta\rangle_{QA'BE} = \sum_z |f(z)\rangle_Q|z\rangle_{A'}|\varphi(z)\rangle_{BE}\,, \tag{18.4}$$

where Q is held by Alice. Applying the decoupling argument to $|\theta\rangle$, we can infer the existence of a isometric operation on Bob's systems $A'B$, which produce Q' such that the QQ' joint system has high fidelity with $|\Phi\rangle_{QQ'}$. □

18.3 Achievability: from reconciliation to extraction

To show achievability, we show how randomness extraction can be reduced to information reconciliation. The setup is the same as using information reconciliation for entanglement distillation; indeed, it is essentially the same argument. Here, however, we need only resort to (13.12), the bound relevant for Version 2 of the uncertainty game.

Proposition 18.2. *Consider an arbitrary CQ state ρ_{AE} with classical A and $|A| = 2^n$ for some integer n. Without loss of generality, we can take A to be diagonal in the Z_A basis. For an arbitrary purification ψ_{ABE} of ρ_{AE}, let $(\breve{g}, \Lambda_B)$ be a $(2^m, \varepsilon)$ information reconciliation protocol for X_A relative to B with a surjective linear function $\breve{g} : \mathbb{Z}_2^n \to \mathbb{Z}_2^m$. Then every surjective function $\bar{f} : \mathbb{Z}_2^n \to (\mathbb{Z}_2^n/\ker\breve{g})^\perp$ is an $(|A|/k, \sqrt{\varepsilon(2-\varepsilon)})$ randomness extractor for Z_A relative to E in ρ_{AE}.*

Proof. Recall the unitary implementation U of the reversible extension of $\breve{g}$ from (17.5). Proceed as there, first applying U_A and then measuring the first m qubits; denote the output by Y and the remaining qubits by M. By assumption, $P_{\text{guess}}(X_M|BY)_\psi \geq 1-\varepsilon$. Therefore, by (13.12) and (10.20), $\delta(\rho_{Z_M E}, \pi_{Z_M}\otimes\rho_E) \leq \sqrt{\varepsilon(2-\varepsilon)}$. Since $Z_M = \bar{f}(Z_A)$ for $\bar{f}$ as described after (17.5), the proof is complete. □

Note that we could adopt the entire result of Proposition 17.2, rather than parts of its proof, and arrive at the same conclusion. Since Alice and Bob are able to create a high-fidelity entangled state in M and M', along with the fact that the protocol does not involve E, the state ρ_{ME} must be similarly close to $\pi_M \otimes \rho_E$ in fidelity. As the construction in Proposition 17.2 makes use of linear hashing, Alice's operations can be interpreted as applying a randomness extractor to Z_A. Observe that the approximation parameter resulting from this argument is precisely the same, $\sqrt{\varepsilon(2-\varepsilon)}$.

Exercise 18.1. Consider the three following CQ states ρ_{ZB} in which Z is random and B is related to X via (a) BSC(p), (b) BEC(q), and (c) PSC(f). Show that extraction in these cases is related to information reconciliation of uniformly random X relative to R with R related to X via (a) PSC($1-2p$), (b) BEC($1-q$), and (c) BSC($\frac{1}{2}(1-f)$).

By the above reduction we can conclude that a particular extractor exists from the existence of an information reconciliation protocol. However, we do not have a means to find it. The same problem plagues our coding results, but in those contexts the problem is perhaps less acute. There we may imagine testing a coding scheme to determine if it is good, but in the cryptographic setting, this is not possible. We cannot ask the eavesdropper to confirm that the key is secret.

Therefore in the cryptographic setting, we usually settle for a *seeded extractor*, in which the particular extractor function is chosen by a *seed* random variable. The reduction above implies that the dual functions of a suitable set of (surjective linear) universal hash functions form a seeded extractor. Averaged over the choice of hash function, the information reconciliation protocol is ε-good, and therefore the randomness extractor is $\sqrt{2\varepsilon}$-good.

A little more is also usually demanded of the extractor, namely that the output key is not just independent of eavesdropper's information E, but also the seed S; that is, the output ω_{KSE} of an ε-good seeded extractor should obey $\delta(\omega_{KSE}, \pi_K \otimes \pi_S \otimes \omega_E) \leq \varepsilon$. This is called a *strong extractor*. Because the seed is classical and uniformly random in ω_{KSE}, the condition is equivalent to $\frac{1}{|S|}\sum_{s\in\mathcal{S}} \delta(\omega_{KE|S=s}, \pi_K \otimes \omega_E) \leq \varepsilon$. This is nothing other than the average of the distinguishability under the choice of seed. The reduction above therefore implies that duals of universal hash functions form strong extractors.

18.4 Extraction from i. i. d. sources

Using the one-shot achievability and converse results, along with Stein's lemma, we can now show that the optimal rate of randomness extraction is indeed given by the conditional entropy of the source.

Proposition 18.3. *For any CQ state ρ_{ZE} with classical Z and any $\varepsilon \in (0,1)$,*

$$\lim_{n\to\infty} \frac{1}{n} \log K_\varepsilon^\star \left(Z^n|E^n\right)_\rho = H(Z|E)_\rho . \tag{18.5}$$

Proof. By Proposition 18.1, (k,ε) randomness extraction is subject to the PPT converse bound (17.2) for A relative to $A'B$ in the state ψ. Then we can follow the converse bound in (17.9) to obtain

$$\frac{1}{n} \log K_\varepsilon^\star (Z^n|E^n)_\rho \leq -H(A|A'B)_\psi . \tag{18.6}$$

Because $|\psi\rangle_{AA'BE}$ is a pure state, $-H(A|A'B)_\psi = H(A|E)_\psi$, which is equal to $H(Z_A|E)_\rho$.

For the achievability bound, by Proposition 18.2 we can appeal to the achievability bound for information reconciliation of X_A relative to side information $A'B$. By the calculations in (17.8) and (17.9) this leads back to $H(Z_A|E)_\rho$. □

Exercise 18.2. Referring back to the discussion of types in Section 12.3.2, construct a seedless randomness extractor for a given i. i. d. source (with no side information).

18.5 Notes and further reading

The quote is from [295], as is the description of the extractor. Randomness extraction relative to classical side information held by an eavesdropper was introduced by Bennett et al. [30]. It is closely related to the wiretap communication model of classical information theory, introduced by Wyner [310]. Optimal rates in the i. i. d. setting were established by Csiszár and Körner [66]. One-shot bounds were given by Impagliazzo, Levin, and Luby [151] and Bennett et al. [28]. For eavesdroppers holding quantum information, optimal rates for the i. i. d. case were given by Devetak and Winter [80] (who implicitly uses the quantum eraser) and a one-shot analysis by Renner and König [242] and Renner [241]. The approach taken here, of relating randomness extraction to the task of information reconciliation via quantum effects such as the uncertainty relation, follows the approach pioneered by Lo and Chau in their analysis of QKD security [193], the full security proof by Shor and Preskill [264], and follow up works by Koashi and Preskill [168], Koashi [166, 167], Hayashi [125], and Renes and Boileau [238], among others. The idea of decoupling goes back to Schumacher and Westmoreland [255]. The sufficiency of hash functions whose dual functions form a universal set was emphasized by Tsurumaru and Hayashi [284].

19 Quantum error correction

> It has often been said that classical error correction is based on making multiple copies and then doing a measurement and majority voting. And both of those things sounded like something that you can't do with quantum information.
>
> Charles H. Bennett

In this chapter, we consider the use of quantum error correction both for transmission of quantum information over noisy quantum channels and for distillation of maximal entanglement from arbitrary bipartite quantum states. Continuing our theme of protocol reductions, we will use the latter to achieve the former.

19.1 Quantum communication: setup and basic properties

19.1.1 Definitions

The goal of quantum communication is of course to simulate the identity channel, as in Figure 19.1. A $(k, \varepsilon)_{\mathrm{wc}}$ protocol for a channel $\mathcal{N}_{B|A}$ consists of an encoding map $\mathcal{E}_{A|Q}$ and a decoding map $\mathcal{D}_{Q|B}$ such that $\delta(\mathcal{I}_Q, \mathcal{D}_{Q|B} \circ \mathcal{N}_{B|A} \circ \mathcal{E}_{A|Q}) \leq \varepsilon$.

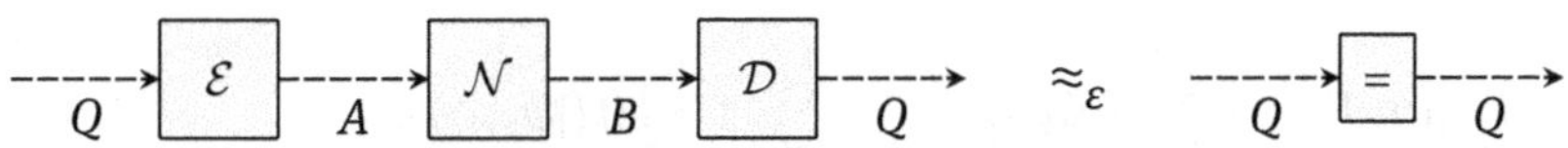

Figure 19.1: A $(k, \varepsilon)_{\mathrm{wc}}$ protocol for quantum communication over the noisy channel $\mathcal{N}_{B|A}$ with $k = |Q|$.

As with classical communication, it is simpler to consider the average case instead of the worst case, where now "average case" means that we attempt to transmit the state $|\Phi\rangle_{QQ'}$. Moreover, it will also prove simpler to work with the fidelity, and therefore we define a (k, ε) protocol for $\mathcal{N}_{B|A}$ under average error to be an encoder and decoder such that

$$F(\Phi_{QQ'}, \mathcal{D}_{M|B} \circ \mathcal{N}_{B|A} \circ \mathcal{E}_{A|M}[\Phi_{QQ'}]) \geq 1 - \varepsilon\,. \tag{19.1}$$

Notice that by (10.19) a $(k, \varepsilon)_{\mathrm{wc}}$ protocol is automatically a (k, ε) protocol. The fidelity requirement can be expressed as $F_{\mathrm{ent}}(\pi_Q, \mathcal{F}_Q) \geq 1 - \varepsilon$ for $\mathcal{F}_Q = \mathcal{D}_{Q|B} \circ \mathcal{N}_{B|A} \circ \mathcal{E}_{A|Q}$ or using the "agreement probability" $P_{\mathrm{agree}}(\mathcal{F}_Q) = F_{\mathrm{ent}}(\pi_Q, \mathcal{F}_Q)^2$ by $P_{\mathrm{agree}}(\mathcal{F}_Q) \geq (1 - \varepsilon)^2$. We denote by $M^\star_\varepsilon(\mathcal{N}_{B|A})$ the largest k for which there exists a (k, ε) quantum coding protocol for $\mathcal{N}_{B|A}$.

https://doi.org/10.1515/9783110570250-019

19.1.2 Reduction of worst case to average case

Analyzing average-case protocols is no real loss of generality. As with classical communication, we can convert an average-case code into a worst-case code by throwing away the worst half of the input. In this way, a (k,ε) protocol can be converted into a $(\frac{k}{2}, \sqrt{8\varepsilon})_{\mathrm{wc}}$ protocol.

Proof. For convenience, define the function $f : |\psi\rangle \to F(|\psi\rangle\langle\psi|, \mathcal{F}[|\psi\rangle\langle\psi|])$ for a given channel $\mathcal{F}$, e. g., $\mathcal{F}_Q$ above. Starting from $\Pi_0 = \mathbb{1}$, iteratively define the projectors $\Pi_j = \Pi_{j-1} - |\psi_j\rangle\langle\psi_j|$ for the minimizer $|\psi_j\rangle$ of f in the support of Π_{j-1}. Thus $|\psi_1\rangle$ is the state with the lowest fidelity, $F_{\mathrm{pure}}(\mathcal{F}) = F(\mathcal{F}[|\psi_1\rangle\langle\psi_1|], |\psi_1\rangle\langle\psi_1|)$, $|\psi_2\rangle$ is the state with the lowest fidelity in the space orthogonal to $|\psi_1\rangle$, and so forth. By construction, the set $\{|\psi_j\rangle\}_{j=1}^k$ is an orthonormal set. Therefore by convexity of the entanglement fidelity (Exercise 10.16) it follows that

$$F_{\mathrm{ent}}(\pi, \mathcal{F}) \le \frac{1}{k}\sum_{j=1}^{k} F_{\mathrm{ent}}(|\psi_j\rangle\langle\psi_j|, \mathcal{F}) = \frac{1}{k}\sum_{j=1}^{k} f(|\psi_j\rangle) . \tag{19.2}$$

Since the sequence of $f(|\psi_j\rangle)$ is nondecreasing by construction, we can further bound the entanglement fidelity by splitting the sum at the middle (supposing k is divisible by 2):

$$F_{\mathrm{ent}}(\pi, \mathcal{F}) \le \frac{1}{k}\sum_{j=1}^{k/2} f(|\psi_{k/2}\rangle) + \frac{1}{k}\sum_{j=k/2+1}^{k} 1 = \tfrac{1}{2}(f(|\psi_{k/2}\rangle) + 1) . \tag{19.3}$$

Prefixing $\mathcal{F}$ with a channel $\mathcal{E}'$ mapping a $\frac{k}{2}$-dimensional space onto support of $\Pi_{k/2}$, the above inequality is just $F_{\mathrm{pure}}(\mathcal{F}\circ\mathcal{E}') \ge 2F_{\mathrm{ent}}(\pi, \mathcal{F}) - 1$. In combination with (10.36), we therefore obtain a lower bound on the channel fidelity of $\mathcal{F}\circ\mathcal{E}'$ in terms of the entanglement fidelity of $\mathcal{F}$ for a maximally entangled input, namely $F(\mathcal{F}\circ\mathcal{E}', \mathcal{I})^2 \ge 2(2F_{\mathrm{ent}}(\pi, \mathcal{F}) - 1)^2 - 1$. Using (10.20) and making some simplifications, it then follows that a (k,ε) protocol can be converted into a $(\frac{k}{2}, \sqrt{8\varepsilon})_{\mathrm{wc}}$ protocol as claimed. □

19.1.3 Isometric encoding suffices

One advantage of focusing on the average case is that for any encoding map $\mathcal{E}_{A|Q}$ of a (k,ε) protocol, it is possible to construct an isometric encoder that performs at least as well as $\mathcal{E}_{A|Q}$, that is, an encoder taking the form $\mathcal{E}_{A|Q} : \rho_Q \mapsto V_{A|Q}\rho_Q V_{A|Q}^*$ for some isometry $V_{A|Q}$.

To see this, suppose $W_{AC|Q}$ is a Stinespring dilation (isometric extension) of the encoding map, which involves the additional output system C. Defining $\sqrt{P(j)}|\phi(j)\rangle_{AQ'} = \langle j|_C W_{AC|Q}|\Phi\rangle_{QQ'}$ for some orthonormal basis $\{|j\rangle_C\}$ of C, with $P(j) \ge 0$ chosen so that $|\phi(j)\rangle$ is normalized, observe that $\sum_j P(j) = \langle\Phi|W^*W|\Phi\rangle = 1$. Furthermore, the marginal state $\phi_{Q'}(j)$ must be maximally mixed, since W does not act on Q'. Hence by the properties of purifications there exists an isometry $V_{A|Q}(j)$ such that $|\phi(j)\rangle_{AQ'} = V_{A|Q}(j)|\Phi\rangle_{QQ'}$. Thus the encoding map applied to $\Phi_{QQ'}$ can be expressed as a convex combination of isometric channels $\mathcal{V}_{A|B}(j)$: $\mathcal{E}_{A|Q}[\Phi_{QQ'}] = \sum_j P(j)\, \mathcal{V}_{A|Q}(j)[\Phi_{QQ'}]$. Now the square of the entanglement fidelity $F_{\text{ent}}(\pi_Q, \mathcal{F}_Q)$ is manifestly linear in the channel: From (10.35) we have $F_{\text{ent}}(\pi_Q, \mathcal{F}_Q)^2 = \operatorname{Tr}[\Phi_{QQ'}\, \mathcal{F}_Q[\Phi_{QQ'}]]$. Therefore at least one of the $\mathcal{V}_{A|Q}(j)$ must lead to a squared entanglement fidelity at least as large as $(1-\varepsilon)^2$.

Exercise 19.1. Modify this argument to show that classical communication from sender to receiver cannot yield better protocols for quantum communication; that is, take the encoder to be a quantum instrument and the decoder to depend on the classical output of the encoder.

19.1.4 Reduction of noisy channel coding to entanglement distillation

Our construction of protocols for quantum communication proceeds via entanglement distillation. Specifically, suppose we have a (k,ε) entanglement distillation protocol for a state ω_{AB} that is the result of transmitting part of $\rho_{AA'}$ through a quantum channel $\mathcal{N}_{B|A}$: $\omega_{AB} = \mathcal{N}_{B|A'}[\rho_{AA'}]$. Then from the compressor $\mathcal{E}_{QY|A}$ and decompressor $\mathcal{D}_{Q'|YB}$ we can construct an encoder/decoder pair of a $(k, 2\varepsilon)$ channel coding protocol for $\mathcal{N}_{B|A}$.

The proof is similar to the case of isometric encoding above using the simple form of the squared entanglement fidelity. In particular, the squared fidelity is an average over the classical information transmitted from A to B. The compressor produces a CQ state $\sigma_{YAQ'} = \mathcal{E}_{YA|Q}[\rho_{AA'}]$, which of course has the form $\sigma_{YAQ'} = \sum_{y\in\mathcal{Y}} P_Y(y)|y\rangle\langle y|_Y \otimes \sigma_{AQ'}(y)$ for $P_Y(y)\sigma_{AQ'}(y) = \mathcal{E}_{Y=y,A|Q}[\rho_{AA'}]$. By normalization, $P_Y \in \operatorname{Prob}(\mathcal{X})$. Since the distillation protocol is ε-good, the squared entanglement fidelity is an average over y:

$$\sum_{y\in\mathcal{Y}} P_Y(y)\ \operatorname{Tr}[\Phi_{QQ'}\, \mathcal{D}_{Q'|Y=yB} \circ \mathcal{N}_{B|A}[\sigma_{AQ'}(y)]] \geq 1-\varepsilon\,. \tag{19.4}$$

Thus there is $y^\star$ for which $\sigma_{AQ'}(y^\star)$ is also ε-good in this sense. Let $\psi_{RAQ'}$ be a purification of $\sigma_{AQ'}(y^\star)$. By monotonicity of the fidelity the marginal Q' satisfies $F(\psi_{Q'}, \pi_{Q'})^2 \geq 1-\varepsilon$. Hence there exists an isometry $V_{RA|Q}$ such that $|\psi\rangle_{RAQ'}$ is ε-close to $V_{RA|Q}|\Phi\rangle_{QQ'}$ in squared fidelity, and therefore $\operatorname{Tr}_R[V_{RA|Q}\Phi_{QQ'}V^*_{RA|Q}]$ is ε-close to $\psi_{AQ'} = \sigma_{AQ'}(y^\star)$.

Define the encoding map of the quantum communication protocol to be $\hat{\mathcal{E}}_{A|Q} = \operatorname{Tr}_R \circ \mathcal{V}_{RA|Q}$ for the channel $\mathcal{V}$ with single Kraus operator V, and the decoding map to be $\hat{\mathcal{D}}_{Q|B} = \mathcal{D}_{Q|BY=y^\star}$. For convenience, let $\mathcal{F}_Q = \hat{\mathcal{D}}_{Q|B} \circ \mathcal{N}_{B|A} \circ \hat{\mathcal{E}}_{A|Q}$ and $\varphi_{QQ'} = \mathcal{D}_{Q|BY=y^\star} \circ \mathcal{N}_{B|A}[\sigma_{AQ'(y^\star)}]$. Then we have $F(\mathcal{F}_Q[\Phi_{QQ'}], \varphi_{QQ'})^2 \geq 1-\varepsilon$ and $F(\varphi_{QQ'}, \Phi_{QQ'}])^2 \geq 1-\varepsilon$. Hence, by the triangle inequality of fidelity in (10.23), $F(\Phi_{QQ'}, \mathcal{F}_Q[\Phi_{QQ'}]) \geq 1-2\varepsilon$.

19.2 CSS codes

Restricting attention to isometric encoding motivates the notion of a quantum error-correcting code, which is simply a subspace of the channel input. A codeword of the code is simply an element of this subspace. A good code for a given channel is one for which any input in the code subspace can be approximately reconstructed from the output of the channel. In this chapter, we will be especially interested in *CSS codes*, named for their inventors Calderbank,[1] Shor, and Steane.[2]

Let us describe a generic CSS code by the form of its encoding isometry in the case of an n-qubit code. To better align with our initial focus on entanglement distillation here, we first describe the *inverse* of the encoding operation. Consider an invertible function $f : \mathbb{Z}_2^n \to \mathbb{Z}_2^n$ and the associated unitary implementation $U_A = \sum_{z\in\mathbb{Z}_2^n} |f(z)\rangle\langle z|_A$ on the n-qubit space $\mathcal{H}_A$. In Section 17.3.1, we defined a similar U, and the only difference now is that U is defined by the action of an invertible function in the amplitude basis. To appreciate the action of U in the conjugate basis, it is useful to work with the matrix representation of f by F, acting to the right on column vectors. Then from (17.5) we already have $U|\tilde{x}\rangle = |\widetilde{Gx}\rangle$ for $G = (F^T)^{-1}$.

Subdivide $\mathbb{Z}_2^n$ into $\mathbb{Z}_2^m \oplus \mathbb{Z}_2^k \oplus \mathbb{Z}_2^\ell$ for some $m, k, \ell \in \mathbb{N}$ such that $m + k + \ell = n$ (e. g., taking the first m elements, the next k, and the final ℓ), and write $f(z) = \check{z} \oplus \bar{z} \oplus \hat{z}$. This defines the functions $\check{f} : \mathbb{Z}_2^n \to \mathbb{Z}_2^m, \bar{f} : \mathbb{Z}_2^n \to \mathbb{Z}_2^k$, and $\hat{f} : \mathbb{Z}_2^n \to \mathbb{Z}_2^\ell$ by $\check{z} = \check{f}(z), \bar{z} = \bar{f}(z)$, and $\hat{z} = \hat{f}(z)$. Decomposing F into a block matrix of three appropriately sized rows $\check{F}$, $\bar{F}$, and $\hat{F}$, we also have $\check{z} = \check{F}z, \bar{z} = \bar{F}z$, and $\hat{z} = \hat{F}z$. Similarly, G can be decomposed into a block matrix of three rows $\check{G}$, $\bar{G}$, and $\hat{G}$, which defines the functions $\check{g}$, $\bar{g}$, and $\hat{g}$ from F via $G = (F^T)^{-1}$. The states $|f(z)\rangle$ form a basis of the state space $\mathcal{H}_A$ for z ranging over a basis of $\mathbb{Z}_2^n$, and this decomposition induces a decomposition of $\mathcal{H}_A$ into $\mathcal{H}_{\check{A}} \otimes \mathcal{H}_{\bar{A}} \otimes \mathcal{H}_{\hat{A}}$. The unitary U may be regarded as a map from the former to the latter. Note that the earlier construction of Section 17.3.1 was simply the case that $\ell = 0$, i. e., of no system $\hat{A}$.

The CSS encoding isometry $V_{A|\bar{A}}$ associated with f and the decomposition of n into k, ℓ, and m is then defined by the action

$$V_{A|\bar{A}}|\psi\rangle_{\bar{A}} = U^*_{\check{A}\bar{A}\hat{A}|A}(|+\rangle_{\check{A}}^{\otimes m} \otimes |\psi\rangle_{\bar{A}} \otimes |0\rangle_{\hat{A}}^{\otimes \ell}). \tag{19.5}$$

Put differently, the CSS code is the subspace of vectors in $\mathcal{H}_A$ such that application of U and measurement of the system $\hat{A}$ (the last ℓ qubits) in the standard basis results in $|0\rangle^{\otimes \ell}$ with certainty, whereas measurement of $\check{A}$ (the first m qubits) results in all $|+\rangle^{\otimes m}$ with certainty.

In general, the outputs of such a measurement procedure are $\hat{f}(z)$ and $\check{g}(x)$, which are known as the *syndromes* of the CSS code. We will be interested in the syndromes as

1 Robert Calderbank, born 1954.

2 Andrew Martin Steane, born 1965.

hash functions for the purposes of entanglement distillation, as foreshadowed in the special case of Section 17.3.1. Related to the syndromes are the *stabilizers* of the code, which we mention in passing. From (19.5), the codewords are evidently eigenstates of the projectors $U^*|+\rangle\langle+|_{\check{A}_i}U$ and $U^*|0\rangle\langle 0|_{\hat{A}_j}U$, where $\check{A}_i$ is the ith qubit of $\check{A}$ and similarly for $\hat{A}_j$, with $i \in \{1,\dots,m\}$ and $j \in \{1,\dots,\ell\}$. Equivalently, the codewords are +1 eigenstates of the stabilizers $U^*(\sigma_x)_{\check{A}_i}U$ and $U^*(\sigma_z)_{\hat{A}_j}U$. The syndromes are the result of measuring the stabilizer operators (associating the outcome 0 with eigenvalue +1 and 1 with −1). In fact, the former stabilizers are simply $X^{\check{G}_{i1}} \otimes X^{\check{G}_{i2}} \otimes \cdots \otimes X^{\check{G}_{in}}$, where $\check{G}_{ij}$ is the (i,j) component of $\check{G}$. Similarly, the latter stabilizers are tensor products of Z operators, one for each row of $\hat{F}$. This holds because measurement of $U^*(\sigma_x)_{\check{A}_i}U$ can be accomplished by applying U and then measuring the ith qubit in the phase basis, which just gives the ith bit of $\check{g}(x)$. But this bit is just the parity of a particular collection of the bits of x, indexed by the position of 1s in the ith row of $\check{G}$. This parity value is precisely the value of the given tensor product of X operators. The same argument holds for the Z-type stabilizers.

Exercise 19.2. Confirm that Z_1Z_2 and Z_2Z_3 are stabilizers of the three-bit quantum repetition code (a CSS code), the subspace spanned by $|000\rangle$ and $|111\rangle$.

The syndromes $\hat{f}(z)$ and $\check{g}(x)$ are each associated with classical linear error-correcting codes, which we denote C_Z and C_X, respectively. Specifically, the corresponding $\mathbb{Z}_2$-valued matrices $\hat{F}$ and $\check{G}$ are the *parity-check matrices*, which annihilate the respective codewords; that is, $C_Z = \{z \in \mathbb{Z}_2^n : \hat{F}z = 0\}$, and similarly for C_X. Observe that $FG^T \doteq FF^{-1} = \mathbb{1}$, which implies $\hat{F}\binom{\bar{G}}{\check{G}}^T = 0$ and $\check{G}\binom{\hat{F}}{\bar{F}}^T = 0$. Therefore a basis for the codewords of C_Z is given by the rows of $\bar{G}$ and $\check{G}$, while a basis for C_X is given by the rows of $\hat{F}$ and $\bar{F}$. Arranged as rows of a matrix, a basis for a linear code is referred to as a *generator matrix*.

Now we are in a position to connect our discussion of CSS codes with the more standard presentation. Usually, a CSS code is defined from a pair of classical error-correcting codes C_1 and C_2, where $C_2 \subseteq C_1$. Then the code is defined by its stabilizers, which are of X or Z type, as we have here. In particular, the Z-type stabilizers are given by the parity check matrix of C_1, while the X-type stabilizer $\check{G}$, corresponding to the X-type stabilizers, is given by the generator matrix of C_2. This is nothing other than the above construction with the parity check matrix $\hat{F}$ of C_1 and the generator matrix $\check{G}$ of C_2.

Exercise 19.3. Confirm the above construction for the Steane code, given by the code C_1 with parity check matrix

$$H = \begin{pmatrix} 1 & 0 & 1 & 0 & 1 & 0 & 1 \\ 0 & 1 & 1 & 0 & 0 & 1 & 1 \\ 0 & 0 & 0 & 1 & 1 & 1 & 1 \end{pmatrix} \tag{19.6}$$

and the subcode C_2 of C_1 codewords with even Hamming weight.

19.3 Quantum coding theorems

Given the difficulties of relating k and ε for classical communication over quantum channels, we may suspect that similar problems arise for quantum communication. This is indeed the case. Again, we will settle for analyzing the asymptotic limit, even though here again we will not be able to derive a single-letter result except in certain cases.

19.3.1 Statement

Each (k, ε) code for $\mathcal{N}_{B|A}^{\otimes n}$ gives an achievable rate $R = \frac{\log k}{n}$. Suppose $M^\star(\mathcal{N}, \varepsilon, n)$ is the largest k such that a (k, ε) code exists for $\mathcal{N}_{B|A}^{\otimes n}$. Just as in the classical case, define

$$C_Q(\mathcal{N}_{B|A}, \varepsilon) := \lim_{n\to\infty} \frac{1}{n} \log M^\star(\mathcal{N}, \varepsilon, n) \quad \text{and} \quad C_Q(\mathcal{N}_{B|A}) := \lim_{\varepsilon\to 0} C(\mathcal{N}_{B|A}, \varepsilon). \tag{19.7}$$

The latter is the quantum capacity, the highest possible rate of arbitrarily reliable communication.

To formulate the quantum noisy channel coding theorem, we first define the *coherent information*

$$Q(\mathcal{N}_{B|A}) := \max_{\rho_{AA'}} -H(A|B)_{\mathcal{N}_{B|A'}[\rho_{AA'}]} . \tag{19.8}$$

Here the maximization is over all states $\rho_{AA'}$, but concavity of the conditional entropy (12.58) implies that the optimal $\rho_{AA'}$ will be a pure state. The quantum noisy channel coding theorem then states that the capacity of any $\mathcal{N}_{B|A}$ is given by the following expression:

$$C_Q(\mathcal{N}_{B|A}) = \lim_{n\to\infty} \frac{1}{n} Q(\mathcal{N}_{B|A}^{\otimes n}) . \tag{19.9}$$

Below we will show that for arbitrary channels, the quantum capacity is achievable by CSS codes.

As with the classical capacity, the expression for the quantum capacity is the regularization of a single-letter quantity. It is also known that regularization is necessary, and we will shortly investigate this point. On the other hand, it follows immediately from concavity of the conditional entropy that the capacity of entanglement-breaking channels is zero, as expected.

Exercise 19.4. Compute the coherent information of a Pauli channel and for the erasure channel.

Owing to the relation between the two protocols, the optimal rate for one-way entanglement distillation takes a form similar to the quantum capacity. Take $E^\star(\rho_{AB}, \varepsilon, n)$

to be the largest k such that a (k,ε) one-way entanglement distillation protocol exists for the state $\rho_{AB}^{\otimes n}$. The optimal distillation rate of arbitrarily good entanglement from ρ_{AB} is then defined as $D_{\rightarrow}(\rho_{AB}) = \lim_{\varepsilon\to 0}\lim_{n\to\infty}\frac{\log E^*(\rho_{AB},\varepsilon,n)}{n}$. The distillation coding theorem states that the optimal rate, the *one-way distillable entanglement*, is given by the regularized expression

$$D_{\rightarrow}(\rho_{AB}) = \lim_{n\to\infty}\frac{1}{n}D_{\rightarrow}^{(1)}(\rho_{AB}^{\otimes n}) \quad \text{for} \tag{19.10}$$

$$D_{\rightarrow}^{(1)}(\rho_{AB}) = \max_{\mathcal{Q}_{XA'|A}} -H(A'|BX)_{\mathcal{Q}_{XA'|A}[\rho_{AB}]}, \tag{19.11}$$

where the optimization is over all quantum instruments $\mathcal{Q}_{XA'|A}$ that output a classical system X and a quantum system A'. Essentially, it is the same regularized negative conditional entropy rate as in the coding theorem, but first an arbitrary quantum instrument can be applied to the state with the classical output communicated to Bob. Although forward communication does not increase the quantum capacity, it can increase the entanglement distillation rate, and we will encounter an example of this in Section 19.5.2.

19.3.2 Converse

The converses to these statements, the upper bounds in (19.9) and (19.10), both rely on the continuity of the conditional entropy. Consider a (k,ε) code for which the state $\omega_{QQ'}$ produced by the protocol satisfies $F(\Phi_{QQ'},\omega_{QQ'}) \geq 1-\varepsilon$. Clearly, $\log k = -H(Q|Q')_{\Phi_{QQ'}}$. Letting $\varepsilon' = \sqrt{2\varepsilon}$, from (10.20) it follows that $\delta(\Phi_{QQ'},\omega_{QQ'}) \leq \varepsilon'$. Now define $f(\varepsilon',k) = 2\varepsilon'\log|k| + (1+\varepsilon')\,h_2(\frac{\varepsilon'}{1+\varepsilon'})$ and then use (12.59), the continuity of conditional entropy, data processing, and isometric encoding to obtain

$$\begin{aligned}\log k - f(\varepsilon',k) &\leq -H(Q|Q')_{\omega_{QQ'}} \leq -H(Q|B)_{\mathcal{N}\circ\mathcal{E}[\Phi_{QQ'}]}\\ &= -H(A|B)_{\mathcal{N}\circ\mathcal{E}[\Phi_{QQ'}]} \leq Q(\mathcal{N}_{B|A}).\end{aligned} \tag{19.12}$$

Now suppose we have a sequence $\{(k_n,\varepsilon_n)\}_n$ of codes such that $\frac{1}{n}\log k_n \to R$ and $\varepsilon_n \to 0$. Applying the above bound, we have $\frac{1}{n}\log k_n \leq \frac{1}{n}Q(\mathcal{N}_{B|A}^{\otimes n})+\frac{1}{n}f(\varepsilon_n',k_n)$. Taking the limit as $n\to\infty$, we only need to show that the second term vanishes:

$$\lim_{n\to\infty}\frac{1}{n}\left(2\varepsilon_n'\log k_n + (1+\varepsilon_n')\,h_2\!\left(\frac{\varepsilon_n'}{1+\varepsilon_n'}\right)\right) = 0. \tag{19.13}$$

The second term in this expression vanishes because the binary entropy is bounded above by 1. Since $k_n \leq |A|^n$, the first term is bounded by $2\varepsilon_n'|A| = 2\sqrt{2\varepsilon_n}|A|$, which tends to zero as $n\to\infty$.

Exercise 19.5. Give a proof of the converse for classical communication over quantum channels using continuity of conditional entropy.

The argument for entanglement distillation is similar. Again, the protocol outputs a state $\omega_{QQ'} = \mathcal{D}_{Q'|XB} \circ \mathcal{E}_{XQ|A}[\rho_{AB}^{\otimes n}]$, which is close to $\Phi_{QQ'}$. By continuity, $\log|Q| = -H(Q|Q')_\Phi \leq -H(Q|Q')_\omega + f(\varepsilon, k)$, and again the second term will vanish in the limit. Meanwhile, by monotonicity under $\mathcal{D}_{Q'|XB}$, $-H(Q|Q')_\omega \leq -H(Q|XB)_{\mathcal{E}_{XQ|A}[\rho_{AB}^{\otimes n}]}$. Maximizing over the instrument $\mathcal{E}_{XQ|A}$ gives the result.

19.4 Achievability

By the reduction of quantum communication to entanglement distillation we need only construct a protocol for the latter. We do so by a further reduction of the entanglement distillation task to a combination of two information reconciliation protocols, one for amplitude information and one for phase information. The two compression functions of the reconciliation protocols are based on a CSS code family, and Alice's part of the protocol is to generate the two compressor outputs by appropriate unitary action and measurement. Given these measurement results, Bob could then accurately predict both amplitude and phase of Alice's remaining systems by suitable measurements on his system. Therefore, by the bipartite uncertainty relation (13.10), from these measurements he can construct a decoder to recover maximal entanglement from his system.

19.4.1 Entanglement distillation protocol

Let us now describe the protocol in more detail. First, fix a state ρ_{AB} shared by Alice and Bob. By isometrically embedding A into a larger space we can take $|A| = 2^n$, i. e., n qubits. In principle, any d^n with prime d would also work. Let $|\psi\rangle_{ABR}$ be a purification of ρ_{AB}, and arbitrarily pick an amplitude basis $\{|z\rangle_A\}_{z\in\mathbb{Z}_2^n}$. The purification can be written as $|\psi\rangle_{ABR} = \sum_{z\in\mathbb{Z}_2^n} |z\rangle_A|\varphi(z)\rangle_{BR}$, where the normalization of the state $|\varphi(z)\rangle_{BR} = {}_A\langle z|\psi\rangle_{ABR}$ encodes its prior probability $P_Z(z)$. By Z we denote the random variable corresponding to the outcome of measurement of A in the amplitude basis. Abusing notation somewhat, denote by ψ_{Z_AB} the CQ state $\sum_{z\in\mathbb{Z}_2^n} |z\rangle\langle z|_A \otimes \varphi_B(z)$. Further, define the state $|\psi'\rangle_{AA'BR} = \sum_{z\in\mathbb{Z}_2^n} |z\rangle_A|z\rangle_{A'}|\varphi(z)\rangle_{BR}$, a purification of ψ_{Z_AB}. By X we denote the random variable corresponding to the outcome of measurement of A in the phase basis defined by $|\tilde{x}\rangle = \frac{1}{\sqrt{2^n}} \sum_{z\in\mathbb{Z}_2^n}(-1)^{x\cdot z}|z\rangle$ for $x \in \mathbb{Z}_2^n$. The CQ state resulting from measurement of A in the conjugate basis is denoted by $\psi'_{X_AA'B}$. Now we can state the precise reduction from entanglement distillation to information reconciliation.

Proposition 19.1. *Suppose $(\hat{f}, \Lambda_B)$ is a $(2^{m_z}, \varepsilon_z)$ information reconciliation protocol for amplitude Z_A relative to B in $\psi_{Z_A B}$ and $(\breve{g}, \Gamma_{A'B})$ is a $(2^{m_x}, \varepsilon_x)$ information reconciliation protocol for phase X_A relative to $A'B$ in $\psi'_{X_A A'B}$ such that $\hat{f} : \mathbb{Z}_2^n \to \mathbb{Z}_2^{m_z}$ and $\breve{g} : \mathbb{Z}_2^n \to \mathbb{Z}_2^{m_x}$ are linear functions satisfying $(\ker \hat{f})^\perp \subseteq \ker \breve{g}$. Then, provided that $n - m_x - m_z > 0$, there exists a $(2^{n-m_x-m_z}, \sqrt{\varepsilon_x} + \sqrt{\varepsilon_z})$ CSS-based entanglement distillation scheme for ρ_{AB} whose encoder is constructed from $\hat{f}$ and $\breve{g}$ and whose decoder is constructed from Λ_B and $\Gamma_{A'B}$.*

Proof. The outputs of the functions $\hat{f}$ and $\breve{g}$ can be generated by applying a unitary U and measuring the appropriate output qubits in the amplitude or phase bases, as described in Section 19.2. Performing these operations and transmitting the measurement results to Bob via a noiseless classical channel is Alice's portion of the entanglement distillation protocol. She will be left with $2^{n-m_z-m_x}$ qubits.

The decoder is constructed as in Section 13.3. In particular, consider the $\bar{A}$ subsystem generated by $U_{\bar{A}\hat{A}\breve{A}}$. Nominally, the decoding measurement results of the information reconciliation protocols are indexed by $x, z \in \mathbb{Z}_2^n$. Since $\bar{X}$ and $\bar{Z}$ are both (linear) functions of X and Z, respectively, the POVM elements can be combined to give POVMs indexed by $\bar{x}$ and $\bar{z}$. Because the compressors are ε-good, these POVMs will be ε-good at predicting $\bar{X}$ and $\bar{Z}$, respectively. Observe that the system A' of $\psi'_{X_A A'B}$ can also be decomposed into $\bar{A}'$, $\hat{A}'$, and $\breve{A}'$, so that the coherent copy of $\bar{Z}$ is present in $\bar{A}'$. Therefore we may apply the entanglement recovery map of (13.27) with systems A and A' therein replaced by systems $\bar{A}$ and $\bar{A}'$ here. The bound (13.26) then implies that the squared fidelity of the output with the maximally entangled state will satisfy $\arccos F \leq \arccos(1 - \varepsilon_z) + \arccos(1 - \varepsilon_x)$. It can be verified that this implies the stated error bound for the squared fidelity. □

19.4.2 Rate calculation

Next, we show the existence of a suitable pair of hash functions $\hat{f}$ and $\breve{g}$ such that the achievability construction of Proposition 16.2 applies to both, giving a relation between the output sizes m_z and m_x on the one hand and the errors ε_z and ε_x on the other. As usual, the argument is that a random choice will be good, and thus there must exist at least one good pair. Using the results of Section 19.2, we choose a random reversible function $f : \mathbb{Z}_2^n \to \mathbb{Z}_2^n$ and construct $\hat{f}$ and $\breve{g}$ from it.

Consider the expected probability $\langle \Pr[Z_A \neq Z'_A \vee X_A \neq X'_A] \rangle_{\hat{f}, \breve{g}}$ that, under the random choice of $\hat{f}$ and $\breve{g}$, one or both of the information reconciliation tasks fail. By the union bound and linearity of expectation,

$$\begin{aligned} \langle \Pr[Z_A \neq Z'_A \vee X_A \neq X'_A] \rangle_{\hat{f}, \breve{g}} &\leq \langle \Pr[Z_A \neq Z'_A] + \Pr[X_A \neq X'_A] \rangle_{\hat{f}, \breve{g}} \\ &= \langle \Pr[Z_A \neq Z'_A] \rangle_{\hat{f}, \breve{g}} + \langle \Pr[X_A \neq X'_A] \rangle_{\hat{f}, \breve{g}} \,. \end{aligned} \tag{19.14}$$

The choice of f delivers precisely the same set of possible functions $\hat{f}$ as just choosing a full-rank set of $\hat{f}$ directly, and the same for $\breve{g}$. Hence Proposition 16.2 applies to each term in (19.14) separately and gives an upper bound on m_z as a function of $\varepsilon_1 = \langle \Pr[Z_A \neq Z'_A] \rangle_{\hat{f}}$ as well as an upper bound on m_x as a function of $\varepsilon_2 = \langle \Pr[X_A \neq X'_A] \rangle_{\breve{g}}$. It follows that there exists a CSS pair $\hat{f}$ and $\breve{g}$ such that $\hat{f}$ yields a $(2^{m_z}, \varepsilon_1 + \varepsilon_2)$ reconciliation protocol for amplitude information and $\breve{g}$ yields a $(2^{m_x}, \varepsilon_1 + \varepsilon_2)$ reconciliation protocol for phase information. The resulting CSS-based entanglement distillation protocol will have an output whose squared fidelity with the maximally entangled state is at least $1 - 2\sqrt{\varepsilon_1 + \varepsilon_2}$ by Proposition 19.1. The output will be $n - m_x - m_z$ qubits in size, where $2^{m_z} \leq \frac{2}{\varepsilon_1} \beta_{1-\frac{\varepsilon_1}{2}}(\psi_{Z_A B}, \mathbb{1}_A \otimes \psi_B) + 1$ and $2^{m_x} \leq \frac{2}{\varepsilon_2} \beta_{1-\frac{\varepsilon_2}{2}}(\psi'_{X_A A' B}, \mathbb{1}_A \otimes \psi'_{A'B}) + 1$ by Proposition 16.2.

The rate expression simplifies in the asymptotic limit for an input state $\rho_{AB}^{\otimes n}$. In this case, we obtain $m_z = H(Z_A|B)_\psi$ and $m_x = H(X_A|A'B)_{\psi'}$. The former is just $m_z = n - D(\psi_{Z_A B}, \pi_A \otimes \psi_B)$, and as we have seen in Section 17.3.2, the latter is simply $n - D(\psi_{AB}, \psi_{Z_A B})$. Therefore, by the relative entropy chain rule (12.23), $n - m_z - m_x = -n + D(\psi_{AB}, \psi_{Z_A B}) + D(\psi_{Z_A B}, \pi_A \otimes \psi_B) = -n + D(\psi_{AB}, \pi_A \otimes \psi_B) = -H(A|B)_\rho$. Hence arbitrarily high quality entanglement distillation is possible at the rate given by the negative conditional entropy of the state.

Given any particular shared state $\rho_{AB}^{\otimes nm}$, Alice is free to perform an arbitrary instrument $\mathcal{Q}_{A'X|A^m}$ identically on blocks of m of her systems and broadcast the value of X to Bob. Then Alice and Bob can execute the entanglement distillation protocol on the n resulting blocks of the resulting state $\omega_{A'B^m X}$, achieving the rate $-\frac{1}{m} H(A'|B^m X)_\omega$ in the asymptotic limit of large n. Therefore the regularized distillation rate given in (19.10) is achievable.

By a similar argument it follows that the channel capacity can be achieved. For m uses of the channel, pick $\psi_{A^m A'^m}$ to be the state such that $\rho_{A^m B^m} = \mathcal{N}_{B|A'}^{\otimes m}[\psi_{A^m A'^m}]$ optimizes the coherent information $Q(\mathcal{N}_{B|A}^{\otimes n})$. Applying the distillation construction to $\rho_{A^m B^m}$ and the reduction from Section 19.1.4 implies the existence of a CSS code with high output fidelity and rate $\frac{1}{m} Q(\mathcal{N}_{B|A}^{\otimes m})$. Taking the large m limit gives the capacity formula (19.9).

19.5 Discussion of the achievability construction

19.5.1 Complementary information

We have established that transmission of the complementary classical amplitude and phase information is sufficient for quantum communication at capacity. We can think of quantum information as sort of being comprised of these two classical pieces, at least for the purposes of information transmission. This is especially clear in the case of a Pauli channel, which describes additive amplitude and phase *errors*. Take the simple case of independent amplitude and phase errors at identical rate p, i. e., the channel $\rho \mapsto (1-p)^2\rho + (1-p)p\,\sigma_x\rho\,\sigma_x + (1-p)p\,\sigma_z\rho\,\sigma_z + p^2\,\sigma_y\rho\,\sigma_y$. This is called the BB84 channel for reasons that will become apparent in the following chapter. The job of $\hat{f}$ in the distillation or coding procedure is to correct amplitude errors, caused by σ_x, while the job of $\check{g}$ is to correct phase errors, caused by σ_z. Correcting each separately will also correct their combination, caused by σ_y. Approaching the quantum communication task in terms of information instead of errors ensures that the above construction applies to arbitrary channels where there is no sensible notion of additive amplitude error or phase error.

It is also important to appreciate that the phase transmission task is formulated assuming that the amplitude transmission task was successful; that is, the pertinent state in the phase information reconciliation task is $\psi'_{X_A A' B}$, not $\psi_{X_A B}$ itself. The system A' is essentially a coherent copy of the amplitude information in A, and hence ψ' encodes the fact that the amplitude reconciliation was successful. The coding argument would indeed go through for phase reconciliation formulated for $\psi_{X_A B}$ by appealing to (13.10) instead of (13.26). However, the subsequent rate calculation would not deliver the coherent information, as having $H(X_A|A'B)$ instead of $H(X_A|B)$ in the expression is crucial in the general case.

Again, a simple illustration is given by Pauli channels, where A' accounts for correlations between amplitude and phase errors. For instance, take the channel with only σ_y errors at rate p. In the construction, we are free to choose the amplitude basis, so we could simply choose the eigenbasis of σ_y such that there are no phase errors; the coherent information is $1 - h_2(p)$. However, we can equally well choose the eigenbasis of σ_z as the amplitude basis, and we will arrive at the same rate expression. Though there are now amplitude and phase errors, they are perfectly correlated, which leads to $H(X_A|A'B) = 0$. Failing to account for these correlations would lower the rate to $1 - 2h_2(p)$.

19.5.2 Error degeneracy

Observe that in the construction, it is not necessary to use reconciliation protocols that enable Bob to reconstruct both X_A and Z_A in their entirety, as we have done. To apply (13.27) and (13.26), we need that $\bar{Z}$ and $\bar{X}$ could be determined by Bob. Recall that the recovery map in (13.27) is constructed by first applying the amplitude information recovery operation and then the phase recovery operation. Thus we could instead base the phase reconciliation on the output of the first stage, i. e., consider reconciliation of $\bar{X}\check{X}$ relative to side information $A'B$, instead of the entire X_A itself. Presumably, this would reduce the necessary m_x and increase the rate beyond the single-letter coherent information. However, at present we have no any way of determining the asymptotic limits for such a protocol.

The difference between focusing on X_A and Z_A versus $\bar{X}$ and $\bar{Z}$ is related to the issue of *error degeneracy*. Return to the case of a Pauli channel. The use of information reconciliation for X_A and Z_A in the construction above means that the precise error pattern of both bit and phase flips will be exactly determined (with high probability). Otherwise, one of the information reconciliation protocols fails. However, determining the error pattern precisely is clearly not necessary, since the construction only relies on predictability of $\bar{X}$ and $\bar{Z}$. The bit flip patterns fall into equivalence classes of errors that lead to the same value of $\bar{Z}$, the same for phase, and it is only necessary to identify the particular class.

A concrete example for the BB84 channel is given by concatenating a repetition code with the random construction above; that is, consider the above construction applied to the channel $\mathcal{N}'_{B_1B_2B_3|A_1} = (\bigotimes_{j=1}^3 \mathcal{N}_{B_j|A_j}) \circ \mathcal{E}_{A_1A_2A_3|A_1}$, where $\mathcal{E}_{A_1A_2A_3|A_1}$ is the encoding operator for the repetition code. The encoder is given by the isometry $W_{A_1A_2A_3}|0\rangle_{A_2}|0\rangle_{A_3}$, where $W = \text{CNOT}_{B_1B_2}\text{CNOT}_{B_1B_3}$, and always outputs states in the span of $|000\rangle$ and $|111\rangle$. Observe that the action of Z_1 in $\bigotimes_{j=1}^3 \mathcal{N}_{B_j|A_j}$ is the same as Z_2 or Z_3 on the repetition code subspace. These three phase errors are degenerate.

If we now follow the general code construction for $\mathcal{N}'$, then the effects of error degeneracy increase the rate beyond the coherent information. For the BB84 channel with error rate p, the coherent information is $1 - 2h_2(p)$, which has a threshold value of $p \approx 0.110028$, but the coherent information of $\mathcal{N}'$ at this value of p is positive, and it remains positive until at least $p \approx 0.111652$. This is a tiny improvement in the threshold, to be sure. Nonetheless, it is sufficient to demonstrate that the coherent information is not additive, since the state $W\pi W^*$ is a possible input state in the $Q(\mathcal{N}^{\otimes 3})$ optimization.

To perform the rate calculation, choose a maximally mixed input for the coherent information, i. e., a maximally entangled input state for the calculation of the negative conditional entropy. Let $\omega_{AB} = \mathcal{N}'_{B|A'}[\Phi_{AA'}]$ be the output state, where B is an abbreviation for $B_1B_2B_3$, while A and A' are single-qubit systems (A_1 and A'_1). It is simpler to compute $-H(A|B)$ for the state $\omega'_{AB} = W_B\omega_{AB}W_B^*$, and since W is unitary, $H(A|B)_\omega = H(A|B)_{\omega'}$. The action of the channel also has a simple form; using the notation $Z_B^v = Z_{B_1}^{v_1} \otimes Z_{B_2}^{v_2} \otimes Z_{B_3}^{v_3}$ as in Section 13.3.3 and $c(j) = (j,j,j) \in \mathbb{Z}_2^3$, we have

$$\omega_{AB} = \frac{1}{2}\sum_{j,k\in\mathbb{Z}_2} |j\rangle\langle k|_A \otimes \sum_{u,v\in\mathbb{Z}_2^3} P_{U,V}(u,v)\, X_B^u Z_B^v |c(j)\rangle\langle c(k)|_B Z_B^v X_B^u, \tag{19.15}$$

where $P_{U,V}$ is the joint distribution of amplitude and phase error patterns, described by the random variables $U, V \in \mathbb{Z}_2^3$, respectively. Due to the form of the channel, $P_{U,V}(u,v) = \prod_{j=1}^3 P(u_j)P(v_j)$, where P is the binary distribution with $P(1) = p$. The errors Z_B^v act degenerately, meaning that $Z_B^v|r(j)\rangle_B = Z_{B_1}^{v_1+v_2+v_3}|r(j)\rangle_B$. Let us abbreviate $v_1+v_2+v_3$ by $s(v)$. Then, because X and Z anticommute and Z_{B_1} commutes with W_B, for $\Pi(k) = |k\rangle\langle k|$, we have

$$\begin{aligned}\omega'_{AB} &= \frac{1}{2}\sum_{j,k\in\mathbb{Z}_2} |j\rangle\langle k|_A \otimes \sum_{u,v\in\mathbb{Z}_2^3} P_{U,V}(u,v)\, Z_{B_1}^{s(v)} W_B |c(j)+u\rangle\langle c(k)+u|_B W_B^* Z_{B_1}^{s(v)} \\ &= \sum_{u,v\in\mathbb{Z}_2^3} P_{U,V}(u,v)\, Z_{B_1}^{s(v)} X_{B_1}^{u_1} \Phi_{AB_1} X_{B_1}^{u_1} Z_{B_1}^{s(v)} \otimes \Pi(u_1+u_2)_{B_2} \otimes \Pi(u_1+u_3)_{B_3}\,.\end{aligned} \tag{19.16}$$

Hence the systems B_2 and B_3, which now store the two syndromes of the repetition code, are classical. It is easy to see that $-H(A|B)_\rho = 1-H(X)_P$ for ρ a Bell-diagonal state with eigenvalues given by the distribution P_X. Here we have the conditional version of the same, so the achievable rate is $1 - H(U_1, s(V)|U_1 + U_2, U_1 + U_3)_P$. Since U and V are independent in $P_{U,V}$, the rate expression simplifies to $1 - H(U_1|U_1 + U_2, U_1 + U_3)_P - H(V_1 + V_2 + V_3)_P$.

The entropies are now not difficult to determine. The former is an average of binary entropies over three cases, corresponding to the syndrome having no values equal to zero, exactly one, or two. The probabilities for these cases are just $q_0 = p^3 + (1-p)^3$, $q_1 = 2p(1-p)$, and $q_2 = p(1-p)$, respectively, and the respective probabilities for $U_1 = 1$ in each case are $r_0 = p^3/(1-3p(1-p))$, $r_1 = p$, and $r_2 = 1-p$. Meanwhile, the probability that $V_1 + V_2 + V_3 = 1$ is just $t = 3(1-p)^2 p + p^3 = p((1-p)^2 + 3p^2)$. Altogether, the rate expression is $1 - h_2(t) - \sum_{j=1} q_j h_2(r_j)$. Setting this equal to zero, we can numerically obtain the above estimate of the threshold value of p.

We can also appreciate the large role degeneracy plays in the overall coding scheme from the results of the calculation. In particular, there is far too little information about the phase errors sacrificed as syndrome information to determine the precise phase error pattern. From the information reconciliation converse, roughly $H(V)_P = h_2(p)$ bits of information per qubit are needed to determine the phase error pattern and $H(U)_P = h_2(p)$ bits per qubit for the amplitude information. In the concatenated coding scheme under consideration, there will be only roughly $\frac{1}{3}h_2(t)$ syndrome bits for phase error correction per qubit, all told, whereas the total number of amplitude syndromes is roughly $\frac{1}{3}(2 + \sum_{j=1} q_j h_2(r_j))$. The additional 2 in this expression comes from the two syndrome bits of the repetition code.

These quantities are shown as a function of p in Figure 19.2, which illustrates that the number of phase syndromes is far too small to be able to determine the exact phase error pattern. Unfortunately, the overhead of the repetition code wipes out essentially all of this advantage, as there are far more syndromes than needed to determine the exact amplitude error pattern. How to make use of the savings degeneracy offers without incurring this kind of overhead is still an open question.

It would be remiss not to mention an even more striking example of nonadditivity of coherent information than that shown above, namely that of *superactivation*. Unlike the case of classical communication, in which zero capacity only occurs when the channel output is the same for every possible input, there are a variety of

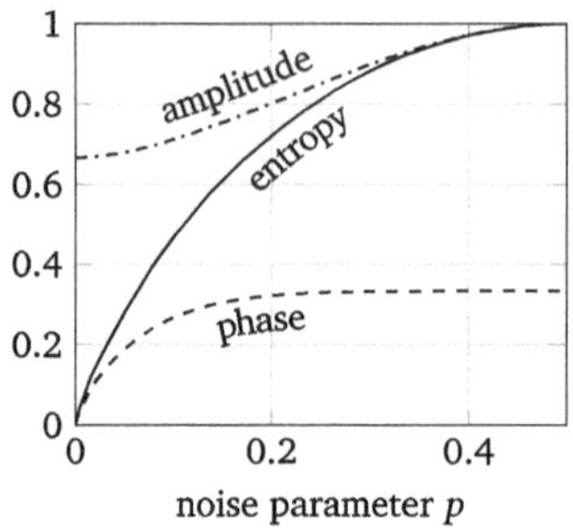

Figure 19.2: Fractional size of amplitude and phase syndromes versus the entropy.

kinds of quantum channels having zero capacity. Amazingly, it turns out that combining different kinds of zero-capacity channels in parallel can lead to a channel with nonzero capacity! There exist qubit-input quantum channels $\mathcal{N}_{B_1|A_1}$ and $\mathcal{N}_{B_2|A_2}$ such that $Q(\mathcal{N}_{B_i|A_i}) = 0$ for $i = 1, 2$ but $Q(\mathcal{N}_{B_1|A_1} \otimes \mathcal{N}_{B_2|A_2}) > 0$. Superactivation calls into question the usefulness of the notion of capacity in the first place, since the ability of a quantum channel to transmit information appears to depend on how it is used with other channels.

19.5.3 Channel degradability

There is at least one class of channels, *degradable channels*, for which degeneracy plays no role and the quantum capacity is equal to the coherent information. A quantum channel $\mathcal{N}_{B|A}$ is degradable if there exists a degrading channel $\mathcal{K}_{R|B}$ such that the complementary channel satisfies $\widehat{\mathcal{N}}_{R|A} = \mathcal{K}_{R|B} \circ \mathcal{N}_{B|A}$. Recall the that complementary channel is not unique, but the only freedom is an isometry on the output R, the inverse of which can then be included into $\mathcal{K}$. An example is the quantum erasure channel with erasure probability q, whose Stinespring dilation maps arbitrary $|\psi\rangle_A$ to $\sqrt{1-q}|\psi\rangle_B|?\rangle_R + \sqrt{q}|?\rangle_B|\psi\rangle_R$. Hence the complement is also an erasure channel but with erasure probability $1-q$. To degrade the former to the latter, just erase with probability $r = (1-2q)/(1-q)$, for which $(1-q)r + q = 1-q$.

Exercise 19.6. Show that dephasing and amplitude damping are degradable.

For degradable channels, the coherent information is additive, i. e., $Q(\mathcal{N}^{\otimes 2}) = 2Q(\mathcal{N})$. This is easily proven. The only thing to show is that $Q(\mathcal{N}^{\otimes 2})$ is upper bounded by $2Q(\mathcal{N})$. First, observe that the conditional entropy $-H(A|B)_\omega$ of the channel output ω_{ABR} for pure state input $\rho_{AA'}$ using the Stinespring dilation of the channel can be expressed as $-H(A|B)_\omega = H(B)_\omega - H(R)_\omega$. Therefore, for the optimal ρ in $Q(\mathcal{N}^{\otimes 2})$, we have $Q(\mathcal{N}^{\otimes 2}) = H(B_1B_2) - H(R_1R_2)$. By the chain rule this is $H(B_1) + H(B_2) - H(R_1) - H(R_2) - I(B_1 :$

$B_2) + I(R_1 : R_2)$. The latter mutual info is smaller than the former by the degradability assumption, so the net contribution of these two terms is negative. Dropping them gives $Q(\mathcal{N}^{\otimes 2}) \le H(B_1) - H(R_1) + H(B_2) - H(R_2) \le 2Q(\mathcal{N})$.

For degradable channels, the optimization in the coherent information is convex, that is, the function $\rho_A \mapsto -H(A|B)_{\mathcal{N}_{B|A}[\rho_{AR}]}$ is concave, and symmetries of the channel simplify finding the optimal input ρ_A. Suppose that $\rho_A = \sum_x P_X(x)\sigma_A(x)$ for states $\sigma_A(x)$ and an arbitrary distribution P_X. Purifying the $\sigma_A(x)$ to $|\varphi(x)\rangle_{AR}$ and calling the associated channel output $|\omega(x)\rangle_{BR}$, let $\theta_{XBR} = \sum_x P_X(x)|x\rangle\langle x|_X \otimes \omega_{BR}(x)$. Degradability implies that $I(X : B)_\theta \ge I(X : R)_\theta$. In terms of entropy and conditional entropy, this inequality is simply $H(B)_\theta - H(R)_\theta \ge H(B|X)_\theta - H(R|X)_\theta$, the concavity statement.

Exercise 19.7. Compute the quantum capacity of the amplitude damping channel.

19.6 Notes and further reading

The quote is from [24]. Long thought to be impossible in principle, quantum error correction was discovered by Shor [262] and Steane [273]. The connection between quantum coding and entanglement distillation was first discussed in [31]. CSS codes were first introduced in [52, 272]. A more standard presentation somewhat similar to ours can be found in Nielsen and Chuang [211] and the lecture notes of Preskill [230]. For an overview, see the edited volume of Lidar and Brun [188].

The formula for the quantum capacity was established in a long series of papers. The coherent information was introduced by Schumacher and Nielsen [253]. The upper bound in terms of the regularized coherent information was established by Barnum, Knill, Nielsen, and Schumacher [11, 12]. See also Allahverdyan and Saakian [5]. Lloyd [192] gave a heuristic argument that the coherent information should be achievable. Shor [263] delivered a lecture outlining a proof, and finally Devetak completed a proof in [77] based on the relation to private communication over quantum channels (which we do not consider here). Several different approximation metrics were considered by various authors, which complicated the initial investigations. An overview of several of them and their relations is given by Kretschmann and Werner's delightful paper [173]. Devetak and Winter [80] proved the optimal rate on entanglement distillation using similar methods.

Several different quantum coding achievability proofs were published in a special issue of Open Systems & Information Dynamics in 2008. Hayden et al. [130] and Klesse [163, 164] follow a decoupling approach, while Horodecki, Lloyd, and Winter [149] also take an approach based on privacy, and Hayden, Shor, and Winter [131] make more direct use of the uncertainty relation.

For the specific case of CSS codes, Hamada [118] showed that the capacity expression can be achieved for Pauli channels. The achievability proof here follows Renes and Boileau [238], correcting an error in the random coding argument. The role of degeneracy was recognized by Divincenzo, Shor, and Smolin [84, 265]. Superactivation was discovered by Smith and Yard [269]. The additivity of the coherent information for degradable channels was discovered by Devetak and Shor [78].

20 Quantum key distribution

> Be aware in all of this of the Heisenberg–Schrödinger Credulity Effect. That effect is that the word "quantum" sucks people's brains out, and otherwise sensible people suffer from impaired reasoning.
>
> Jon Callas

20.1 Private communication over public channels

20.1.1 Encryption

Finally, we turn to the cryptographic possibilities of quantum information processing, in particular, the task of quantum key distribution (QKD). QKD is a means to the ultimate end of private communication by our two parties Alice and Bob over a public communication channel. "Public" refers to the fact that Alice and Bob themselves do not control the channel and therefore cannot assume anything about how it actually operates. This includes the possibility that someone (Eve) is eavesdropping on their communication.

By itself this task is clearly impossible if Alice and Bob only communicate using classical information carriers. Nothing in principle prohibits Eve simply copying all of the transmitted information, though, depending on how it is encoded, it may be difficult in practice. In the language of resource simulation: There is no protocol that enables Alice and Bob to simulate an ideal private classical channel using only a public classical channel.

The solution to this problem is to add another resource, a *secret key* shared by Alice and Bob, and then use an *encryption* protocol. Suppose that Alice would like to transmit message M to Bob over the public channel and that they already share a secret key unknown to Eve. We describe the key as a pair of completely correlated random variables K and K' with uniform marginal distribution. The encryption protocol is specified by an encryption function $f : \mathcal{M} \times \mathcal{K} \to \mathcal{M}$ for Alice and decryption function $g : \mathcal{M} \times \mathcal{K} \to \mathcal{M}$ for Bob such that $g(f(m,k),k) = m$ for all $k \in \mathcal{K}$ and $m \in \mathcal{M}$. Alice encrypts the message $M = m$, called the *plaintext*, with the particular value of the key $K = k$ by computing the *ciphertext* $c = f(m,k)$. She then transmits the ciphertext over the public channel to Bob, who computes $m' = g(c,k)$ to obtain the original plaintext message.

20.1.2 Information-theoretic security

Before discussing a specific choice of f and g, we should first formalize the security statement we would like to have in terms of resource simulation. The available real

https://doi.org/10.1515/9783110570250-020

resources are the secret key and the classical channel, which we assume to be noiseless. Eve has no access to the key, but does have access the classical channel and may obtain information from it. Let us also assume that Eve may not alter messages on the classical channel, i. e., she is a purely passive observer. Meanwhile, an ideal private communication resource would seem to simply transmit Alice's message unchanged to Bob while outputting nothing to Eve.

However, this ideal resource is too simplistic. What we really require is that Eve obtain no *information* about the transmitted message, not no output whatsoever. For one thing, if the eavesdropper literally receives no output in the ideal case, but does in the real case, then distinguishing the real and ideal cases is trivial. Perhaps more importantly, it is information that we are concerned with, not the information carriers per se.

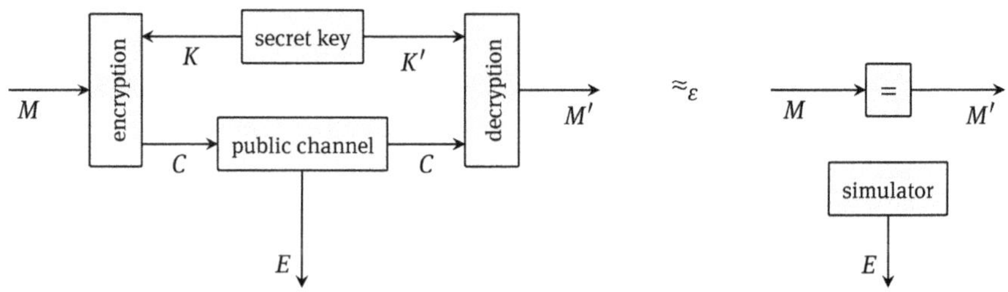

Figure 20.1: Information-theoretic security of an encryption scheme for private communication over public channels. The simulator outputs an approximation to the distribution Eve obtains from the public channel.

We can formalize the requirement that Eve obtains no information about the message by including a *simulator* with the ideal resource. The simulator simply gives Eve some output independent of the private message, and the particular output can be chosen to match as closely as possible whatever Eve obtains in the real case. Now it is sensible to compare the real resources and protocol on the one hand with the ideal resource and simulator on the other. This is depicted in Figure 20.1.

More precisely, denoting by $\mathcal{I}_{M'|M}$ the ideal channel from Alice to Bob and by $\mathcal{N}_{M'E|M}$ the actual action of the protocol and resources, the approximate resource simulation statement is that $\min_{P_E} \delta(\mathcal{N}_{M'E|M}, P_E \otimes \mathcal{I}_{M'|M}) \leq \varepsilon$, where the minimization is over probability distributions P_E, which describe the output of the simulator. By minimizing we only require that a single simulator output P_E exists, which makes the real and ideal cases nearly identical, not that all simulator output distributions will do so. If there exists a simulator satisfying the bound, then the protocol is essentially indistinguishable from one that employs the ideal resource, and hence the protocol is information-theoretically secure.

20.1.3 One-time pad

An example of a protocol meeting the information-theoretic security definition with $\varepsilon = 0$ for $|\mathcal{K}| = |\mathcal{M}|$ is the *one-time pad*. For simplicity, take $\mathcal{M} = \mathcal{K} = \mathbb{Z}_2^n$. Alice's encryption function f is simply $f(m,k) = m + k$, i. e., the bitwise modulo-two sum of the two inputs. Bob's function g is precisely the same operation, since this will return Alice's input m for all k. Meanwhile, Eve's output E is just the ciphertext, i. e., in the worst case, she simply copies all the information transmitted from Alice to Bob. The simulator distribution P_E is uniformly random, since the ciphertext will be uniformly random given a uniformly random key. With this choice, $\delta(\mathcal{N}_{M'E|M}, P_E \otimes \mathcal{I}_{M'|M}) = 0$.

The one-time pad has an important drawback: The length of the key is as long as the message. This is necessarily the case, as was first shown by Shannon. The argument is straightforward, and we give a version in the same spirit as the bounds of Chapter 8. Ideal security has the feature that the probability of guessing the message M is independent of whether we have the ciphertext C or not: $P_{\text{guess}}(M|C) = P_{\text{guess}}(M)$. On the other hand, Bob can ideally recover M given C and the key K, meaning that $P_{\text{guess}}(M|CK) = 1$. Since $P_{\text{guess}}(M|CK) \leq |K| P_{\text{guess}}(M|C) = |K| P_{\text{guess}}(M)$ by (11.10), we have $|K| \geq 1/P_{\text{guess}}(M)$. For a uniformly distributed message, this gives $|K| \geq |M|$. The following shows that the key length is very stable to increasing approximation error ε. In particular, $\varepsilon \approx 1/|M|$ is required to reduce the key length $\log |K|$ by one bit.

Exercise 20.1. Show that an ε-good protocol for private communication of a uniformly distributed message M requires a key of length $|K| \geq |M| \frac{1-\varepsilon}{|M|\varepsilon+1}$.

Via an entropic argument, we can infer that the one-time pad is essentially the only information-theoretically secure protocol with $|K| = |M|$.

Exercise 20.2. Show that in terms of entropy, the two security conditions above imply that $H(K) - H(M) = I(K : C) + H(K|MC) \geq 0$.

Hence, assuming that M is uniformly distributed, having $|K| = |M|$ requires that the key be independent of the ciphertext but be a deterministic function of the ciphertext and the message.

It is important to stress that the above is a formalization of information-theoretic security, not computational security, which is possibly more familiar to readers. In information-theoretic security, there are no constraints or assumptions on what operations Eve can perform on her data. They can take as much computational time or memory as necessary. Security holds because there is simply no useful information for Eve to work with in the first place. Computational security, by contrast, is predicated on Eve's limited ability to extract the information about the plaintext contained in the ciphertext.

20.2 Key distribution

In the context of information-theoretic security, the one-time pad reduces the problem of private communication to obtaining the secret key in the first place. However, unless Alice and Bob have exchanged secret keys in the past, it would appear that the bound on the key length presents a logical Catch-22, as the only available solution to the problem of private communication seems to require at least that much private communication to establish the key. Amazingly, via the BB84 protocol (as well as other quantum key distribution schemes) quantum mechanics allows us to escape this conclusion by offering a means for Alice and Bob to detect how much information Eve has acquired.

20.2.1 Real and ideal resources

Recall from Chapter 1 that the BB84 protocol involves transmitting randomly chosen quantum systems via an insecure channel, measuring them, and then creating a key from the classical information specifying which state was sent and which measurement was obtained. Before describing the specifics of the protocol and analyzing its security, we first need to reconsider the real and ideal resources, as there are two additional subtleties beyond the above discussion of private communication. There we made the assumption that Eve is merely a passive observer, which is too simplistic. Eve has two attacks at her disposal in any real-world cryptosystem that any protocol must deal with. The first is an *impersonation attack* on classical communication between Alice and Bob, in which Eve simply pretends to be Bob to Alice, and Alice to Bob. Since Alice and Bob are in physically distant locations, there is no immediate way to know with whom they are actually communicating. The second is a *jamming attack* on the quantum channel, in which Eve simply does not allow quantum information transmission from Alice to Bob.

Due to the possibility of a jamming attack, it is not possible for any real resources to emulate an ideal resource that always outputs a secret key. Therefore we must alter the ideal resource slightly. The BB84 protocol is designed to abort when Alice and Bob notice that the quantum channel is not working or is too noisy. Besides the key outputs, the ideal resource must therefore have an additional dedicated output for Alice and Bob that indicates whether the key was successfully created. Because Eve can choose to jam the physical quantum channel, the ideal resource has an *input* for Eve.

In particular, we model the ideal resource as follows. Eve has a binary input D, while Alice and Bob have length-ℓ key outputs K and K', respectively. Alice and Bob also have access to an additional binary "flag" output F, which indicates whether the key output is good. (Nominally, Alice has access to F and Bob has access to F', but F' always equals F.) Input $D = 1$ results in successful ideal key creation in the key outputs, and this is indicated by $F = 1$ in the auxiliary flag output. The input $D = 0$ corresponds to the situation in which the protocol aborts, which is indicated by $F = 0$. The key values are unimportant when $F = 0$, but for concreteness, let them be $K = K' = 0$. Note that the output key length is fixed, and not variable, so that technically there is a different ideal resource for every ℓ. This will simplify the security analysis later.

The real resources enable Eve to jam the quantum channel, but of course the actual description is not as simple as just a binary input as in the ideal resource. The mismatch between what kinds of actions Eve can perform in the real case versus the ideal case is handled by the simulator. In the actual scenario, Eve can choose the quantum channel through which Alice communicates with Bob. Perhaps more correctly, there are two channels, one from Alice to Eve and another from Eve to Bob. Eve is free to do whatever she likes with the output of the first and the input of the second. However, this ultimately results in some quantum channel from Alice to Bob, with Eve keeping the purification.

The possibility of an impersonation attack requires additional real-world resources for Alice and Bob. In particular, they require a means of *authenticating* their communication. This prevents Eve from tampering with their messages but does not prevent her from reading them. Hence we must assume that Eve obtains a copy of all classical information exchanged in the protocol.

Authentication can be simulated by a standard classical channel and a secret key by appending a short hash of the message to the message as a tag. The particular hash function is chosen by the key, which enables the receiver to authenticate the message by comparing the received tag to the hash of the received message. Crucially, the shared key can be much shorter than the message. We will not go into further details of authentication and instead simply assume that Alice and Bob have an authentic classical channel at their disposal. Nonetheless, it is important to appreciate the need for at least a short key in any quantum key distribution scheme and that the resulting protocol is more technically key *expansion* rather than key distribution.

20.2.2 Approximate simulation

Now let us more precisely state the approximate simulation requirement. First, consider the real state of affairs. Eve selects some quantum channel $\mathcal{N}_{BE|A}$ through which Alice communicates to Bob. The protocol makes use of this quantum channel and an authentic classical channel, ultimately leading to an output $\omega_{KK'FPE}$, where K and K' are the key outputs, F is the flag output, while E is the purification of the action of the quantum channel, and P denotes all public communication between Alice and Bob. Note that E is the only quantum system in the output state.

On the ideal side the simulator makes use of the ideal key resource in some manner, so that the ultimate output state $\theta_{KK'FPE}$ is as indistinguishable from $\omega_{KK'FPE}$ as possible. Formally, an (ℓ, ε) QKD protocol is a protocol with output key length ℓ such that for every channel $\mathcal{N}_{BE|A}$, there exists a simulator leading to output $\theta_{KK'FPE}$ for which

$$\delta(\omega_{KK'FPE}, \theta_{KK'FPE}) \leq \varepsilon\,. \tag{20.1}$$

This is depicted in Figure 20.2. Note that the simulator is free to depend on Eve's choice of the channel $\mathcal{N}_{BE|A}$. That the condition must hold for all channels is of course the major difference to the coding tasks we considered previously, where the channel or state resource was simply fixed.

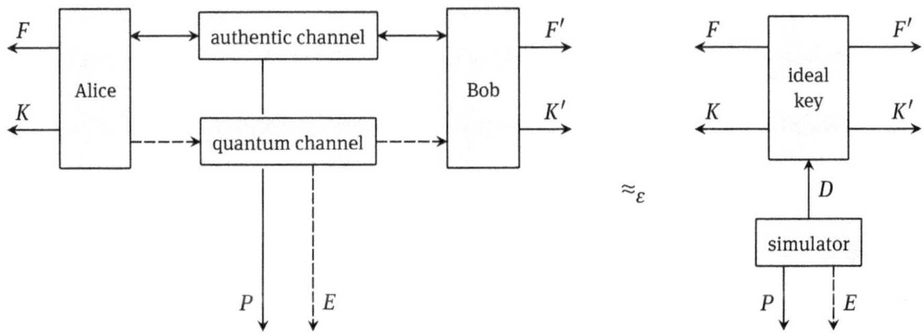

Figure 20.2: An (ℓ, ε) protocol for quantum key distribution. The protocol is secure with approximation parameter ε if for all quantum channels, there exists a simulator such that the distinguishability is less than ε. The key length $|K| = |K'| = \ell$.

A simple choice of simulator is that it just executes the QKD protocol itself for the specific $\mathcal{N}_{BE|A}$. It triggers key creation from the ideal key resource using $D = 1$ when the simulated protocol terminates successfully and sets $D = 0$ when it aborts. Thus the probability $p = \Pr[F = 1]$ that a key is created is precisely the same in the real and ideal cases. The simulator hands Eve the outputs P and E from the simulated $\omega_{KK'FPE}$. The output state in the ideal case for this simulator is therefore

$$\theta_{KK'FPE} = p\tau_{KK'} \otimes |1\rangle\langle 1|_F \otimes \omega_{PE|F=1} + (1-p)|00\rangle\langle 00|_{KK'} \otimes |0\rangle\langle 0|_F \otimes \omega_{PE|F=0}\,, \tag{20.2}$$

where $\tau_{KK'} = \frac{1}{|K|}\sum_k |k,k\rangle\langle k,k|_{KK'}$ is the quantum state of an ideal secret key. Note that the real output has a similar form with precisely the same second term. Hence the security condition reduces to $p\delta(\omega_{KK'PE|F=1}, \tau_{KK'} \otimes \omega_{PE|F=1}) \leq \varepsilon$. Using the triangle inequality, we can further decompose the quantity to be bounded into two parts, *correctness* and *secrecy*:

$$\begin{aligned} &p\,\delta(\omega_{KK'PE|F=1}, \tau_{KK'} \otimes \omega_{PE|F=1}) \\ &\quad \leq p\Pr[K \neq K'|F=1]_\omega + p\delta(\omega_{KPE|F=1}, \pi_K \otimes \omega_{PE|F=1})\,. \end{aligned} \tag{20.3}$$

Exercise 20.3. Show (20.3).

Basing the simulator on the actual protocol allows for the possibility of trivially secure protocols, those that never attempt to output a secret key in the first place! To make a claim of *robustness* of a QKD protocol, we should specify an eavesdropper attack, i. e., a specific channel that we expect from the design of the hardware implementing the protocol, and show that the protocol achieves a good security parameter and a small probability of not creating a key. The chosen attack is sometimes called the “honest implementation”, since in this case, Eve is not particularly malicious.

20.3 The BB84 protocol

The broad idea of the full BB84 protocol is to first generate raw keys by preparing, transmitting, and measuring qubits in the amplitude or phase basis, as described in Chapter 1, and then use the observed rate of errors to perform information reconciliation and privacy amplification to generate the final output key. Let us now precisely specify the protocol. It is determined by a long list $(n, n_x, n_z, s, t, \ell)$ of integer parameters and real-valued parameters $q \in [0,1]$ and $\delta \in (0, \frac{1-q}{2})$.

The information reconciliation step utilizes a $(2^s, \varepsilon_{\mathrm{ir}})$ syndrome-based information reconciliation protocol designed to recover from i. i. d. additive binary noise of rate q, i. e., from $\mathrm{BSC}(q)^{\otimes n}$. This protocol consists of an invertible linear function $f_{\mathrm{ir}} : \mathbb{Z}_2^{n_z} \to \mathbb{Z}_2^{n_z}$, the first s bits of which are the syndrome, and a syndrome decoding function $f_{\mathrm{dec}} : \mathbb{Z}_2^s \to \mathbb{Z}_2^{n_z}$. The output of f_{dec} is the estimated error pattern. A further set of invertible linear functions are used to verify that information reconciliation succeeded. These are specified by a two-argument function $f_{\mathrm{chk}} : \mathbb{Z}_2^{n_z-s} \times \mathbb{Z}_2^c \to \mathbb{Z}_2^{n_z-s}$, where the second argument picks an invertible function of the first argument, the set of which is such that the first t bits of the output yields a universal hash family. These output bits are called the *checksum*. The privacy amplification step, meanwhile, utilizes a seeded extractor $f_{\mathrm{ext}} : \mathbb{Z}_2^{n_z-s-t} \times \mathbb{Z}_2^r \to \mathbb{Z}_2^\ell$, where the second input is the seed, and the function is linear in its first input.

With these definitions in hand, the precise steps of the protocol are as follows. Steps 4–6 are depicted in Figure 20.3.

1. *State preparation.* Alice randomly generates two n-bit strings $Y = (Y_1, Y_2, \dots, Y_n)$ and $W = (W_1, W_2, \dots, W_n)$, and then prepares n qubits in the pure state $|\psi\rangle = \bigotimes_{j=1}^n e^{-i\frac{\pi}{4}(2y_j+t_j)\sigma_y}|0^n\rangle$; that is, the value of W_j chooses the amplitude ($W_j = 0$) or phase ($W_j = 1$) basis, and the value of Y_j chooses the particular basis element of the jth qubit. She transmits the n qubits to Bob via the quantum channel.
2. *Measurement.* Bob randomly generates an n-bit string W' and for each $j \in \{1, \dots, n\}$, measures the jth qubit of the output of the channel in the amplitude basis for $W_j' = 0$ or the phase basis if $W_j' = 1$. He stores the measurement results in an n-bit string Y'.

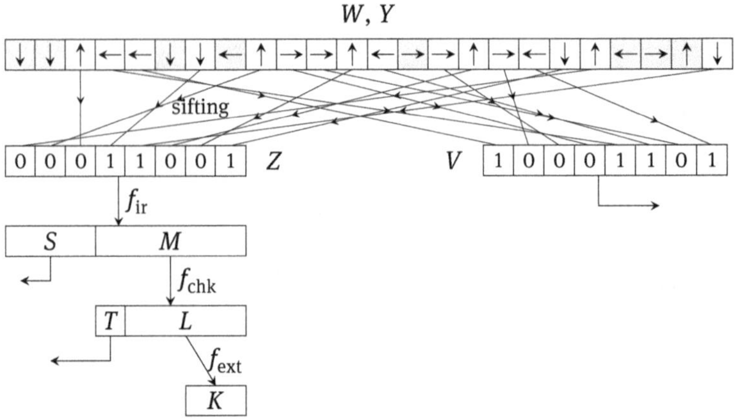

Figure 20.3: Alice's classical processing steps in BB84. The sifting step is a random assignment of Ys with $W = 0$ to Z and Ys with $W = 1$ to V. The arrows from S, T, V indicate that these are announced to Bob (and known to Eve). Gray shaded entries in the W, Y row indicate rounds in which $W_j \neq W'_j$. The dependence of f_{chk} and f_{ext} on the random variables C and R is not depicted.

These are the only parts of the protocol that involve the quantum channel. The remaining steps only make use of the authenticated classical channel.

3. *Randomization setup.* Alice and Bob agree on random bit strings $C \in \mathbb{Z}_2^c$ and $R \in \mathbb{Z}_2^r$.
4. *Sifting.* Alice announces W. Bob finds a set of n_z indices for which $W_j = W'_j = 0$ and a set of n_x indices for which $W_j = W'_j = 1$. He randomly orders each into the sequences I_Z and I_X, respectively, and announces both. If there are insufficiently many matching indices, then he sets $F = 0$, and Alice and Bob abort the protocol. Alice and Bob keep the substrings $Z = Y|_{I_Z}$ and $Z' = Y'|_{I_Z}$ of Y and Y' with indices in I_Z, respectively. Similarly, they keep the substrings $V = Y|_{I_X}$ and $V' = Y'|_{I_X}$ with indices in I_X, respectively. The former are the raw keys, and the latter are the verification bits.
5. *Information reconciliation.* Alice computes $S \oplus M = f_{\text{ir}}(Z)$ and $T \oplus L = f_{\text{chk}}(M, C)$ and announces the syndrome S and the checksum T. Bob reconciles Z' to Z by first computing $S' \oplus M' = f_{\text{ir}}(Z')$ and then $\hat{S} \oplus \hat{M} = f_{\text{ir}}(Z' + f_{\text{dec}}(S + S'))$. He tests if reconciliation was successful by computing $T' \oplus L' = f_{\text{chk}}(\hat{M}, C)$ and comparing T' with T. He informs Alice if the checksums match or not. If not, then Alice and Bob set $F = 0$ and abort the protocol.
6. *Privacy amplification.* Alice announces V and Bob determines if $|V+V'| \leq n_x(q+\delta)$. If the inequality is not satisfied, then he sets $F = 0$, and they abort the protocol. If the inequality is satisfied, then he sets $F = 1$ and informs Alice. She computes $K = f_{\text{ext}}(L, R)$, while he computes $K' = f_{\text{ext}}(L', R)$. The protocol terminates.

Before embarking on a sketch of why the protocol is secure, first note that we can alter the initial quantum phase of the protocol to use entanglement. Instead of step 1 above, Alice executes the following alternative:

1′. *Entanglement preparation.* Alice generates n maximally entangled qubit pairs $|\Phi\rangle_{AB}$ and transmit systems B to Bob. She then generates a random n-bit string T and measures the jth qubit of her remaining systems in the amplitude or phase basis according to the value of T. She records the measurement result in an n-bit string Y.

In other words, Alice can make use of steering to achieve the same result as directly preparing the ensemble of the four qubit states. The remainder of the protocol proceeds as before. The two protocols look identical from Eve's point of view, and therefore if the latter is secure, then so is the former.

20.4 Security and robustness analysis

Now we turn to the analysis of security and robustness. For the latter, let us fix the honest implementation to be the i. i. d. depolarizing channel with depolarization probability $2q$. (The factor of 2 will simplify several calculations.) We first examine the implications of the final steps of the protocol for security and robustness and then collect the results into a relation between ℓ and ε in terms of the parameters of the protocol.

20.4.1 Sifting

The sifting step does not affect the security parameter, only the robustness probability. It fails if there are too few amplitude basis matches or too few phase basis matches. By the union bound the probability the protocol aborts is upper bounded by the sum of the probabilities of these events individually.

The probability of an amplitude basis match for a single qubit is just $1/4$. Thus the entire sequence of matches in n attempts is described by a sequence of i. i. d. binary-valued random variables with individual probability $1/4$ of taking the value 1. Then by the Hoeffding bound the probability of fewer than n_z matches in n attempts is smaller than $\exp(-\frac{(n-4n_z)^2}{8n})$. Therefore, for every possible eavesdropper attack, the probability the sifting stage aborts is no larger than $\exp(-\frac{(n-4n_z)^2}{8n}) + \exp(-\frac{(n-4n_x)^2}{8n})$. Choosing $n > 4\max(n_x, n_z)$ suffices to ensure exponentially small (in n) probability of failure in the sifting stage.

20.4.2 Information reconciliation

The information reconciliation step affects both correctness and robustness. Correctness for every possible channel $\mathcal{N}_{BE|A}$ is ensured by the checksum test. Since the checksum is computed by universal hashing, the probability that $T = T'$ even though $M \neq \hat{M}$ is less than 2^{-t}. Because the final key is a deterministic function of M through the choice of C and R, we have

$$\Pr[K \neq K' \wedge T = T'] \leq \Pr[M \neq \hat{M} \wedge T = T'] \leq \Pr[T = T'|M \neq \hat{M}] \leq \frac{1}{2^t}. \qquad (20.4)$$

Since $F = 1$ implies $T = T'$, $\Pr[K \neq K' \wedge F = 1] \leq \Pr[K \neq K' \wedge T = T']$. Therefore we have an upper bound for the correctness condition in (20.3).

The above ignores $\Pr[M \neq \hat{M}]$, just bounding it by 1. In the honest implementation, the probability that amplitude inputs to the channel are flipped is just q. Thus using a $(2^s, \varepsilon_z)$ reconciliation protocol for this noise rate will ensure that $\Pr[M \neq \hat{M}] \leq \varepsilon_z$, further reducing the probability that the protocol aborts.

20.4.3 Privacy amplification

The privacy amplification step has implications for the secrecy of the key and, of course, robustness. Let us first remark on the latter, as the former is considerably more involved. For the honest implementation, the probability of a mismatch in the phase basis is again q, as just discussed for the amplitude basis. Thus the probability that there are more than $n_x(q+\delta)$ mismatches is less than $\exp(-2n_x\delta^2)$ by the Hoeffding bound.

As for secrecy, it is first important to observe that we do not require the distinguishability $\delta(\omega_{KPE|F=1}, \pi_K \otimes \omega_{PE|F=1})$ to be small for every attack to ensure secrecy, which is fortunate, as it is not possible to do so. Consider an intercept-resend attack in which Eve measures every qubit in the amplitude basis (as opposed to the attack considered in Chapter 1, where she chose the basis at random). This will result in completely uncorrelated strings V and V', and a very small chance of passing the $V + V' \leq n_x(q + \delta)$ test. However, when the test does pass, Eve knows the key with certainty. The randomness extractor cannot ensure secrecy in this case. Instead, secrecy rests on the fact that the test is very unlikely to pass, via the prefactor p in the secrecy condition in (20.3).

Let us deal with this subtlety right away, as it simplifies the logical argument going forward. Divide the attacks into those for which the probability p_{pa} of passing the test in the privacy amplification step is larger than some threshold $\eta \in (0, 1)$ and those for which it is not. In the latter case, we have $p\delta(\omega_{KPE|F=1}, \pi_K \otimes \omega_{PE|F=1}) \leq \eta$, since $p \leq p_{\text{pa}}$ and the distinguishability is less than 1. Then, for the remaining cases, we only need to show that $\delta(\omega_{KPE|F=1}, \pi_K \otimes \omega_{PE|F=1}) \leq f(\eta/p_{\text{pa}})$ for some function f such that $p_{\mathrm{pa}} f(\eta/p_{\mathrm{pa}}) \to 0$ as η and p_{pa} go to zero. The result will be the secrecy bound $p\delta(\omega_{KPE|F=1}, \pi_K \otimes \omega_{PE|F=1}) \leq \max(\eta, p_{\mathrm{pa}} f(\eta/p_{\mathrm{pa}}))$. Though the prospects of incorporating p_{pa} into the analysis of the randomness extractor may seem ominous here at the outset, this quantity will actually appear quite naturally.

The broad idea to establish extractor secrecy is to show that the test of σ_x basis outcomes gives Alice and Bob sufficient information about phase observables associated with the amplitude values L and L' to perform phase information reconciliation. As in Proposition 18.2, this will imply that the output key is independent of the information held by the eavesdropper.

Observe that in the entanglement-based version of the protocol, Alice and Bob can execute all operations on the n bits coherently, only measuring when necessary to exchange classical information. In the sifting step, they simply form two sets of qubits, one destined to actually be measured in the amplitude basis and the other destined to be measured in the phase basis. The functions computed in the information reconciliation step can be implemented coherently, as in Sections 18.3 and 19.4. Alice can apply unitary implementations of f_{ir} and f_{chk} and only then measure some of the qubits in the amplitude basis to generate S and T. Measurement of the remaining $n_z - s - t$ qubits can be deferred for the moment.

The same holds for Bob's operations, though these are slightly more complicated. Note that the correction step can be simplified using linearity of f_{ir}. Letting $S'' \oplus M'' = f_{\mathrm{ir}}(f_{\mathrm{dec}}(S+S'))$, it follows that $\hat{M} = M' + M''$. Therefore Bob's information reconciliation steps can be accomplished by first applying the unitary implementation of f_{ir}, then measuring the first s qubits in the amplitude basis to generate S', computing M'' from the final $n_z - s$ bits of $f_{\mathrm{ir}}(f_{\mathrm{dec}}(S+S'))$, and finally applying the bit flip pattern specified by M'' to the remaining $n_z - s$ qubits. To complete the information reconciliation step, he subsequently coherently applies f_{chk} and measures the first t qubits.

Thus, immediately prior to the privacy amplification stage of the protocol, we can imagine that Alice and Bob still hold $n_z - s - t$ qubits, ostensibly to be measured in the amplitude basis, and n_x qubits to be measured in the phase basis. Suppose now that the first set of qubits were instead measured in the phase basis, generating the $(n_z - s - t)$-bit random variables $\bar{X}$ and $\bar{X}'$. Due to the form of the actual amplitude information reconciliation operations in Step 5, these quantities are linear functions of X and X', the n_z-bit strings that would result from Alice and Bob measuring their qubits prior to Step 5. This is immediately clear for $\bar{X}$ and X, on Alice's side, given the discussion in Section 19.2. It also holds for $\bar{X}'$ and X' since the only additional operation on Bob's qubits is that σ_x is applied to some of them, which does not affect phase basis measurements.

Now we make use of the relation between randomness extraction of amplitude information and information reconciliation of phase information. Specifically, if we can construct a $(2^{n_z-s-t-\ell}, 2\varepsilon_x)$ information reconciliation protocol consisting of a family of surjective linear compression functions for $\bar{X}$ relative to $\bar{X}'$ for some ε_x, then by the discussion in Section 18.2 it follows that their dual functions yield a $(2^\ell, 2\sqrt{\varepsilon_x})$ strong randomness extractor for L relative to all systems held by Eve. (The awkward-looking factor of two in the error will be useful later.) Hence $\delta(\omega_{KPE|F=1}, \pi_K \otimes \omega_{PE|F=1}) \le 2\sqrt{\varepsilon_x}$.

So our focus shifts to information reconciliation of $\bar{X}$ relative to $\bar{X}'$. These random variables have some distribution $P_{\bar{X}\bar{X}'}$ depending on the particulars of the channel $\mathcal{N}_{BE|A}$ chosen by Eve. From Proposition 16.8 we know that $2\varepsilon_x$ is an achievable error in information reconciliation provided $2^{n_z-s-t-\ell} \le \frac{1}{\varepsilon_x}\beta_{1-\varepsilon_x}(P_{\bar{X}\bar{X}'}, \mathbb{1}_{\bar{X}} \times P_{\bar{X}'}) + 1$. This bound gives a constraint between the secrecy parameter and the key length. Since $\bar{X}$ and $\bar{X}'$ are functions of X and X', by (20.5) below we can further bound the right-hand side to obtain $2^{n_z-s-t-\ell} \le \frac{1}{\varepsilon_x}\beta_{1-\varepsilon_x}(P_{XX'}, \mathbb{1}_X \times P_{X'}) + 1$.

Exercise 20.4. Show that for any three random variables X, Y, and Z such that $Y = f(X)$ for some function f, we have the following inequality for all $\alpha \in [0,1]$:

$$\beta_\alpha(P_{YZ}, \mathbb{1}_Y \times P_Z) \le \beta_\alpha(P_{XZ}, \mathbb{1}_X \times P_Z)\,. \tag{20.5}$$

Hint: Construct a map from P_{YZ} to P_{XZ} and then use monotonicity and Exercise 9.14.

Next, define $U = X + X'$, the difference between the two binary strings. The deterministic conditional sum map $(x, x') \mapsto (x, x + x')$ transforms $P_{XX'}$ to P_{XU}, whereas $\mathbb{1}_X P_{X'}$ becomes the unnormalized distribution with probability $P_{X'}(x + u)$ at (x, u). Marginalization over X results in P_U for the former and simply $\mathbb{1}_U$ for the latter. Since the conditional sum and marginalization are classical channels, monotonicity of β_α implies $2^{n_z-s-t-\ell} \le \frac{1}{\varepsilon_x}\beta_{1-\varepsilon_x}(P_U, \mathbb{1}_U) + 1$.

By design Alice and Bob directly test the properties of U in the protocol, and this gives a means to further bound the hypothesis testing quantity. The strings $X \oplus V$ and $X' \oplus V'$ are phase measurement results from all the qubits that are kept after the sifting stage, and the division into X and V, respectively, X' and V', is done randomly. Thus V is a random sample of the entire set of Alice's phase measurement results, as is Bob's random sample V'. This implies that the fraction of mismatches between X and X', i. e., the number of 1s in U, will not be very different from the fraction of mismatches observed in $V + V'$.

The test ensures that no more than $n_x(q + \delta)$ mismatches are observed in $V + V'$. Recall that p_{pa} denotes the probability that this test passes. Given that the test passed, it turns out that the probability of more than $n_z(q + 2\delta)$ 1s in U is smaller than η/p_{pa}, where

$$\eta := \exp\left(-2\delta^2\left(\frac{n_z n_x^2}{(n_x + n_z)^2}\right)\right). \tag{20.6}$$

We will return to why this is the case below.

This bound is only useful when $\eta < p_{\mathrm{pa}}$, so we use the parameter η to split the set of attacks in two, as described at the beginning of this subsection. Hence we have also used the same letter 'η'. Now we can focus on the case $\eta/p_{\mathrm{pa}} < 1$, where it is permissible to set $\varepsilon_x = \eta/p_{\mathrm{pa}}$. This gives the extractor secrecy bound $\delta(\omega_{KPE|F=1}, \pi_K \otimes$

$\omega_{PE|F=1}) \leq 2\sqrt{\eta/p_{\mathrm{pa}}}$ corresponding to $f(x) = 2\sqrt{x}$. Hence the overall secrecy bound is $\max(\eta, 2p_{\mathrm{pa}}\sqrt{\eta/p_{\mathrm{pa}}}) \leq 2\sqrt{\eta}$.

Next, we turn to the implications of this choice of ε_x on the key length ℓ. The event that the number of 1s in U is no larger than $n_z(q+2\delta)$ has probability $1-\varepsilon_x$ and is thus a feasible hypothesis test in $\beta_{1-\varepsilon_x}(P_U, \mathbb{1}_U)$. Making use of the exercise to follow, we therefore have

$$\beta_{1-\varepsilon_x}(P_U, \mathbb{1}_U) \leq \sum_{j=0}^{\lfloor n_z(q+2\delta)\rfloor} \binom{n_z}{j} \leq 2^{n_z h_2(q+2\delta)}. \tag{20.7}$$

Hence there exists a $2\sqrt{\varepsilon_x}$-good randomness extractor for ℓ satisfying $2^{n_z-s-t-\ell} \leq \frac{p_{\mathrm{pa}}}{\eta} 2^{n_z h_2(q+2\delta)} \leq \frac{1}{\eta} 2^{n_z h_2(q+2\delta)}$. Taking the worst case for ℓ gives a restriction on the allowable values of ℓ in terms of the other parameters.

Exercise 20.5. Show that $\sum_{j=0}^{\lfloor np\rfloor}\binom{n}{j} \leq 2^{nh_2(p)}$ for any $p \in (0,1)$. *Hint: Start from the binomial expansion of* $1 = (p + (1-p))^n$.

It remains to show the sampling bound for η in (20.6). For this purpose, let us recycle Z and now use it as the random variable $Z = U \oplus (V + V')$ and define $m = n_x + n_z$. Let $\mu(z)$ be the empirical mean of a string $z \in \mathbb{Z}_2^m$, i. e., $\mu(z) = \frac{1}{m}\sum_{i=1}^m z_i$, and let $\mu_1(z)$ and $\mu_2(z)$ be the empirical means of the first n_z entries and the last n_x entries, respectively. Therefore $m\mu(z) = n_z\mu_1(z) + n_x\mu_2(z)$. Then consider the probability $\Pr[\mathsf{A} \wedge \mathsf{B}]$ of events A: $\mu_1(Z) \geq q + 2\delta$ and B: $\mu_2(Z) \leq q + \delta$. Note that $\Pr[\mathsf{B}] = p_{\mathrm{pa}}$ and the bound we are looking for is $\Pr[\mathsf{A}|\mathsf{B}] \leq \eta/p_{\mathrm{pa}}$.

Together, these two events imply $\mathsf{C} : \mu_1(Z) \geq \mu_2(Z) + \delta$, and so $\Pr[\mathsf{C}] \geq \Pr[\mathsf{A} \wedge \mathsf{B}]$. However, C is equivalent to $\mu_1(Z) \geq \mu(Z) + \frac{n_x}{m}\delta$. Since the order of entries in Z was ultimately random to begin with, for any given z, the event C describes randomly sampling, without replacement, $m\mu_1(z) + n_x\delta$ marked items in n_z tries from a population of $m\mu(z)$ marked items in m total. As luck would have it, though nominally formulated for sampling with replacement (the i. i. d. case), the Hoeffding bound also applies to sampling without replacement. Therefore $\Pr[\mathsf{C}]$ satisfies

$$\begin{aligned}\Pr[\mathsf{C}] &= \sum_{z\in\mathbb{Z}_2^m} P_Z(z)\Pr\left[\mu_1(z) \geq \mu(z) + \frac{n_x}{m}\delta\right] \\ &\leq \sum_{z\in\mathbb{Z}_2^m} P_Z(z)\exp\left(-2n_z\left(\frac{n_x}{m}\delta\right)^2\right) = \eta.\end{aligned} \tag{20.8}$$

Hence $\Pr[\mathsf{A} \wedge \mathsf{B}] \leq \eta$, and the proof is complete.

20.4.4 Security and robustness statement

Putting everything together, we have established that BB84 is an (ℓ, ε) QKD protocol, for which

$$\varepsilon \leq 2^{-t} + 2\exp\left(-\delta^2\left(\frac{n_z n_x^2}{(n_x+n_z)^2}\right)\right) \quad \text{and} \tag{20.9}$$

$$\ell \leq n_z - n_z h(q+2\delta) - s - t - \frac{2}{\ln 2}\delta^2 \frac{n_z n_x^2}{(n_x+n_z)^2}\,. \tag{20.10}$$

Furthermore, in the honest implementation the protocol will abort with probability no larger than

$$\exp\left(-\frac{(n-4n_z)^2}{8n}\right) + \exp\left(-\frac{(n-4n_x)^2}{8n}\right) + \varepsilon_z 2^{-t} + \exp(-2n_x\delta^2)\,. \tag{20.11}$$

We can get a clearer picture by making a specific choice of parameters in terms of n_z and considering the case of large n_z. In particular, set $n > 4n_z$, $n_x = n_z^{2/3}$, $\delta^2 = 1/\log n_z$, and $t = n_z^{1/2}$. We know from Section 16.4 that $s = n_z(h_2(q) - \nu)$ is sufficient to ensure that ε_z is small in n_z. Then the security parameter will decay as $\exp(-n_z^{1/3}/\log n_z)$, while the key rate ℓ/n_z will approach $1 - 2h_2(q)$.

Interestingly, this rate is precisely the coherent information of the Pauli channel with independent amplitude and phase errors of identical rate q, the BB84 channel. Although we specified the depolarizing channel as the honest implementation, the analysis also holds for this channel. It is to be expected that the protocol performs the same for depolarization as for the BB84 channel, since it does not make use of any correlations between amplitude and phase errors. (Doing so would effectively require modeling the eavesdropper attack.) The key rate in terms of the number of qubits n will have an additional factor of at most $1/4$.

20.5 Discussion of the security proof

The interplay between amplitude and phase information plays a crucial role both in our construction of quantum error-correcting codes in Chapter 19 and here in the security proof of BB84, highlighting the close relationship between these two tasks. Indeed, we are effectively utilizing a CSS code in the information reconciliation and privacy amplification steps above. Note that the compatibility of the hashing functions is automatically ensured here because f_{ext} acts on the L output of f_{chk}, and not directly on the raw key Z. We did not do this for quantum error correction, as remarked in Section 19.5.2, because in the general case, it is not apparent how to recover the asymptotic statement. Here we are able to employ (20.5), which does not hold when the conditioning system Z there is quantum, because we reduced the calculation to one involving only classical random variables. This approach would work to recover the quantum capacity of Pauli channels, but a different argument is needed for general channels.

We can now more fully appreciate that, as remarked in the Introduction, leakage of amplitude information in a quantum information processing protocol manifests itself as corruption of phase information. This is the uncertainty principle at work, as manifest in Version 2 of the uncertainty game, and was of course the basis for our construction of (amplitude) randomness extractors from phase information reconciliation protocols. Version 1 of the game ties this phenomenon back to preserving quantum information, as this task can be accomplished, as we have done it, by preserving amplitude and phase information. It is gratifying to comprehend the vital role played by the uncertainty principle in quantum information processing.

20.6 Notes and further reading

The quote is from [53]. The one-time pad goes back at least to Miller [206] in 1882. Vernam and Mauborgne [291] developed an electrical implementation. See Bellovin [21] for more on the history of the one-time pad. Shannon's argument that the key must be as long as the message appears in [259].

The approach of defining approximate cryptographic security by distinguishability to the ideal resource was developed by Maurer and Renner [201] and termed *abstract cryptography* by them. Portmann and Renner [227, 228] give a detailed treatment of its application to QKD. We follow the latter discussion, as well as that of Tomamichel and Leverrier [282]. See also the recent book by Wolf [308].

Ekert [91] proposed a QKD scheme based on Bell inequalities and utilizing entanglement. Bennett, Brassard, and Mermin [29] showed that a similar entanglement-based formulation of BB84 is possible, which nearly all proofs make use of in some way or another. There are too many QKD security proofs to list here; see the review by Portmann and Renner [228]. We are broadly following the approach by Shor and Preskill [264], in which the secrecy of the key is ensured by the complementary coding task of correcting phase errors, but making the argument in the style of [228, 282].

A The postulates of quantum mechanics

Here we recount one of the standard approaches to quantum mechanics, the axioms of Dirac [83] and von Neumann [293].

1. *Observables:*
   Any physical property of a system that can be measured is an *observable*, and all observables are represented by Hermitian linear operators acting on some Hilbert space $\mathcal{H}$. The possible values of the observable are the eigenvalues of the operator.
2. *States:*
   Complete descriptions of physical systems are called *states*, and states of an isolated physical system are represented by a normalized vector $|\phi\rangle \in \mathcal{H}$. Vectors differing only by phase represent the same state, e. g., $|\phi\rangle$ and $e^{i\theta}|\phi\rangle$ for $\theta \in \mathbb{R}$.
3. *Measurements:*
   The measurement of an observable X yields an eigenvalue x. The *Born rule* gives the probability of observing outcome x for a system in state $|\phi\rangle \in \mathcal{H}$:

   $$P_X(x) = \mathrm{Tr}\big[\Pi(x)|\phi\rangle\langle\phi|\big], \tag{A.1}$$

   where $\Pi(x)$ is the operator that projects onto the subspace of eigenvectors with eigenvalue x. More generally, we need not regard the projection operators as being associated with a particular observable. It is enough to specify a measurement by a set of projectors $\{\Pi(x)\}_{x=1}^{n}$ such that $\Pi(x)\Pi(x') = 0$ for $x \neq x'$ and $\sum_{x=1}^{n}\Pi(x) = \mathbb{1}$. According to the *projection postulate*, the state $|\phi_x'\rangle$ of the system after the measurement, given the outcome x, is

   $$|\phi_x'\rangle = \frac{1}{\sqrt{P_X(x)}}\Pi(x)|\phi\rangle\,. \tag{A.2}$$

4. *Dynamics:*
   Dynamical evolution of an isolated physical system over any fixed time interval $[t_0, t_1]$ is represented by some unitary operator U determined from the Hamiltonian of the system by the Schrödinger equation. In the *Schrödinger picture*, U maps the states $|\phi\rangle \in \mathcal{H}$ at time t_0 to the states $|\phi'\rangle = U|\phi\rangle$ at time t_1. In the *Heisenberg picture*, U maps observables O at time t_0 to observables $O' = U^*OU$ at time t_1.
5. *Composition:*
   For two physical systems with state spaces $\mathcal{H}_A$ and $\mathcal{H}_B$, the state space of the product system is isomorphic to $\mathcal{H}_A \otimes \mathcal{H}_B$.

The *measurement problem* arises if we ask for a description of measurement as a physical process subject to the dynamical laws. How can Schrödinger dynamics give rise to a measurement process having a particular outcome? Or how could it even lead an initial state $|\psi\rangle$ to the collection of possible outcomes?

https://doi.org/10.1515/9783110570250-021

B Vectors and operators

B.1 Linear operators

Consider the vector space $\mathcal{H} = \mathbb{C}^d$ for some finite dimension $\dim(\mathcal{H}) = d < \infty$. We denote the usual inner product of two vectors $x, y \in \mathcal{H}$ by $\langle x, y \rangle := \sum_{j=1}^{d} x_j^* y_j$. Here x_j and y_j are the components of x and y, and λ^* denotes the complex conjugate of any $\lambda \in \mathbb{C}$. Observe that $\langle x, y \rangle^* = \langle y, x \rangle$. The inner product gives rise to the usual norm $\| \cdot \| : \mathcal{H} \to \mathbb{R}_+$ defined by $\|v\| := \sqrt{\langle v, v \rangle}$.

In quantum mechanics, we need to also consider inner product spaces of infinite dimension, in particular, *Hilbert* spaces, e. g., to describe the free particle. There one must address issues of convergence, which are trivial in the finite case. Nonetheless, in quantum information theory, it is common to refer to $\mathcal{H}$ as Hilbert space. This is somewhat unwarranted from the mathematical point of view, and we will mostly refrain from doing so here.

We denote the set of linear maps, or operators, from $\mathcal{H}$ to a possibly different space $\mathcal{H}'$ by $\mathrm{Lin}(\mathcal{H}, \mathcal{H}')$, and when $\mathcal{H}' = \mathcal{H}$, we just write $\mathrm{Lin}(\mathcal{H})$. Of course, these are vector spaces, too. The identity operator, which maps any vector $v \in \mathcal{H}$ to itself, is denoted by $\mathbb{1}$. By S^* we denote the *adjoint* of an operator $S \in \mathrm{Lin}(\mathcal{H}, \mathcal{H}')$, which is the unique operator in $\mathrm{Lin}(\mathcal{H}', \mathcal{H})$ such that

$$\langle v', Sv \rangle = \langle S^* v', v \rangle \tag{B.1}$$

for all $v \in \mathcal{H}$ and $v' \in \mathcal{H}'$. We have $(S^*)^* = S$ for any $S \in \mathrm{Lin}(\mathcal{H}, \mathcal{H}')$.

The operators $S \in \mathrm{Lin}(\mathcal{H})$ for which $S = S^*$ are Hermitian (or self-adjoint), and the set of these, denoted $\mathrm{Herm}(\mathcal{H})$, forms a real vector space. Normal operators are those for which $SS^* = S^*S$. They include unitary operators U, for which $U^*U = UU^* = \mathbb{1}$. A unitary operator U is an *isometry* since it preserves the inner product between vectors: $\langle Ux, Uy \rangle = \langle x, U^*Uy \rangle = \langle x, y \rangle$. More generally, $V \in \mathrm{Lin}(\mathcal{H}, \mathcal{H}')$ is an isometry when $V^*V = \mathbb{1}_{\mathcal{H}}$. It is called a *partial isometry* when it acts as an isometry on the orthogonal complement of its kernel, its *support*. Observe that V can only be an isometry if $\dim(\mathcal{H}') \geq \dim(\mathcal{H})$ to preserve the inner product between basis vectors. By including a kernel, partial isometries avoid this constraint. The operators $S \in \mathrm{Lin}(\mathcal{H})$ for which $S^2 = S$ are projection operators, in particular, orthogonal projection operators when $S \in \mathrm{Herm}(\mathcal{H})$. We will not consider nonorthogonal projection operators and just refer to orthogonal projections as "projection operators".

B.2 Dirac notation

Dirac notation for vectors and operators simplifies many calculations and is widely used in quantum information theory. It is a more elaborate version of the representation of vectors by column vectors and action of linear maps by matrix multiplication

https://doi.org/10.1515/9783110570250-022

familiar from linear algebra. Formally, we associate any vector $v \in \mathcal{H}$ with the linear map $|v\rangle \in \mathrm{Lin}(\mathbb{C}, \mathcal{H})$ defined by

$$|v\rangle : z \mapsto zv \tag{B.2}$$

for $z \in \mathbb{C}$. This mapping is referred to as "ket", and we can think of ket as promoting a vector v to a linear map in this sense.

Dual to $\mathcal{H}$ is the vector space $\mathcal{H}^*$ of all linear functions from $\mathcal{H}$ to $\mathbb{C}$, i. e., $\mathcal{H}^* = \mathrm{Lin}(\mathcal{H}, \mathbb{C})$. Then the action of a dual vector $\omega \in \mathcal{H}^*$ on $|v\rangle$ is just the composition of the two maps. We essentially never deal with dual vectors directly, only as adjoints of elements of $\mathcal{H}$ using the inner product on $\mathcal{H}$. The adjoint of $|v\rangle$, denoted $\langle v|$ and called a "bra", is the linear functional defined by

$$\langle v| : u \mapsto \langle v, u\rangle \tag{B.3}$$

for $u \in \mathcal{H}$. Note that although $\langle v|$ is an element of $\mathcal{H}^*$, its label v is (or specifies) an element of $\mathcal{H}$. Nonetheless, the *Riesz*[1] *representation theorem* states that every element of the dual space is of the form given in (B.3).

Using this notation, the composition $\langle u| \circ |v\rangle$ of a bra $\langle u| \in \mathrm{Lin}(\mathcal{H}, \mathbb{C})$ with a ket $|v\rangle \in \mathrm{Lin}(\mathbb{C}, \mathcal{H})$ results in an element of $\mathrm{Lin}(\mathbb{C}, \mathbb{C})$, which can be identified with $\mathbb{C}$. It follows immediately from the above definitions that for all $u, v \in \mathcal{H}$, $\langle u| \circ |v\rangle = \langle u, v\rangle$. Thus, in the following, we will omit the $\circ$ and denote the scalar product by $\langle u|v\rangle$.

Conversely, the composition $|v\rangle \circ \langle u|$ (the outer product of v and u, also sometimes called a dyadic) is an element of $\mathrm{Lin}(\mathcal{H})$ or of $\mathrm{Lin}(\mathcal{H}, \mathcal{H}')$ when $\mathcal{H}' \neq \mathcal{H}$ and $u \in \mathcal{H}$, $v \in \mathcal{H}'$. We will just denote this by $|v\rangle\langle u|$. Its adjoint is $|u\rangle\langle v| \in \mathrm{Lin}(\mathcal{H}', \mathcal{H})$. This follows because acting with $|v\rangle\langle u|$ on $|x\rangle \in \mathcal{H}$ and taking the inner product of the result with $|y\rangle \in \mathcal{H}'$ gives $\langle y|v\rangle\langle u|x\rangle$, which is the same as the inner product of $|x\rangle$ with the result of applying $|u\rangle\langle v|$ to $|y\rangle$.

Any map $S \in \mathrm{Lin}(\mathcal{H}, \mathcal{H}')$ can be written as a linear combination of such outer products,

$$S = \sum_i |u_i\rangle\langle v_i| \tag{B.4}$$

for some families of vectors $\{u_i \in \mathcal{H}'\}_i$ and $\{v_i \in \mathcal{H}\}_i$. For example, for any orthonormal basis $\{b_i\}$ of $\mathcal{H}$, the identity $\mathbb{1} \in \mathrm{Lin}(\mathcal{H})$ can be written as

$$\mathbb{1} = \sum_{i=0}^{d-1} |b_i\rangle\langle b_i| . \tag{B.5}$$

1 Frigyes Riesz, 1880–1956.

This is the *completeness relation* of the basis vectors. For general S, we can choose $|v_i\rangle$ to be an orthonormal basis for the support of S and then set $|u_i\rangle = S|v_i\rangle$, which will form a basis for its image.

The *trace* of a linear map $S \in \mathrm{Lin}(\mathcal{H})$ is the function $\mathrm{Tr} : \mathrm{Lin}(\mathcal{H}) \to \mathbb{C}$ defined by extending the action $\mathrm{Tr}[|v\rangle\langle u|] = \langle u|v\rangle$ linearly using expansion (B.4). Thus, for S as there, $\mathrm{Tr}[S] = \sum_i \langle v_i|u_i\rangle$. The trace is a linear map from the vector space $\mathrm{Lin}(\mathcal{H})$ to $\mathbb{C}$, the set of which we denote as $\mathrm{Map}(\mathcal{H}, \mathbb{C})$. It is cyclic, meaning that $\mathrm{Tr}[ST] = \mathrm{Tr}[TS]$ for all $S \in \mathrm{Lin}(\mathcal{H}, \mathcal{H}')$, $T \in \mathrm{Lin}(\mathcal{H}', \mathcal{H})$. Furthermore, since $\mathrm{Tr}[|u\rangle\langle v|] = \mathrm{Tr}[|v\rangle\langle u|]^*$, the trace is compatible with the adjoint in that $\mathrm{Tr}[S^*] = \mathrm{Tr}[S]^*$.

Exercise B.1. Show that $\mathrm{Tr}[S] = \sum_i \langle b_i|S|b_i\rangle$ for all $S \in \mathrm{Lin}(\mathcal{H})$ and an arbitrary orthonormal basis $\{b_i\}$ of $\mathcal{H}$.

B.3 Matrix representations

Given an orthonormal basis $\{|b_k\rangle\}_{k=1}^d$, we can associate a matrix (S_{jk}) with any operator $S \in \mathrm{Lin}(\mathcal{H})$ with components $S_{jk} = \langle b_j|S|b_k\rangle$. We have chosen j to be the row index and k the column index, so that a composition of operators such as $S \circ T$ corresponds to the product of the corresponding matrices. Moreover, the map $|b_j\rangle\langle b_k|$ is represented by the matrix with a single 1 in the jth row and kth column.

The representation of an operator by a matrix is not unique, but depends on the choice of basis. One way to see this is to use the completeness relation (B.5) to write

$$S = \mathbb{1}\, S\, \mathbb{1} = \sum_{j,k} |b_j\rangle\langle b_j|S|b_k\rangle\langle b_k| = \sum_{j,k} S_{j,k} |b_j\rangle\langle b_k| . \tag{B.6}$$

Now the basis dependence is plain to see. Matrix representations can be given for more general operators $S \in \mathrm{Lin}(\mathcal{H}, \mathcal{H}')$ by the same technique:

$$S = \mathbb{1}_{\mathcal{H}'}\, S\, \mathbb{1}_{\mathcal{H}} = \sum_{j,k} |b_j'\rangle\langle b_j'|S|b_k\rangle\langle b_k| = \sum_{j,k} S_{j,k} |b_j'\rangle\langle b_k| . \tag{B.7}$$

In Dirac notation, $|v\rangle$ is itself an operator, meaning we can apply the above method to this case. As the input space is one-dimensional, we drop the associated basis vector and simply write

$$|v\rangle = \sum_j v_j |b_j\rangle . \tag{B.8}$$

According to the above convention, the representation of $|v\rangle$ is automatically a column vector, as it is the column index (which would take only one value) that has been omitted. Similarly, the representation of $\langle v|$ is automatically a row vector. This is a consequence of representing operators by matrices acting to their right, such that operator composition $S \circ T$ is the multiplication $(S_{jk})(T_{k\ell})$. If, as would plausibly be more sensible, composition were written in the same order as it is read left to right, i. e., with $T \circ S$ denoting first T then S, we would end up using matrices acting to their left and row vectors for kets. Either choice works, but consistency in that choice is certainly helpful.

We expect the representation of the adjoint of an operator to be the conjugate transpose of the matrix, but let us verify that this is indeed the case. The defining property of the adjoint is (B.1) or $\langle u|Sv\rangle = \langle S^*u|v\rangle$ in Dirac notation. By not identifying $|v\rangle$ with elements in $\mathcal{H}$, unlike what is usually done in Dirac notation, the expression on the right-hand side is still sensible. In terms of matrix representatives, reading the above from right to left, we have

$$\sum_j (S^*u)_j^* v_j = \sum_k u_k^*(Sv)_k = \sum_{jk} u_k^* S_{kj} v_j = \sum_{jk} (S_{kj}^* u_k)^* v_j. \tag{B.9}$$

Thus the jth component of S^*u is $\sum_k S_{kj}^* u_k$, so it must be that $(S^*)_{jk} = (S_{kj})^*$, as intended.

Exercise B.2. Show that $\mathrm{Herm}(\mathbb{C}^d)$ is a real vector space of dimension d^2.

B.4 Tensor products

The tensor product of two vector spaces is essentially just the product compatible with linearity. Let us first give a more abstract definition and properties before examining how it concretely works with Dirac notation.

The most straightforward product of two arbitrary vector spaces $\mathcal{H}_A$ and $\mathcal{H}_B$ is their Cartesian product $\mathcal{H}_A \times \mathcal{H}_B$, the set of all ordered pairs (u, v) with $u \in \mathcal{H}_A$ and $v \in \mathcal{H}_B$. This can be made into a vector space by including all formal linear combinations, e. g., $a(u, v) + b(u', v')$ for all choices of $a, b \in \mathbb{C}$. This is called the free vector space generated by $\mathcal{H}_A \times \mathcal{H}_B$. However, doing so does not respect the linearity of $\mathcal{H}_A$ or $\mathcal{H}_B$, since the Cartesian product is just a product of sets; that is, if $u = u_0 + u_1$, then $(u, v) \neq (u_0, v) + (u_1, v)$. The idea behind the *tensor product* is to enforce this sort of linearity on the free vector space. There are four combinations of vectors, which we would expect to vanish by linearity:

$$\begin{aligned}
&(u, v) + (u', v) - (u + u', v), \\
&(u, v) + (u, v') - (u, v + v'), \\
&a(u, v) - (au, v), \quad a \in \mathbb{C}, \\
&a(u, v) - (u, av), \quad a \in \mathbb{C}.
\end{aligned} \tag{B.10}$$

These vectors define an equivalence relation on the free vector space in that we can consider two elements of that space to be equivalent if they differ by some vector of the form in (B.10). These equivalence classes themselves form a vector space, and the resulting vector space is precisely the tensor product $\mathcal{H}_A \otimes \mathcal{H}_B$.

Since the construction enforces linearity of the products of vectors, we may consider the tensor product to be the space spanned by products of basis elements of each space. This is precisely how we work with $\mathbb{R}^n$ or $\mathbb{C}^n$ using the *standard basis* formed by the vectors $(1, 0, \dots)$, $(0, 1, 0, \dots)$, and so forth.

In Dirac notation, with bases $\{|b_j\rangle_A\}_{j=1}^{d_A}$ and $\{|b'_k\rangle_B\}_{k=1}^{d_B}$ for $\mathcal{H}_A$ and $\mathcal{H}_B$, respectively, a set basis vectors of the tensor product is $\{|b_j\rangle_A \otimes |b'_k\rangle_B\}$. Furthermore, the inner product on $\mathcal{H}_A \otimes \mathcal{H}_B$ is defined by the linear extension of

$$\langle u \otimes v, u' \otimes v' \rangle = \langle u|u' \rangle \langle v|v' \rangle . \tag{B.11}$$

The tensor product $S \otimes T$ of two linear maps $S \in \mathrm{Lin}(\mathcal{H}_A, \mathcal{H}_{A'})$ and $T \in \mathrm{Lin}(\mathcal{H}_B, \mathcal{H}_{B'})$ is the element of $\mathrm{Lin}(\mathcal{H}_A \otimes \mathcal{H}_B, \mathcal{H}_{A'} \otimes \mathcal{H}_{B'})$ defined by the action

$$(S \otimes T) : (u \otimes v) \mapsto (Su) \otimes (Tv) \tag{B.12}$$

for $u \in \mathcal{H}_A$ and $v \in \mathcal{H}_B$. In fact, linear combinations of tensor products suffice to span all of $\mathrm{Lin}(\mathcal{H}_A \otimes \mathcal{H}_{A'}, \mathcal{H}_B \otimes \mathcal{H}_{B'})$, which is the statement that

$$\mathrm{Lin}(\mathcal{H}_A, \mathcal{H}_{A'}) \otimes \mathrm{Lin}(\mathcal{H}_B, \mathcal{H}_{B'}) \simeq \mathrm{Lin}(\mathcal{H}_A \otimes \mathcal{H}_B, \mathcal{H}_{A'} \otimes \mathcal{H}_{B'}) . \tag{B.13}$$

This can be easily seen using expansions as in (B.4).

From the form of (B.12) we immediately have the possibility of tracing out part of a linear map as follows. For an element of $\mathrm{Lin}(\mathcal{H}_A \otimes \mathcal{H}_B)$ of the form $S_A \otimes T_B$ with $S_A \in \mathrm{Lin}(\mathcal{H}_A)$ and $T_B \in \mathrm{Lin}(\mathcal{H}_B)$, define the partial trace over the B factor as the map $\mathrm{Tr}_B : \mathrm{Lin}(\mathcal{H}_A \otimes \mathcal{H}_B) \to \mathrm{Lin}(\mathcal{H}_A)$ having the action

$$\mathrm{Tr}_B : S_A \otimes T_B \mapsto \mathrm{Tr}[T_B]\, S_A . \tag{B.14}$$

Then we can linearly extend the action to an arbitrary maps in $\mathrm{Lin}(\mathcal{H}_A \otimes \mathcal{H}_B)$. Here we have used subscripts to label which vector spaces the operators act on, a notation we will frequently use.

Similarly to the trace operation, the partial trace Tr_B is linear and commutes with the operation of taking the adjoint. Furthermore, it commutes with the left and right multiplication with an operator of the form $T_A \otimes \mathbb{1}_B$, where $T_A \in \mathrm{Lin}(\mathcal{H}_A)$; that is, for any operator $S_{AB} \in \mathrm{Lin}(\mathcal{H}_A \otimes \mathcal{H}_B)$,

$$\mathrm{Tr}_B[S_{AB}(T_A \otimes \mathbb{1}_B)] = \mathrm{Tr}_B[S_{AB}]T_A \quad \text{and} \quad \mathrm{Tr}_B[(T_A \otimes \mathbb{1}_B)S_{AB}] = T_A\, \mathrm{Tr}_B[S_{AB}] . \tag{B.15}$$

We will also make use of the property that the trace on a bipartite system can be decomposed into partial traces on the individual subsystems. Specifically, we have

$\mathrm{Tr}[S_{AB}] = \mathrm{Tr}[\mathrm{Tr}_B[S_{AB}]]$, or, for an operator $S_{ABC} \in \mathrm{Lin}(\mathcal{H}_A \otimes \mathcal{H}_B \otimes \mathcal{H}_C)$,

$$\mathrm{Tr}_{AB}[S_{ABC}] = \mathrm{Tr}_A[\mathrm{Tr}_B[S_{ABC}]] . \tag{B.16}$$

As it is a linear map between two inner product spaces, we may ask for the adjoint of Tr_B. By definition it must satisfy

$$\langle S_A, \mathrm{Tr}_B[T_{AB}]\rangle = \langle \mathrm{Tr}_B^*[S_A], T_{AB}\rangle . \tag{B.17}$$

From this we can easily see that the only choice that works for arbitrary S_A and T_{AB} is the mapping $\mathrm{Tr}_B^* : S_A \mapsto S_A \otimes \mathbb{1}_B$.

B.5 Positive operators

An operator $S \in \mathrm{Lin}(\mathcal{H})$ is *positive* (or positive semidefinite), denoted $S \geq 0$, if $\langle v|S|v\rangle \geq 0$ for all $v \in \mathcal{H}$. Positive definite corresponds to the strict inequality $S > 0$. Note that we should ostensibly refer the latter as positive and the former as nonnegative, but we will always take "positive" to mean "positive semidefinite" as our uses for positive definite operators are limited. Positivity gives rise to a partial ordering of operators by taking $S \geq T$ to mean $S - T \geq 0$. The simple definition of positivity is enough to imply some very important properties.

Exercise B.3. Show that for any $M \in \mathrm{Lin}(\mathcal{H}, \mathcal{H}')$, the operators M^*M and MM^* are both positive and that $MSM^* \geq 0$ if $S \geq 0$, where $S \in \mathrm{Lin}(\mathcal{H})$.

Positive operators are necessarily Hermitian, at least when $\mathcal{H}$ is complex.

Lemma B.1. *If $S \in \mathrm{Lin}(\mathcal{H})$ is positive, then it is Hermitian.*

Proof. For positive S, we have $\langle x, (S - S^*)x\rangle = \langle x, Sx\rangle - \langle x, S^*x\rangle = \langle x, Sx\rangle - \langle Sx, x\rangle = \langle x, Sx\rangle - \langle x, Sx\rangle^* = 0$. Thus it is enough to show that an arbitrary $M \in \mathrm{Lin}(\mathcal{H})$ is zero if $\langle x, Mx\rangle = 0$ for all $x \in \mathcal{H}$. To this end, consider arbitrary $x, y \in \mathcal{H}$ for which $\langle x, Mx\rangle = 0$, $\langle y, My\rangle = 0$, $\langle x + y, M(x + y)\rangle = 0$, and $\langle x + iy, M(x + iy)\rangle = 0$ by assumption. Together these imply $\langle x, My\rangle = 0$ for all $x, y \in \mathcal{H}$. Hence My is orthogonal to all of $\mathcal{H}$, which is only possible if $M = 0$. □

The positivity condition is quite restrictive. An important property is the following, which states that $\langle u, Su\rangle = 0$ implies that u is in the kernel of S.

Lemma B.2. *For positive semidefinite $S \in \mathrm{Lin}(\mathcal{H})$, if $\langle u, Su\rangle = 0$ for some $u \in \mathcal{H}$, then $\langle v, Su\rangle = 0$ for all $v \in \mathcal{H}$.*

Exercise B.4. Prove the lemma. *Hint: Apply positivity of S to $w = su + v$ and $w' = su + iv$ for $s \in \mathbb{R}$.*

By the spectral theorem it immediately follows that positive operators have positive eigenvalues, and in light of the above, we could have made this the definition. However, it is often more useful to directly apply our basic definition than to use the spectral theorem, for instance, in Exercise B.3. In any case the trace of a positive operator S is necessarily positive and can only vanish if $S = 0$. The square root of a positive operator S can be defined by taking the square root of the eigenvalues. For eigendecomposition $S = \sum_{j=1}^{d} \lambda_j |b_j\rangle\langle b_j|$, where $\lambda_j \geq 0$ are the eigenvalues, and $\{|b_j\rangle\}$ is the orthonormal eigenbasis of S, the square root $S^{1/2}$ is simply $S^{1/2} = \sum_{j=1}^{d} \sqrt{\lambda_j} |b_j\rangle\langle b_j|$. We sometimes denote this operator as $\sqrt{S}$.

Another restrictive property of positive operators, which turns out to be very useful, is the following:

Lemma B.3. *Suppose S and T are arbitrary positive operators in* $\mathrm{Lin}(\mathcal{H})$. *Then* $\mathrm{Tr}[ST] \geq 0$, *and equality holds only if $ST = 0$, i. e., S and T have disjoint supports.*

Proof. For the first statement, notice that $\mathrm{Tr}[ST] = \mathrm{Tr}[S^{1/2}TS^{1/2}]$ by the properties of the trace and positivity of S. Positivity of T then implies positivity of $S^{1/2}TS^{1/2}$. This implies that the trace must be positive and only vanishes if $S^{1/2}TS^{1/2} = 0$. It must therefore be that $S^{1/2}|b_j\rangle = 0$ for all eigenvectors $|b_j\rangle$ of T with strictly positive eigenvalues. Hence $\mathrm{Tr}[ST] = 0$ implies $ST = 0$. □

Exercise B.5. For $S, T \in \mathrm{Lin}(\mathcal{H})$, show that $T \geq 0$ if $\mathrm{Tr}[ST] \geq 0$ for all $S \geq 0$.

The set of all positive semidefinite operators in $\mathrm{Lin}(\mathcal{H})$ forms a convex cone in $\mathrm{Herm}(\mathcal{H})$, that is, λS is positive semidefinite if S is when $\lambda \geq 0$, and a convex combination of positive semidefinite operators is also positive semidefinite. This cone is a closed set. Consider the function $\lambda_{\min}(S) = \min\{\langle v, Sv\rangle : \|v\| = 1, v \in \mathcal{H}\}$. It is continuous since $|\lambda_{\min}(S+T) - \lambda_{\min}(S)| \leq \|T\|$ for $\|T\| = \max\{\|Tv\| : \|v\| \leq 1, v \in \mathcal{H}\}$. Then, because the set of positive semidefinite operators is just the continuous preimage of the closed set $[0, \infty)$, it must be closed.

Exercise B.6. Prove the continuity statement.

B.6 Operator decompositions

Several operator decompositions play an important role in the formalism of quantum information theory. Foremost is the *spectral decomposition* of a normal operator into eigenvalues and eigenvectors, which we just encountered. More precisely, for any normal $S \in \mathrm{Lin}(\mathcal{H})$, there exist an orthonormal basis $\{|\phi_j\rangle\}$ and complex eigenvalues λ_j such that

$$S = \sum_j \lambda_j |\phi_j\rangle\langle\phi_j| . \tag{B.18}$$

For a fixed basis $\{|b_j\rangle\}$, (B.18) is equivalent to the existence of a unitary U such that $S = UDU^*$, where D is the diagonal matrix in the basis $\{|b_j\rangle\}$ with entries λ_j. In particular, $U = \sum_j |\phi_j\rangle\langle b_j|$.

The spectral decomposition can be used to give a meaning to applying a function $f : \mathbb{C} \to \mathbb{C}$ to a normal operator S by taking the action

$$f : S \mapsto \sum_j f(\lambda_j)|\phi_j\rangle\langle\phi_j| . \tag{B.19}$$

This is sometimes called the functional calculus. A particularly important operator function is the square root for positive operators, which we just defined above. Positive operators can have other "square roots" though, any M such that $S = M^*M$.

All other possibilities are determined by the *singular value decomposition* of an arbitrary $M \in \mathrm{Lin}(\mathcal{H}, \mathcal{H}')$. The singular values s_j of M are simply the square roots of the eigenvalues of $M^*M \in \mathrm{Lin}(\mathcal{H})$. Since M^*M is positive, we can use the square root defined by the spectral decomposition, and the number n of nonzero singular values is no larger than the dimension of $\mathcal{H}$.

Now, given bases $\{|b_j\rangle\}$ and $\{|b_j'\rangle\}$ for $\mathcal{H}$ and $\mathcal{H}'$, there exist isometries $U \in \mathrm{Lin}(\mathbb{C}^n, \mathcal{H}')$ and $V \in \mathrm{Lin}(\mathbb{C}^n, \mathcal{H})$ and a diagonal operator $D \in \mathrm{Lin}(\mathbb{C}^n)$ with entries $s_j > 0$ such that

$$M = UDV^* . \tag{B.20}$$

This is the singular value decomposition. Note that if we think of M as a collection of column vectors, i. e., $M = \sum_k |\varphi_k\rangle\langle b_k|$, then from the form of the singular value decomposition n must be the dimension of their span, the column rank of M. Similarly, regarded as a collection of row vectors $M = \sum_k |k\rangle\langle\theta_k|$, the same argument implies that n must be the row rank of M, which accords with the fact that these two ranks are always equal.

Lemma B.4. *Every $M \in \mathrm{Lin}(\mathcal{H}, \mathcal{H}')$ has a singular value decomposition as in* (B.20).

Proof. The proof proceeds by applying the spectral decomposition to M^*M. Since M^*M is positive, there exists a unitary $\hat{V} \in \mathrm{Lin}(\mathcal{H})$ such that $M^*M = \hat{V}\hat{D}\hat{V}^*$, where $\hat{D} \in \mathrm{Lin}(\mathcal{H})$ is diagonal with nonnegative entries s_j^2. Let $\mathcal{H}_n$ be the subspace of $\mathcal{H}$ corresponding to nonzero s_j; this subspace is isomorphic to $\mathbb{C}^n$. Furthermore, define $V \in \mathrm{Lin}(\mathcal{H}_n, \mathcal{H})$ to have the same action as $\hat{V}$ restricted to inputs in $\mathcal{H}_n$. By construction, V is an isometry, i. e., $V^*V = \mathbb{1}_{\mathcal{H}_n}$, and moreover the operator $VV^* = \Pi_{\mathcal{H}_n}$, the projector of $\mathcal{H}$ onto $\mathcal{H}_n$. If we order the basis in which $\hat{D}$ is diagonal so that the nonzero singular values appear first, then the matrix representative of V is just the first n columns of that of $\hat{V}$. Since D is invertible, we can set $U = MVD^{-1} \in \mathrm{Lin}(\mathcal{H}_n, \mathcal{H}')$ to get $UDV^* = M\Pi_{\mathcal{H}_n}$. However, $M\Pi_{\mathcal{H}_n}$ must be equal to M, since, otherwise, $\mathcal{H}_n$ would not contain all the nonzero eigenvalues of M^*M. Furthermore, U is an isometry, since

$U^*U = D^{-1}V^*M^*MVD^{-1} = D^{-1}V^*\hat{V}\hat{D}\hat{V}^*VD^{-1} = D^{-1}V^*VD^2V^*VD^{-1} = \mathbb{1}_{\mathcal{H}_n}$, completing the proof. □

Observe that $MM^* = UD^2U^*$, so M^*M and MM^* have the same nonzero eigenvalues. Therefore the number of singular values is less than the lesser of $\dim(\mathcal{H})$ and $\dim(\mathcal{H}')$.

Returning to the question of possible square roots of $S \geq 0$, the proof above shows that V diagonalizes S in that $S = VD^2V^*$. Therefore $D = V^*\sqrt{S}\,V$, and M can be written $M = UV^*\sqrt{S}$. This is precisely the *polar decomposition* of M. Namely, for any $M \in \mathrm{Lin}(\mathcal{H}, \mathcal{H}')$, there exists an isometry $W \in \mathrm{Lin}(\mathcal{H}, \mathcal{H}')$ such that

$$M = W\sqrt{M^*M} = \sqrt{MM^*}\,W\,. \tag{B.21}$$

In particular, $W = UV^*$ satisfies the two equations. In analogy with the case of complex numbers, we define the *absolute value* $|M| = \sqrt{M^*M}$, so that $M = W|M|$. Since $|M|$ and W do not necessarily commute, we have to make a choice for this ordering or the other. Note that either choice $\sqrt{M^*M}$ or $\sqrt{MM^*}$ agrees with the absolute value of a Hermitian operator using the functional calculus, and in this case, $W = \mathbb{1}$.

The singular value decomposition also allows us to define a *pseudoinverse* M^+ for an arbitrary matrix M as $M^+ = VD^{-1}U^*$. This fulfills the properties that $MM^+M = M$ and $M^+MM^+ = M^+$, so that M^+ acts as an inverse on M, whereas MM^+ is the projection onto the range of M (i. e., the column space), and M^+M is the projection onto the support of M (i. e., the row space). For a positive semidefinite matrix M, the pseudoinverse is just the usual inverse on the support of M.

B.7 Inner products and norms of operators

A simple inner product for the space $\mathrm{Lin}(\mathcal{H}, \mathcal{H}')$ can be defined as follows using the trace:

$$\langle S, T\rangle := \mathrm{Tr}[S^*T]\,. \tag{B.22}$$

This is the *Hilbert–Schmidt inner product*, and the induced norm $\|S\|_2 := \sqrt{\langle S, S\rangle}$ is the *Hilbert–Schmidt* or *Frobenius*[2] *norm*. This inner product is what we obtain by treating using the usual vector inner product for the matrix representatives of S and T in the following sense. Choosing a basis $\{|b_k\rangle\}$ for $\mathcal{H}$ and $\{|b_j'\rangle\}$ for $\mathcal{H}'$, suppose $S = \sum_{jk} S_{jk}|b_j'\rangle\langle b_k|$ and $T = \sum_{jk} T_{jk}|b_j'\rangle\langle b_k|$. Then $\langle S, T\rangle = \sum_{jk} S_{jk}^* T_{jk}$. The induced norm satisfies the following defining properties of a norm: $\|S\|_2 \geq 0$ with equality iff $S = 0$

2 Ferdinand Georg Frobenius, 1849–1917.

(positivity), $\|\alpha S\|_2 = |\alpha|\|S\|_2$ for $\alpha \in \mathbb{C}$ (scalability), and $\|S+T\|_2 \le \|S\|_2+\|T\|_2$ (triangle inequality). Naturally, the Hilbert–Schmidt inner product satisfies the Cauchy–Schwarz inequality $|\langle S, T\rangle| \le \|S\|_2\|T\|_2$.

Two other norms besides the Hilbert–Schmidt norm show up frequently in quantum information theory, though we meet only one of them in this book. This is the *trace norm*, sometimes called the *nuclear norm*, defined by

$$\|S\|_1 := \mathrm{Tr}[\sqrt{S^*S}]\,. \tag{B.23}$$

Given the results of the previous section, it follows that $\|S\|_1 = \mathrm{Tr}[|S|]$, i. e., the trace norm is the sum of the singular values of S. The other is the *operator norm* $\|S\|_\infty := \max\{\|Sv\| : \|v\| = 1, v \in \mathcal{H}\}$, which we implicitly used in the previous section. It is also known as the infinity norm, as it turns out that $\|S\|_\infty = \max\{\sigma_j\}$, where σ_j are the singular values of S. The following lemma provides a useful characterization of the trace norm on $\mathrm{Lin}(\mathcal{H})$.

Lemma B.5. *For any $S \in \mathrm{Lin}(\mathcal{H})$, $\|S\|_1 = \max\{|\mathrm{Tr}[SU]| : U$ unitary on $\mathcal{H}\}$.*

Proof. Let $S = |S|V$ be the polar decomposition of S. Then equality holds for $U = V^*$. To show that the maximization cannot be larger than the trace norm, we employ the Cauchy–Schwarz inequality. Specifically, for every unitary U, we have

$$\begin{aligned}\big|\mathrm{Tr}[SU]\big| &= \big|\mathrm{Tr}[|S|VU]\big| = \big|\mathrm{Tr}[\sqrt{|S|}\sqrt{|S|}VU]\big| \\ &\le \sqrt{\mathrm{Tr}[|S|]\,\mathrm{Tr}[U^*V^*|S|VU]} = \mathrm{Tr}[|S|]\,,\end{aligned} \tag{B.24}$$

and therefore $U = V^*$ is the optimal choice. □

B.8 The Schur complement

A useful tool in characterizing operator positivity is the *Schur complement*. Consider an $(n+m)\times(n+m)$ block matrix of the form

$$S = \begin{pmatrix} A & B \\ B^* & C \end{pmatrix}, \tag{B.25}$$

where $A \in \mathbb{C}^{n\times n}$, $B \in \mathbb{C}^{n\times m}$, and $C \in \mathbb{C}^{m\times m}$, for arbitrary n and m. The Schur complement of C in S is $A - BC^+B^*$, and, similarly, the Schur complement of A in S is $C - B^*A^+B$, where C^+ and A^+ are the pseudoinverses of C and A from the end of Section B.6. Positive semidefiniteness of S is related to positive semidefiniteness of the Schur complements:

Lemma B.6 (Schur complement). *For an arbitrary matrix S as in* (B.25), *$S \ge 0$ iff $A \ge 0$, $C - B^*A^+B \ge 0$, and $(\mathbb{1} - AA^+)B = 0$.*

Proof. Start with the "only if" statement. Clearly, the A block must be positive if S is, so the first condition is fulfilled. To establish the second condition, let $K = \begin{pmatrix} \mathbb{1} & 0 \\ -B^*A^+ & \mathbb{1} \end{pmatrix}$. Since A is square, $A^+ = (A^+)^*$, and hence $K^* = \begin{pmatrix} \mathbb{1} & -A^+B \\ 0 & \mathbb{1} \end{pmatrix}$. Now define

$$S' = \begin{pmatrix} A & (\mathbb{1} - AA^+)B \\ ((\mathbb{1} - AA^+)B)^* & C - B^*A^+B \end{pmatrix} . \tag{B.26}$$

Direct calculation shows that $S' = KSK^*$, so $S' \geq 0$ by Exercise B.3. Hence the second condition is satisfied. The third condition states that S' is block diagonal. Since S is positive, we can write $S = MM^*$ for some $(n + m) \times k$ matrix M, where k is at least the rank of S. Take $M = \sqrt{S}$ or the columns of M to be the eigenvectors of S, normalized, for instance, by the square root of the corresponding eigenvalues. Partitioning M as $M = \begin{pmatrix} X \\ Y \end{pmatrix}$, it follows that $A = XX^*$, $C = YY^*$, and $B = XY^*$. Using the polar decomposition to write $X = PW$ for $P = \sqrt{A}$ and some isometry $W \in \mathrm{Lin}(\mathbb{C}^k, \mathbb{C}^{n+m})$, it follows that $B = \sqrt{A}WY^* = AP^+WY^* = AR$ for $R = P^+WY^*$. Hence $AA^+B = AA^+AR = AR = B$, i. e., the third condition holds.

The "if" statement is now simple. Start from S', which by the third condition is block diagonal, and therefore positive by the first and second conditions. Letting $K' = \begin{pmatrix} \mathbb{1} & 0 \\ B^*A^+ & \mathbb{1} \end{pmatrix}$, it follows that $K'S'K'^* \geq 0$, and direct calculation shows that $S = K'S'K'^*$. □

B.9 Operator monotonicity and convexity

Since we can apply functions f to operators via (B.19), we can also extend the notions of monotonicity and concavity of f to the domain of positive operators by using the partial ordering of operators under positivity. A function $f : \mathbb{R}_+ \to \mathbb{R}$ is called *operator monotone* if $A \geq B$ implies $f(A) \geq f(B)$ for all Hermitian A and B, and it is called *operator convex* if $f(\lambda A + (1 - \lambda)B) \leq \lambda f(A) + (1 - \lambda)f(B)$ for all $\lambda \in [0, 1]$. Reversing the inequalities for the function, f is *operator antimonotone* when $-f$ is operator monotone and *operator concave* when $-f$ is operator convex. These are stricter conditions than when working over the reals; even the square function is not operator monotone. This is easily seen by the example

$$A = \begin{pmatrix} 1 & 1 \\ 1 & 1 \end{pmatrix} \quad \text{and} \quad B = \begin{pmatrix} 2 & 1 \\ 1 & 1 \end{pmatrix} . \tag{B.27}$$

Nevertheless, it is easy to show operator convexity in this case.

Exercise B.7. Show that $f : t \mapsto t^2$ is operator convex.

An important function for our purposes is the square root, which is both monotone and concave.

Lemma B.7. *The function $f : t \mapsto \sqrt{t}$ is operator monotone.*

Proof. We show that for positive semidefinite A and Hermitian B, $A^2 \geq B^2$ implies $A \geq B$, and the proof is an application of the Cauchy–Schwarz inequality. For any vector $|\lambda\rangle$, we have

$$\left|\langle\lambda|AB|\lambda\rangle\right|^2 \leq \langle\lambda|A^2|\lambda\rangle\langle\lambda|B^2|\lambda\rangle \leq \langle\lambda|A^2|\lambda\rangle^2 . \tag{B.28}$$

Now take $|\lambda\rangle$ to be an eigenvector of $A - B$ with eigenvalue λ. Then $\langle\lambda|AB|\lambda\rangle = \langle\lambda|A(A - \lambda\mathbb{1})|\lambda\rangle = \langle\lambda|A^2|\lambda\rangle - \lambda\langle\lambda|A|\lambda\rangle$, and therefore by (B.28) it follows that

$$\left|\langle\lambda|A^2|\lambda\rangle - \lambda\langle\lambda|A|\lambda\rangle\right| \leq \langle\lambda|A^2|\lambda\rangle . \tag{B.29}$$

But $\langle\lambda|A^2|\lambda\rangle \geq 0$, so this can only hold if $\lambda\langle\lambda|A|\lambda\rangle \geq 0$. Since $A \geq 0$, it follows that $\lambda \geq 0$ and hence $A \geq B$. □

Lemma B.8. *The function $f : t \mapsto \sqrt{t}$ is operator concave.*

Proof. Since the square root function is operator monotone, we just need to show

$$\lambda A + (1-\lambda)B \geq \left(\lambda\sqrt{A} + (1-\lambda)\sqrt{B}\right)^2 \tag{B.30}$$

for all $\lambda \in [0,1]$. Expanding the right-hand side gives

$$\lambda A + (1-\lambda)B \geq \lambda^2 A + (1-\lambda)^2 B + \lambda(1-\lambda)(\sqrt{A}\sqrt{B} + \sqrt{B}\sqrt{A}) , \tag{B.31}$$

which is equivalent to $A + B \geq (\sqrt{A}\sqrt{B} + \sqrt{B}\sqrt{A})$. This inequality is just $(\sqrt{A} - \sqrt{B})^2 \geq 0$. □

Meanwhile, anti-monotonicity holds for the inverse.

Lemma B.9. *The function $f : t \mapsto t^{-1}$ is operator antimonotone.*

Proof. For $A > 0$ and $B \geq 0$, the operator $C = A^{-1/2}BA^{-1/2}$ satisfies $C \geq 0$ by Exercise B.3. Therefore it follows that $(\mathbb{1} + C)^{-1} \leq \mathbb{1}$. Now observe that $A^{-1} - (A+B)^{-1} = A^{-1/2}(\mathbb{1} - (\mathbb{1} + C)^{-1})A^{-1/2}$ using $(XYX)^{-1} = X^{-1}Y^{-1}X^{-1}$ for $X, Y \geq 0$. Using the previous inequality in this equality gives $A^{-1} \geq (A+B)^{-1}$, which completes the proof. □

We can also show joint concavity of the operator *geometric mean*. Recall that the geometric mean of two real numbers a and b is just $c = \sqrt{ab}$. This c is the largest possible value such that the matrix $\left(\begin{smallmatrix} a & c \\ c & b \end{smallmatrix}\right)$ is positive semidefinite, since positivity of the determinant requires $ab - c^2 \geq 0$. For two positive definite operators A and B, we define the geometric mean in the same way and denote it as $A\#B$. Specifically, $A\#B$ is the largest (in the operator ordering sense) positive C such that

$$\begin{pmatrix} A & C \\ C & B \end{pmatrix} \geq 0 . \tag{B.32}$$

By definition we have $B\#A = A\#B$. The variational formulation immediately implies that the function $(A, B) \mapsto A\#B$ is jointly concave.

Lemma B.10 (Joint concavity of the geometric mean). *For arbitrary positive definite operators A_0, B_0, A_1, and B_1 on* $\mathrm{Lin}(\mathcal{H})$ *and any $t \in [0,1]$, let $A = tA_0 + (1-t)A_1$ and $B = tB_0 + (1-t)B_1$. Then*

$$A\#B \geq tA_0\#B_0 + (1-t)A_1\#B_1 . \tag{B.33}$$

Proof. For $C_0 = A_0\#B_0$ and $C_1 = A_1\#B_1$, let $C = tC_0 + (1-t)C_1$. Then we have

$$\begin{pmatrix} A & C \\ C & B \end{pmatrix} = t \begin{pmatrix} A_0 & C_0 \\ C_0 & B_0 \end{pmatrix} + (1-t) \begin{pmatrix} A_1 & C_1 \\ C_1 & B_1 \end{pmatrix} . \tag{B.34}$$

Since the convex combination is positive, it follows that $A\#B \geq C$. □

The geometric mean has a closed-form expression, albeit somewhat ungainly:

$$A\#B = A^{1/2}(A^{-1/2}BA^{-1/2})^{1/2}A^{1/2} . \tag{B.35}$$

This follows at once using the properties of Schur complement discussed above. Positivity of the block matrix in the definition implies $B \geq CA^{-1}C$, and therefore $A^{-1/2}BA^{-1/2} \geq (A^{-1/2}CA^{-1/2})^2$. Monotonicity of the square root gives the desired form.

B.10 Notes and further reading

There are a multitude of books on linear algebra. A few favorites are by Strang [277], Körner [177], and Axler [9]. Dirac notation was introduced in [82]. For more on the Schur complement, see the volume edited by Zhang [315]. Carlen [54] gives an excellent overview of operator monotonicity, concavity, and convexity. Readers interested in exploring matrix analysis in further detail should consult Bhatia [38] and the two volumes by Horn and Johnson [147, 148].

C Semidefinite programs

C.1 General form

We start with the simpler case of linear programs. A linear program (LP) is an optimization of a linear function over a set of real variables defined by linear constraints. The general form of an LP is

$$\begin{aligned} &\underset{x}{\text{infimum}} \quad a \cdot x \\ &\text{subject to} \quad Lx \geq b\,, \quad x \geq 0\,, \quad x \in \mathbb{R}^n\,, \end{aligned} \tag{C.1}$$

with $a \in \mathbb{R}^n$, $b \in \mathbb{R}^m$, $L \in \mathbb{R}^{m \times n}$. The optimization is thus completely defined by the three *parameters* (L, a, b). The constraints are given by a set of m inequalities to the *variable* vector $x \in \mathbb{R}^n$. The inequality is pointwise: $(Lx)_j \geq b_j$ for the jth component of Lx and b, respectively, for all j. Equality conditions can be specified by pairs of inequalities, e. g., $x_0 = b_0$ by $x_0 \geq b_0$ and $-x_0 \geq -b_0$ for a given component.

If the resulting feasible set is closed, it forms a polytope. Since the objective function is linear, the optimal value will be found at one of the extreme points of the feasible set—one of the vertices of the polytope. If the feasible set is open, then the optimal value may be finite and again occurs at one of the vertices, or the optimal value may be unbounded. We write infimum instead of minimum to handle the case of an empty feasible set, i. e., when there exists no x such that $Lx \geq b$ or when the problem is unbounded. In the former case the value of the optimization is $+\infty$, and $-\infty$ in the latter.

Semidefinite programs (SDPs) have a completely analogous structure, but the variables are Hermitian matrices, and the notion of positivity is positive semidefiniteness. More precisely, the objective function uses the Hilbert–Schmidt inner product, and the positivity constraint is implemented using a linear map on Hermitian operators:

$$\begin{aligned} &\underset{X}{\text{infimum}} \quad \mathrm{Tr}[AX] \\ &\text{subject to} \quad \mathcal{L}[X] \geq B\,, \quad X \geq 0\,, \quad X \in \mathrm{Herm}(n)\,, \end{aligned} \tag{C.2}$$

with $A \in \mathrm{Herm}(n)$, $B \in \mathrm{Herm}(m)$, $\mathcal{L} : \mathrm{Herm}(n) \to \mathrm{Herm}(m)$. Observe that, in our presentation, SDPs are to LPs as quantum theory is to probability theory. As with LPs, the optimization problem is specified by the three parameters $(\mathcal{L}, A, B)$. Let us denote the optimal value in (C.2) by $f(\mathcal{L}, A, B)$ and call it the *primal optimization*. A Hermiticity-preserving superoperator turns out to be one whose Choi operator is Hermitian.

Exercise C.1. Show that a superoperator that maps Hermitian operators to Hermitian operators has a Hermitian Choi operator. *Hint: First show that such a map $\mathcal{E}$ satisfies $\mathcal{E}[X]^* = \mathcal{E}[X^*]$ for arbitrary X by decomposing $X = \frac{1}{2}(X + X^*) + i\frac{1}{2i}(X - X^*)$ and using linearity. It is also helpful to note that the partial trace is compatible with the adjoint in that* $\mathrm{Tr}_B[S_{AB}]^* = \mathrm{Tr}_B[S_{AB}^*]$.

https://doi.org/10.1515/9783110570250-023

The above form (C.2) is sometimes called the *inequality* form, since the constraint involving $\mathcal{L}$ is given as an inequality. Often, SDPs are defined in *equality form*, where this constraint is an equality. The two forms can be converted into each other using *slack variables*. To convert $\mathcal{L}[X] \geq B$ into equality form, for instance, simply invent a new variable X' and demand that $X' = \mathcal{L}[X] - B$ and $X' \geq 0$. The constraints now become $\mathcal{L}[X] - X' = B$, $X \geq 0$, and $X' \geq 0$.

C.2 Duality

The *dual optimization* gives lower bound on the value of the primal. To derive the dual optimization, consider $Y \in \mathrm{Herm}(m)$. First, observe that for $Y \geq 0$, we have

$$\mathrm{Tr}\big[\mathcal{L}[X]Y\big] \geq \mathrm{Tr}[BY]\,. \tag{C.3}$$

If we ensure that for all positive semidefinite $X \in \mathrm{Herm}(n)$,

$$\mathrm{Tr}[AX] \geq \mathrm{Tr}\big[\mathcal{L}(X)Y\big] \tag{C.4}$$

by requiring $\mathcal{L}^*[Y] \leq A$, then

$$\mathrm{Tr}[AX] \geq \mathrm{Tr}[BY]\,. \tag{C.5}$$

Neither of the constraints $Y \geq 0$ or $\mathcal{L}^*[Y] \leq A$ is necessary to have $\mathrm{Tr}[AX] \geq \mathrm{Tr}[BY]$; they are in general only sufficient. Nonetheless, any Y satisfying these conditions gives a lower bound on the objective function of $f(\mathcal{L}, A, B)$ for all feasible X. The tightest lower bound is given by the dual optimization

$$\begin{aligned} f^{\dagger}(\mathcal{L}, A, B) = \underset{Y}{\text{supremum}} \quad & \mathrm{Tr}[BY] \\ \text{subject to} \quad & \mathcal{L}^*[Y] \leq A\,, \quad Y \geq 0\,, \quad Y \in \mathrm{Herm}(m)\,. \end{aligned} \tag{C.6}$$

Again, the optimization might be infeasible, which now leads to a value of $= -\infty$, finite, or unbounded, i. e., $+\infty$.

Exercise C.2. Show that the dual of the dual is the primal. Therefore either can be used as the original optimization.

Exercise C.3. Show that the dual of (C.1) is

$$\begin{aligned} \underset{y}{\text{supremum}} \quad & b \cdot y\,, \\ \text{subject to} \quad & L^T y \leq a\,, \quad y \geq 0\,, \quad y \in \mathbb{R}^m\,. \end{aligned} \tag{C.7}$$

What happens to equality constraints in the dual?

By construction, $f^\dagger(\mathcal{L}, A, B) \leq f(\mathcal{L}, A, B)$, which is the statement of *weak duality*. The difference $f(\mathcal{L}, A, B) - f^\dagger(\mathcal{L}, A, B)$ is called the *duality gap*. *Strong duality* holds when the primal and dual are equal, i. e., when the duality gap is zero. From the construction of the dual it is easy to see that the following two conditions are each sufficient for zero duality gap.

Proposition C.1 (Zero duality gap). *For an arbitrary SDP, the following are each sufficient for* $f(\mathcal{L}, A, B) = f^\dagger(\mathcal{L}, A, B)$:

1. *(Equal objective functions) If there exist feasible X and Y such that* $\mathrm{Tr}[AX] = \mathrm{Tr}[BY]$, *then X and Y are optimizers and* $f(\mathcal{L}, A, B) = f^\dagger(\mathcal{L}, A, B)$.
2. *(Complementary slackness) If X, Y are feasible and satisfy the conditions*

$$\mathrm{Tr}\big[(\mathcal{L}[X] - B)Y\big] = 0 \quad \textit{and} \quad \mathrm{Tr}\big[(\mathcal{L}^*[Y] - A)X\big] = 0\,, \tag{C.8}$$

then X and Y are optimizers, and $f(\mathcal{L}, A, B) = f^\dagger(\mathcal{L}, A, B)$.

The complementary slackness conditions are just the equality versions of (C.3) and (C.4), which then implies equality of the objective functions. Because both factors in the trace of each slackness condition are positive, they imply the stronger form

$$(\mathcal{L}[X] - B)Y = 0 \quad \text{and} \quad (\mathcal{L}^*[Y] - A)X = 0\,. \tag{C.9}$$

The name complementary slackness refers to the fact that there cannot be “slack” in both the constraint and the dual variable. Either the constraint is *binding* in that it is satisfied with equality (perhaps, only on a subspace), the dual variable is zero (on a subspace), or both. A positive dual variable implies the corresponding constraint is binding, whereas slack in the constraint implies that the dual variable is zero. However, a zero dual variable does not imply slack in the constraint. This occurs when a constraint is redundant.

Exercise C.4. Consider the linear program specified by

$$A = -\begin{pmatrix} 3 & 0 & 2 \\ 3 & 3 & 1 \\ -3 & 0 & 1 \\ 0 & -3 & 2 \end{pmatrix}, \quad b = (-9, -12, 0, 0)\,, \quad c = (0, 0, -1)\,. \tag{C.10}$$

Show that all primal constraints are binding at the optimum, but, nonetheless, some of the dual variables may vanish. Can we drop any of the constraints and still obtain the same optimal value? Is the dual optimizer unique?

It turns out that the duality gap of every linear program is either zero or infinite. Moreover, the latter happens only when both the primal and dual are infeasible. So there are three possibilities for LPs:

1. Both primal and dual are infeasible (infinite duality gap, infinite optimum),

2. One feasible and the other unbounded (zero duality gap, infinite optimum), or
3. Both are feasible (zero duality gap, finite optimum).

Hence, strong duality holds if *either* of the primal or dual is feasible, and the optimal value is finite if *both* are feasible. Unboundedness of the primal (dual) implies infeasibility of the dual (primal).

Exercise C.5. For which of the following linear programs does strong duality hold? For which is the optimal value finite?

1. $A = \left(\begin{smallmatrix} 1 & -1 \\ -1 & 1 \end{smallmatrix}\right)$, $b = (2, 1)$, $c = (1, -2)$,
2. $A = \left(\begin{smallmatrix} 1 & -1 \\ -1 & 1 \end{smallmatrix}\right)$, $b = (-2, 1)$, $c = (1, -2)$,
3. $A = \left(\begin{smallmatrix} 1 & -1 \\ -1 & 1 \end{smallmatrix}\right)$, $b = (-2, 1)$, $c = (1, 2)$.

In contrast, the duality gap for SDPs can be finite, as we will see below. A useful condition for strong duality is known as Slater's condition, which makes use of *strict feasibility*. This simply means that all inequalities describing the feasible region must be strictly satisfied (all equality conditions are stated as such, not as pairs of inequalities). In particular, variables that are constrained to be positive semidefinite in general must be strictly positive to be strictly feasible.

Proposition C.2 (Slater conditions for strong duality). *For an arbitrary semidefinite program specified by* $(\mathcal{L}, A, B)$,

1. *if the primal is feasible and the dual strictly feasible, then* $f(\mathcal{L}, A, B) = f^{\dagger}(\mathcal{L}, A, B)$, *and there exists a primal optimizer* $X^{\star}$, *and*
2. *if the dual is feasible and primal strictly feasible, then* $f(\mathcal{L}, A, B) = f^{\dagger}(\mathcal{L}, A, B)$, *and there exists a dual optimizer* $Y^{\star}$.

When strong duality is known to hold, it follows that the complementary slackness conditions (C.8) must also hold. These conditions are often extremely useful in constructing the optimal variables. The following example shows that strong duality does not always hold for SDPs.

Example C.1. Consider the primal SDP $f(\mathcal{L}, A, B)$ for $X \in \mathrm{Herm}(3)$ with $A = \frac{1}{2}\left(\begin{smallmatrix} 0 & 1 & 0 \\ 1 & 0 & 0 \\ 0 & 0 & 0 \end{smallmatrix}\right)$, $B = \mathrm{diag}(0, 0, -1)$, and

$$\mathcal{L}[X] = \begin{pmatrix} 0 & X_{12} & 0 \\ X_{21} & X_{22} & 0 \\ 0 & 0 & \frac{1}{2}(X_{12} + X_{21}) \end{pmatrix}. \tag{C.11}$$

Notice that the upper left 2×2 submatrix of the constraint $\mathcal{L}[X] \geq B$ amounts to $\left(\begin{smallmatrix} 0 & X_{12} \\ X_{21} & X_{22} \end{smallmatrix}\right) \geq 0$. By Lemma B.2, $X_{12} = X_{21} = 0$, since the diagonal entry in the first row is zero. A zero on the diagonal of a positive semidefinite matrix implies that the entire row and column must also be zero. This can also be seen by using the Schur complement. The optimization is feasible; for instance, $X = 0$ is feasible. Therefore the optimal value is finite and satisfies $f(\mathcal{L}, A, B) = 0$.

Now we can derive the dual. Computing the adjoint of $\mathcal{L}$, we find $\mathrm{Tr}[\mathcal{L}[X]Y] = X_{12}Y_{21} + X_{21}Y_{12} + X_{22}Y_{22} + \frac{1}{2}(X_{12} + X_{21})Y_{33}$, and therefore

$$\mathcal{L}^*[Y] = \begin{pmatrix} 0 & Y_{12} + \frac{1}{2}Y_{33} & 0 \\ Y_{21} + \frac{1}{2}Y_{33} & Y_{22} & 0 \\ 0 & 0 & 0 \end{pmatrix}. \tag{C.12}$$

The task is to find the supremum of the objective function $-Y_{33}$ subject to $\mathcal{L}[Y] \le A$ and $Y \ge 0$. Consider the 22 component in the two constraints. From the latter, $Y_{22} \ge 0$, while the opposite must hold in the former. Hence $Y_{22} = 0$, which in turn implies $Y_{12} = Y_{21} = 0$. All that remains of the $\mathcal{L}[Y] \le A$ constraint is $Y_{33}\sigma_x \le \sigma_x$, which implies $Y_{33} = 1$. Any choice of Y with $Y_{2j} = Y_{j2} = 0$, $Y_{33} = 1$, and Y_{11} and Y_{13} satisfying $Y_{11} - |Y_{13}|^2 \ge 0$ satisfies all the constraints. Therefore $f^\dagger(\mathcal{L}, A, B) = -1 \ne f(\mathcal{L}, A, B)$.

In this particular example, we can appreciate that strong duality does not hold because the constraints $Y \ge 0$ and $\mathcal{L}^*[Y] \le A$ are not necessary for $\mathrm{Tr}[AX] \ge \mathrm{Tr}[BY]$. For instance, $Y = \mathrm{diag}(1, 0, 0)$ is certainly positive, does not satisfy $\mathcal{L}^*[Y] \le A$, and yet gives $\mathrm{Tr}[BY] = 0$. On the other hand, $Y = \frac{1}{2}\left(\begin{smallmatrix} 0 & 1 & 0 \\ 1 & -1 & 0 \\ 0 & 0 & 0 \end{smallmatrix}\right)$ is not positive but satisfies $\mathcal{L}^*[Y] \le A$ and $\mathrm{Tr}[BY] = 0$. Moreover, although there exist primal and dual optimizers, e. g., $X = 0$ and $Y = \mathrm{diag}(0, 0, 1)$, there are no strictly feasible primal or dual variables. The 2×2 constraint in the primal, which is ultimately forced to be diagonal, is necessarily satisfied with equality in the 11 component. Similarly, with $Y_{22} = 0$ in the dual, the 22 component of $Y \ge 0$ is necessarily satisfied with equality.

It is also possible for strong duality to hold, but the optimal value of the primal or the dual fail to be achieved by any feasible variable. Put differently, it can happen that only one of the conditions in Proposition C.2 is fulfilled. The following example also makes use of Lemma B.2.

Example C.2. Consider the SDP for $X \in \mathrm{Herm}(2)$ with

$$A = \begin{pmatrix} 1 & 0 \\ 0 & 0 \end{pmatrix}, \quad B = -\begin{pmatrix} 0 & 1 \\ 1 & 0 \end{pmatrix}, \quad \text{and} \quad \mathcal{L}[X] = \begin{pmatrix} X_{11} & 0 \\ 0 & X_{22} \end{pmatrix}. \tag{C.13}$$

The dual is $f^\dagger(\mathcal{L}, A, B) = \sup\{-Y_{21} - Y_{12} : Y_{11} \le 1, Y_{22} \le 0, Y \ge 0\}$. The latter two constraints imply $Y_{12} = Y_{21} = 0$, whereas Y_{11} is constrained to the unit interval. The objective function is zero, and the dual is feasible.

Meanwhile, the primal is strictly feasible: $X = 2\mathbb{1}$ gives $2\mathbb{1} = \mathcal{L}[X] > B = \sigma_x$. By Slater's condition strong duality therefore holds. The primal optimization is $f(\mathcal{L}, A, B) = \inf\{X_{11} : \left(\begin{smallmatrix} X_{11} & 1 \\ 1 & X_{22} \end{smallmatrix}\right) \ge 0, X \ge 0\}$. Nevertheless, again by Lemma B.2, the infimum $X_{11} = 0$ cannot be attained. Alternatively, in this case the conditions imply $X_{11} \ge 0$, $X_{22} \ge 0$, and $X_{11}X_{22} - 1 \ge 0$, i. e., $X_{11} = 0$ is not a feasible choice.

C.3 Notes and further reading

In this chapter, we only scratch the surface of the very important topic of linear programming, semidefinite programming, and convex optimization in general. For an overview of convex optimization, including semidefinite programming, see Boyd and Vanderberghe [44, 289] for an applied approach and more mathematical treatments by van Tiel [288], Rockafellar [244], and especially Barvinok [13]. The duality derivation (for LPs) goes back to von Neumann [294] in 1947; Dantzig [68] provides interesting historical details. Slater's condition, which is formulated for general convex problems, appeared just a few years later [267].

Bibliography

[1] D. Aharonov, A. Kitaev, and N. Nisan, “Quantum Circuits with Mixed States”, Proceedings of the Thirtieth Annual ACM Symposium on Theory of Computing (1998), pp. 20–30.

[2] P. M. Alberti, “A note on the transition probability over C*-algebras”, Letters in Mathematical Physics **7**, 25–32 (1983).

[3] P. M. Alberti and A. Uhlmann, “Stochastic linear maps and transition probability”, Letters in Mathematical Physics **7**, 107–112 (1983).

[4] R. Alicki and M. Fannes, “Continuity of quantum conditional information”, Journal of Physics A: Mathematical and General **37**, L55–L57 (2004).

[5] A. E. Allahverdyan and D. B. Saakian, “Converse coding theorems for quantums source and noisy channel”, arXiv:quant-ph/9702034 (1997).

[6] T. Ando, Concavity of certain maps on positive definite matrices and applications to Hadamard products. Linear Algebra and its Applications **26**, 203–241 (1979).

[7] H. Araki and E. H. Lieb, “Entropy inequalities”, Communications in Mathematical Physics **18**, 160–170 (1970).

[8] H. Araki and G. A. Raggio, “A remark on transition probability”, Letters in Mathematical Physics **6**, 237–240 (1982).

[9] S. Axler, *Linear Algebra Done Right* (Springer, New York, 2014).

[10] H. Barnum and E. Knill, “Reversing quantum dynamics with near-optimal quantum and classical fidelity”, Journal of Mathematical Physics **43**, 2097–2106 (2002).

[11] H. Barnum, E. Knill, and M. A. Nielsen, “On quantum fidelities and channel capacities”, IEEE Transactions on Information Theory **46**, 1317–1329 (2000).

[12] H. Barnum, M. A. Nielsen, and B. Schumacher, “Information transmission through a noisy quantum channel”, Physical Review A **57**, 4153 (1998).

[13] A. Barvinok, *A Course in Convexity* (American Mathematical Society, Providence, RI, 2002).

[14] W. Beckner, “Inequalities in Fourier analysis”, The Annals of Mathematics, Second Series **102**, 159–182 (1975).

[15] W. Beckner, “Inequalities in Fourier analysis on R^n”, Proceedings of the National Academy of Sciences of the United States of America **72**, 638–641 (1975).

[16] S. Beigi and A. Gohari, “Quantum achievability proof via collision relative entropy”, IEEE Transactions on Information Theory **60**, 7980–7986 (2014).

[17] V. P. Belavkin, “Optimal multiple quantum statistical hypothesis testing”, Stochastics **1**, 315 (1975).

[18] V. P. Belavkin and P. Staszewski, “C*-algebraic generalization of relative entropy and entropy”, Annales de l’I. H. P. Physique théorique **37**, 51–58 (1982).

[19] J. S. Bell, “On the Einstein–Podolsky–Rosen paradox”, Physics **1**, 195 (1964).

[20] J. S. Bell, “On the problem of hidden variables in quantum mechanics”, Reviews of Modern Physics **38**, 447 (1966).

[21] S. M. Bellovin, “Frank Miller: inventor of the one-time pad”, Cryptologia **35**, 203–222 (2011).

[22] C. H. Bennett, “The thermodynamics of computation—a review”, International Journal of Theoretical Physics **21**, 905–940 (1982).

[23] C. H. Bennett, “Demons, engines, and the second law”, Scientific American **257**, 108–117 (1987).

[24] C. H. Bennett, “Quantum information: qubits and quantum error correction”, International Journal of Theoretical Physics **42**, 153–176 (2003).

[25] C. H. Bennett, H. J. Bernstein, S. Popescu, and B. Schumacher, “Concentrating partial entanglement by local operations”, Physical Review A **53**, 2046 (1996).

https://doi.org/10.1515/9783110570250-024

[26] C. H. Bennett and G. Brassard, "Quantum Cryptography: Public Key Distribution and Coin Tossing", Proceedings of International Conference on Computers Systems and Signal Processing (1984), pp. 175–179. Reprinted in Theoretical Computer Science **560**, 7–11 (2014).

[27] C. H. Bennett, G. Brassard, C. Crépeau, R. Jozsa, A. Peres, and W. K. Wootters, Teleporting an unknown quantum state via dual classical and Einstein–Podolsky–Rosen channels. Physical Review Letters **70**, 1895 (1993).

[28] C. H. Bennett, G. Brassard, C. Crépeau, and U. M. Maurer, "Generalized Privacy Amplification", Proceedings of the 1994 IEEE International Symposium on Information Theory (1994).

[29] C. H. Bennett, G. Brassard, and N. D. Mermin, "Quantum cryptography without Bell's theorem", Physical Review Letters **68**, 557 (1992).

[30] C. H. Bennett, G. Brassard, and J.-M. Robert, "Privacy amplification by public discussion", SIAM Journal on Computing **17**, 210–229 (1988).

[31] C. H. Bennett, D. P. DiVincenzo, J. A. Smolin, and W. K. Wootters, "Mixed-state entanglement and quantum error correction", Physical Review A **54**, 3824 (1996).

[32] C. H. Bennett and R. Landauer, "The fundamental physical limits of computation", Scientific American **253**, 48–56 (1985).

[33] C. H. Bennett and S. J. Wiesner, "Communication via one- and two-particle operators on Einstein–Podolsky–Rosen states", Physical Review Letters **69**, 2881 (1992).

[34] M. Berta, M. Christandl, R. Colbeck, and J. M. Renes, R. Renner, "The uncertainty principle in the presence of quantum memory", Nature Physics **6**, 659–662 (2010).

[35] M. Berta, P. J. Coles, and S. Wehner, "Entanglement-assisted guessing of complementary measurement outcomes", Physical Review A **90**, 062127 (2014).

[36] R. Bertlmann and A. Zeilinger, eds., *Quantum [Un]speakables II, The Frontiers Collection* (Springer, Cham, Switzerland, 2017).

[37] R. A. Bertlmann and A. Zeilinger, *Quantum [Un]speakables* (Springer, Berlin, 2002).

[38] R. Bhatia, *Matrix Analysis*, Vol. 169, Graduate Texts in Mathematics (Springer, New York, 1996).

[39] I. Białynicki-Birula and J. Mycielski, "Uncertainty relations for information entropy in wave mechanics", Communications in Mathematical Physics **44**, 129–132 (1975).

[40] I. Bjelaković and R. Siegmund-Schultze, "Quantum Stein's lemma revisited, inequalities for quantum entropies, and a concavity theorem of Lieb", arXiv:quant-ph/0307170v2 (2012).

[41] N. Bohr, "The quantum postulate and the recent development of atomic theory", Nature **121**, 580–590 (1928).

[42] G. Boole, *The Mathematical Analysis of Logic* (Macmillan, Barclay, & Macmillan, Cambridge, and George Bell, London, 1847). Reprinted by Cambridge University Press (Cambridge, 2009).

[43] S. Boucheron, G. Lugosi, and P. Massart, *Concentration Inequalities: A Nonasymptotic Theory of Independence* (Oxford University Press, Oxford, 2013).

[44] S. Boyd and L. Vandenberghe, *Convex Optimization* (Cambridge University Press, Cambridge, 2004).

[45] F. Brandão and N. Datta, "One-shot rates for entanglement manipulation under non-entangling maps", IEEE Transactions on Information Theory **57**, 1754–1760 (2011).

[46] G. Brassard, "Brief History of Quantum Cryptography: A Personal Perspective", Proceedings of the IEEE Information Theory Workshop on Theory and Practice in Information-Theoretic Security, 2005 (2005), pp. 19–23.

[47] L. Brillouin, "Maxwell's demon cannot operate: information and entropy. I", Journal of Applied Physics **22**, 334–337 (1951).

[48] J. Bub, "Von Neumann's Theory of Quantum Measurement", *John von Neumann and the Foundations of Quantum Physics*, Vol. 8, edited by M. Rédei and M. Stöltzner, Vienna Circle Institute Yearbook [2000] (Springer, Dordrecht 2001), pp. 63–74.

[49] F. Buscemi and N. Datta, "Distilling entanglement from arbitrary resources", Journal of Mathematical Physics **51**, 102201 (2010).

[50] P. Busch and C. Shilladay, "Complementarity and uncertainty in Mach–Zehnder interferometry and beyond", Physics Reports **435**, 1–31 (2006).

[51] V. Bush, "Instrumental analysis", Bulletin of the American Mathematical Society **42**, 649–669 (1936).

[52] A. R. Calderbank and P. W. Shor, "Good quantum error-correcting codes exist", Physical Review A **54**, 1098 (1996).

[53] J. Callas, Curious RNG Stalemate [Was: Use of Cpunks], cypherpunks@cpunks.org (2013).

[54] E. Carlen, Trace Inequalities and Quantum Entropy: An Introductory Course, *Contemporary Mathematics*, Vol. 529, edited by R. Sims and D. Ueltschi (American Mathematical Society, Providence, RI, 2010), pp. 73–140.

[55] R. Chandler, *Raymond Chandler speaking*, edited by D. Gardiner, K. S. Walker (University of California Press, Berkeley, 1997).

[56] H. Chernoff, "A measure of asymptotic efficiency for tests of a hypothesis based on the sum of observations", The Annals of Mathematical Statistics **23**, 493–507 (1952).

[57] E. C. Cherry, "A history of the theory of information", Proceedings of the IEE—Part III: Radio and Communication Engineering **98**, 383–393 (1951).

[58] A. M. Childs, J. Preskill, and J. Renes, "Quantum information and precision measurement", Journal of Modern Optics **47**, 155–176 (2000).

[59] M.-D. Choi, "Completely positive linear maps on complex matrices", Linear Algebra and its Applications **10**, 285–290 (1975).

[60] M. Christandl and A. Winter, "Uncertainty, monogamy, and locking of quantum correlations", IEEE Transactions on Information Theory **51**, 3159–3165 (2005).

[61] J. F. Clauser, M. A. Horne, A. Shimony, and R. A. Holt, "Proposed experiment to test local hidden-variable theories", Physical Review Letters **23**, 880 (1969).

[62] R. Clausius, "Ueber die Anwendung des Satzes von der Aequivalenz der Verwandlungen auf die innere Arbeit", Annalen der Physik **192**, 73–112 (1862). Trans. by T. A. Hirst as "On the application of the theorem of the equivalence of transformations to interior work", *The mechanical theory of heat: with its applications to the steam-engine and to the physical properties of bodies* (J. Van Voorst, London, 1867), pp. 215–250.

[63] P. J. Coles, M. Berta, M. Tomamichel, and S. Wehner, "Entropic uncertainty relations and their applications", Reviews of Modern Physics **89**, 015002 (2017).

[64] T. M. Cover and J. A. Thomas, *Elements of Information Theory*, Second edition (Wiley-Interscience, Hoboken, NJ, 2006).

[65] I. Csiszár, "Information-type measures of difference of probability distributions and indirect observations", Studia Scientiarum Mathematicarum Hungarica **2**, 229–318 (1967).

[66] I. Csiszár and J. Körner, "Broadcast channels with confidential messages", IEEE Transactions on Information Theory **24**, 339–348 (1978).

[67] A. I. Dale, "Bayes or Laplace? An examination of the origin and early applications of Bayes' theorem", Archive for History of Exact Sciences **27**, 23–47 (1982).

[68] G. B. Dantzig and M. N. Thapa, *Linear Programming*, Springer Series in Operations Research (Springer, New York, 1997).

[69] N. Datta and F. Leditzky, Second-order asymptotics for source coding, dense coding, and pure-state entanglement conversions. IEEE Transactions on Information Theory **61**, 582–608 (2015).

[70] E. B. Davies, *Quantum Theory of Open Systems* (Academic Press, London, 1976).

[71] E. B. Davies and J. T. Lewis, "An operational approach to quantum probability", Communications in Mathematical Physics **17**, 239–260 (1970).

[72] A. S. Davis, "Markov chains as random input automata", The American Mathematical Monthly **68**, 264–267 (1961).

[73] C. Davis, "Various averaging operations onto subalgebras", Illinois Journal of Mathematics **3**, 538–553 (1959).

[74] B. de Finetti, "La prévision : ses lois logiques, ses sources subjectives", Annales de l'institut Henri Poincaré 7, 1–68 (1937). Trans. by H. E. Kyburg as "Foresight: Its Logical Laws, Its Subjective Sources", *Studies in Subjective Probability*, edited by H. E. Kyburg and H. E. K. Smokler (Robert E. Krieger Publishing Company, Huntington, NY, 1980), pp. 53–118.

[75] J. de Pillis, "Linear transformations which preserve Hermitian and positive semidefinite operators", Pacific Journal of Mathematics **23**, 129–137 (1967).

[76] D. Deutsch, "Uncertainty in quantum measurements", Physical Review Letters **50**, 631 (1983).

[77] I. Devetak, "The private classical capacity and quantum capacity of a quantum channel", IEEE Transactions on Information Theory **51**, 44–55 (2005).

[78] I. Devetak and P. W. Shor, The capacity of a quantum channel for simultaneous transmission of classical and quantum information. Communications in Mathematical Physics **256**, 287–303 (2005).

[79] I. Devetak and A. Winter, "Classical data compression with quantum side information", Physical Review A **68**, 042301 (2003).

[80] I. Devetak and A. Winter, "Distillation of secret key and entanglement from quantum states", Proceedings of the Royal Society A **461**, 207–235 (2005).

[81] D. Dieks, "Communication by EPR devices", Physics Letters A **92**, 271–272 (1982).

[82] P. A. M. Dirac, "A new notation for quantum mechanics", Mathematical Proceedings of the Cambridge Philosophical Society **35**, 416–418 (1939).

[83] P. A. M. Dirac, *The Principles of Quantum Mechanics*, Fourth edition (Oxford University Press, Oxford, 1967).

[84] D. P. DiVincenzo, P. W. Shor, and J. A. Smolin, "Quantum-channel capacity of very noisy channels", Physical Review A **57**, 830 (1998).

[85] F. Dupuis, O. Fawzi, and S. Wehner, "Entanglement sampling and applications", IEEE Transactions on Information Theory **61**, 1093–1112 (2015).

[86] F. Dupuis, L. Krämer, P. Faist, J. M. Renes, R. Renner, and "Generalized Entropies", Proceedings of the XVIIth International Congress on Mathematical Physics (2013), pp. 134–153.

[87] R. Durrett, *Probability: Theory and Examples* (Cambridge University Press, Cambridge, 2010).

[88] A. Einstein, Über die Entwicklung unserer Anschauungen über das Wesen und die Konstitution der Strahlung, Physikalische Zeitschrift **10**, 817–825 (1909). Reprinted in "ÜBER DIE ENTWICKLUNG UNSERER ANSCHAUUNGEN ÜBER DAS WESEN UND DIE KONSTITUTION DER STRAHLUNG", *The Collected Papers of Albert Einstein*, Vol. 2: The Swiss Years: Writings, 1900–1909, edited by J. Stachel (Princeton University Press, Princeton, 1989), pp. 563–583. Translated by A. Beck as "On the Development of Our Views Concerning the Nature and Constitution of Radiation" in the accompanying English translation supplement, pp. 379–394.

[89] A. Einstein, "Maxwell's Influence on the Development of the Conception of Physical Reality", *James Clerk Maxwell: A Commemoration Volume 1831–1931* (Cambridge University Press, Cambridge, 1931), pp. 66–73.

[90] A. Einstein, B. Podolsky, and N. Rosen, Can quantum-mechanical description of physical reality be considered complete? Physical Review **47**, 777 (1935).

[91] A. K. Ekert, "Quantum cryptography based on Bell's theorem", Physical Review Letters **67**, 661 (1991).

[92] Y. Eldar, A. Megretski, and G. C. Verghese, "Designing optimal quantum detectors via semidefinite programming", IEEE Transactions on Information Theory **49**, 1007–1012 (2003).

[93] R. W. Emerson, *Essays & Lectures*, J. Porte (Library of America, New York, 1983).
[94] B.-G. Englert, "Remarks on some basic issues in quantum mechanics", Zeitschrift für Naturforschung A **54**, 11–32 (1999).
[95] H. Everett, "The theory of the universal wavefunction", PhD thesis, Princeton University (1957).
[96] K. Fang, X. Wang, M. Tomamichel, and R. Duan, "Non-asymptotic entanglement distillation", IEEE Transactions on Information Theory **65**, 6454–6465 (2019).
[97] R. M. Fano, *Transmission of Information: A Statistical Theory of Communications* (MIT Press, Cambridge, 1961).
[98] U. Fano, "Liouville Representation of Quantum Mechanics with Application to Relaxation Processes", *Lectures on the Many-body Problems*, Vol. 2, edited by E. R. Caianiello (Academic Press, New York, 1964), pp. 217–239.
[99] R. P. Feynman, *The Character of Physical Law* (MIT Press, Cambridge, 1967).
[100] R. P. Feynman, R. B. Leighton, and M. L. Sands, *The Feynman Lectures on Physics, Vol. 3: Quantum Mechanics* (Addison-Wesley, Reading, MA, 1963).
[101] A. S. Fletcher, P. W. Shor, and M. Z. Win, "Optimum quantum error recovery using semidefinite programming", Physical Review A **75**, 012338 (2007).
[102] R. L. Frank and E. H. Lieb, "Monotonicity of a relative Rényi entropy", Journal of Mathematical Physics **54**, 122201 (2013).
[103] C. Fuchs and J. van de Graaf, "Cryptographic distinguishability measures for quantum-mechanical states", IEEE Transactions on Information Theory **45**, 1216–1227 (1999).
[104] C. A. Fuchs and C. M. Caves, "Mathematical techniques for quantum communication theory", Open Systems & Information Dynamics **3**, 345–356 (1995).
[105] D. Gabor, "Communication theory and physics", Philosophical Magazine Series 7 **41**, 1161–1187 (1950).
[106] D. Gabor, "Light and Information", *Progress in Optics*, Vol. 1, edited by E. Wolf (North-Holland Publishing Co., Amsterdam, 1961), pp. 109–153.
[107] A. Gelman and C. R. Shalizi, "Philosophy and the practice of Bayesian statistics: *Philosophy and the practice of Bayesian statistics*", British Journal of Mathematical and Statistical Psychology **66**, 8–38 (2013).
[108] J. W. Gibbs, *Elementary Principles in Statistical Mechanics* (Charles Scribner's sons, New York, 1902). Reprinted by Cambridge University Press (Cambridge, 2010).
[109] A. Gilchrist, N. K. Langford, and M. A. Nielsen, "Distance measures to compare real and ideal quantum processes", Physical Review A **71**, 062310 (2005).
[110] V. Giovannetti, R. García-Patrón, N. J. Cerf, and A. S. Holevo, "Ultimate classical communication rates of quantum optical channels", Nature Photonics **8**, 796–800 (2014).
[111] N. Gisin, "Stochastic quantum dynamics and relativity", Helvetica Physica Acta **62**, 363 (1989).
[112] S. Givant and P. Halmos, *Introduction to Boolean Algebras*, Undergraduate Texts in Mathematics (Springer, New York, 2009).
[113] A. Gut, *An Intermediate Course in Probability*, Second edition, Springer Texts in Statistics (Springer, Dordrecht, 2009).
[114] R. Haag and D. Kastler, "An algebraic approach to quantum field theory", Journal of Mathematical Physics **5**, 848 (1964).
[115] I. Hacking, *An Introduction to Probability and Inductive Logic* (Cambridge University Press, Cambridge, 2001).
[116] M. J. W. Hall, "Information exclusion principle for complementary observables", Physical Review Letters **74**, 3307 (1995).
[117] P. R. Halmos, "The foundations of probability", The American Mathematical Monthly **51**, 493–510 (1944).

[118] M. Hamada, "Information rates achievable with algebraic codes on quantum discrete memoryless channels", IEEE Transactions on Information Theory **51**, 4263–4277 (2005).

[119] T. Han and S. Verdu, "Approximation theory of output statistics", IEEE Transactions on Information Theory **39**, 752–772 (1993).

[120] T. S. Han, "An information-spectrum approach to source coding theorems with a fidelity criterion", IEEE Transactions on Information Theory **43**, 1145–1164 (1997).

[121] T. S. Han, *Information-Spectrum Method in Information Theory*, Vol. 50, Stochastic Modelling and Applied Probability (Springer, Berlin, 2002).

[122] M. B. Hastings, "Superadditivity of communication capacity using entangled inputs", Nature Physics **5**, 255–257 (2009).

[123] P. Hausladen and W. K. Wootters, "A 'pretty good' measurement for distinguishing quantum states", Journal of Modern Optics **41**, 2385 (1994).

[124] M. Hayashi, Optimal sequence of quantum measurements in the sense of Stein's lemma in quantum hypothesis testing. Journal of Physics A: Mathematical and General **35**, 10759–10773 (2002).

[125] M. Hayashi, "Practical evaluation of security for quantum key distribution", Physical Review A **74**, 022307 (2006).

[126] M. Hayashi, Second-order asymptotics in fixed-length source coding and intrinsic randomness. IEEE Transactions on Information Theory **54**, 4619–4637 (2008).

[127] M. Hayashi, "Information spectrum approach to second-order coding rate in channel coding", IEEE Transactions on Information Theory **55**, 4947–4966 (2009).

[128] M. Hayashi, *Quantum Information Theory: Mathematical Foundation*, Second edition, Graduate Texts in Physics (Springer, Berlin, 2017).

[129] M. Hayashi and H. Nagaoka, "General formulas for capacity of classical-quantum channels", IEEE Transactions on Information Theory **49**, 1753–1768 (2003).

[130] P. Hayden, M. Horodecki, A. Winter, and J. Yard, "A decoupling approach to the quantum capacity", Open Systems & Information Dynamics **15**, 7–19 (2008).

[131] P. Hayden, P. W. Shor, and A. Winter, "Random quantum codes from Gaussian ensembles and an uncertainty relation", Open Systems & Information Dynamics **15**, 71–89 (2008).

[132] W. Heisenberg, "Über den anschaulichen Inhalt der quantentheoretischen Kinematik und Mechanik", Zeitschrift für Physik **43**, 172–198 (1927). Trans. by J. A. Wheeler and W. H. Zurek as "The physical content of quantum kinematics and mechanics", *Quantum theory and measurement* (Princeton University Press, Princeton, 1981), pp. 62–84.

[133] W. Heisenberg, Letter to Wolfgang Pauli, Feb. 23, 1927. Published in A. Hermann, K. v. Meyenn, and V. F. Weisskopf, eds., *Wolfgang Pauli: Wissenschaftlicher Briefwechse mit Bohr, Einstein, Heisenberg u. a. Band I: 1919–1929*, Vol. 2, Sources in the History of Mathematics and Physical Sciences (Springer New York, 1979).

[134] W. Heisenberg, *Der Teil und das Ganze: Gespräche im Umkreis der Atomphysik* (R. Piper, München, 1969). Trans. by A. J. Pomerans as Physics and beyond: encounters and conversations (Harper & Row, New York, 1972).

[135] K.-E. Hellwig and K. Kraus, "Pure operations and measurements", Communications in Mathematical Physics **11**, 214–220 (1969).

[136] K.-E. Hellwig and K. Kraus, "Operations and measurements. II", Communications in Mathematical Physics **16**, 142–147 (1970).

[137] C. W. Helstrom, "Detection theory and quantum mechanics", Information and Control **10**, 254–291 (1967).

[138] C. W. Helstrom, Quantum Detection and Estimation Theory, Vol. 123, Mathematics in Science and Engineering (Academic Press, London, 1976).

[139] F. Hiai and D. Petz, "The proper formula for relative entropy and its asymptotics in quantum probability", Communications in Mathematical Physics **143**, 99–114 (1991).

[140] I. I. Hirschman, "A note on entropy", American Journal of Mathematics **79**, 152–156 (1957).
[141] W. Hoeffding, "Probability inequalities for sums of bounded random variables", Journal of the American Statistical Association **58**, 13–30 (1963).
[142] A. Holevo, "The capacity of the quantum channel with general signal states", IEEE Transactions on Information Theory **44**, 269–273 (1998).
[143] A. S. Holevo, "An analog of the theory of statistical decisions in noncommutative probability theory", Trudy Moskovskogo Matematicheskogo Obshchestva (Transactions of the Moscow Mathematical Society) **26**, 133–149 (1972).
[144] A. S. Holevo, Bounds for the quantity of information transmitted by a quantum communication channel. Problemy Peredachi Informatsii (Problems of Information Transmission) **9**, 3–11 (1973).
[145] A. S. Holevo, *Probabilistic and Statistical Aspects of Quantum Theory*, Lecture Notes (Scuola Normale Superiore) (Edizioni della Normale, Pisa, 2011).
[146] A. S. Holevo, *Quantum Systems, Channels, Information: A Mathematical Introduction*, Second edition (De Gruyter, Berlin, 2019).
[147] R. A. Horn and C. R. Johnson, *Topics in Matrix Analysis* (Cambridge University Press, Cambridge, 1991).
[148] R. A. Horn and C. R. Johnson, *Matrix Analysis*, Second edition (Cambridge University Press, Cambridge, 2013).
[149] M. Horodecki, S. Lloyd, and A. Winter, "Quantum coding theorem from privacy and distinguishability", Open Systems & Information Dynamics **15**, 47–69 (2008).
[150] L. P. Hughston, R. Jozsa, and W. K. Wootters, "A complete classification of quantum ensembles having a given density matrix", Physics Letters A **183**, 14–18 (1993).
[151] R. Impagliazzo, L. A. Levin, and M. Luby, "Pseudo-Random Generation from One-Way Functions", Proceedings of the Twenty-First Annual ACM Symposium on Theory of Computing (1989), pp. 12–24.
[152] A. Jamiołkowski, "Linear transformations which preserve trace and positive semidefiniteness of operators", Reports on Mathematical Physics **3**, 275–278 (1972).
[153] J. Jauch and C. Piron, "Generalized localizability", Helvetica Physica Acta **40**, 559–570 (1967).
[154] E. T. Jaynes, "Information theory and statistical mechanics. II", Physical Review **108**, 171 (1957).
[155] E. T. Jaynes, *Probability Theory: The Logic of Science* (Cambridge University Press, Cambridge, 2003).
[156] D. H. Johnson, "Statistical signal processing", Lecture notes (2017).
[157] R. Jozsa, "Fidelity for mixed quantum states", Journal of Modern Optics **41**, 2315 (1994).
[158] V. Katariya and M. M. Wilde, "Geometric distinguishability measures limit quantum channel estimation and discrimination", Quantum Information Processing **20**, 78 (2021).
[159] J. H. B. Kemperman, "On the optimum rate of transmitting information", The Annals of Mathematical Statistics **40**, 2156–2177 (1969).
[160] N. Killoran, "Entanglement quantification and quantum benchmarking of optical communication devices", PhD thesis, (University of Waterloo, 2012).
[161] A. Y. Kitaev, "Quantum computations: algorithms and error correction", Russian Mathematical Surveys **52**, 1191–1249 (1997).
[162] O. Klein, "Zur quantenmechanischen Begründung des zweiten Hauptsatzes der Wärmelehre", Zeitschrift für Physik **72**, 767–775 (1931).
[163] R. Klesse, "Approximate quantum error correction, random codes, and quantum channel capacity", Physical Review A **75**, 062315 (2007).
[164] R. Klesse, "A random coding based proof for the quantum coding theorem", Open Systems & Information Dynamics **15**, 24–45 (2008).

[165] R. R. Kline, *The Cybernetics Moment: Or Why We Call Our Age the Information Age*, New Studies in American Intellectual and Cultural History (Johns Hopkins University Press, Baltimore, 2015).
[166] M. Koashi, Simple security proof of quantum key distribution via uncertainty principle, arXiv:quant-ph/0505108 (2005).
[167] M. Koashi, Complementarity, distillable secret key, and distillable entanglement, arXiv:0704.3661 [quant-ph] (2007).
[168] M. Koashi and J. Preskill, "Secure quantum key distribution with an uncharacterized source", Physical Review Letters **90**, 057902 (2003).
[169] S. Kochen and E. Specker, "The problem of hidden variables in quantum mechanics", Indiana University Mathematics Journal **17**, 59–87 (1967).
[170] R. L. Kosut and D. A. Lidar, "Quantum error correction via convex optimization", Quantum Information Processing **8**, 443–459 (2009).
[171] K. Kraus, *States, Effects, and Operations: Fundamental Notions of Quantum Theory*, edited by A. Böhm, J. D. Dollard, and W. H. Wootters Lecture Notes in Physics, Vol. 190, Lecture Notes in Physics (Springer, Berlin, 1983).
[172] K. Kraus, "Complementary observables and uncertainty relations", Physical Review D **35**, 3070 (1987).
[173] D. Kretschmann and R. F. Werner, "Tema con variazioni: quantum channel capacity", New Journal of Physics **6**, 26 (2004).
[174] S. Kullback, "A lower bound for discrimination information in terms of variation (Corresp.)", IEEE Transactions on Information Theory **13**, 126–127 (1967).
[175] S. Kullback and R. A. Leibler, "On information and sufficiency", The Annals of Mathematical Statistics **22**, 79–86 (1951).
[176] R. König, R. Renner, and C. Schaffner, "The operational meaning of min- and max-entropy", IEEE Transactions on Information Theory **55**, 4337–4347 (2009).
[177] T. W. Körner, *Vectors, Pure and Applied: A General Introduction to Linear Algebra* (Cambridge University Press, Cambridge, 2013).
[178] J. Ladyman, S. Presnell, and A. J. Short, "The use of the information-theoretic entropy in thermodynamics", Studies in History and Philosophy of Science Part B: Studies in History and Philosophy of Modern Physics **39**, 315–324 (2008).
[179] J. Ladyman, S. Presnell, A. J. Short, and B. Groisman, "The connection between logical and thermodynamic irreversibility", Studies in History and Philosophy of Science Part B: Studies in History and Philosophy of Modern Physics **38**, 58–79 (2007).
[180] L. Landau, "Das Dämpfungsproblem in der Wellenmechanik", Zeitschrift für Physik **45**, 430–441 (1927).
[181] R. Landauer, "Irreversibility and heat generation in the computing process", IBM Journal of Research and Development **5**, 183 (1961).
[182] R. Landauer, "Information is Physical", Physics Today **44**, 23 (1991).
[183] P.-S. Laplace, *Théorie Analytique Des Probabilités*, Second edition (Mme Ve Courcier, Paris, 1814).
[184] P.-S. Laplace, "Sur l'application Du Calcul Des Probabilités à La Philosophie Naturelle", Connaissance Des Temps Ou Des Mouvements Célestes, à l'usage Des Astronomes et Des Navigateurs: Pour l'an 1818 (Mme Ve Courcier, Paris, 1815), pp. 361–377.
[185] P. S. Laplace, "Mémoire sur la probabilité des causes par les événements", Memoires de l'Academie royale des Sciences de Paris (Savants étrangers) **6**, 621–656 (1774).
[186] M. S. Leifer and R. W. Spekkens, Towards a formulation of quantum theory as a causally neutral theory of Bayesian inference. Physical Review A **88**, 052130 (2013).
[187] K. Li, "Second-order asymptotics for quantum hypothesis testing", The Annals of Statistics **42**, 171–189 (2014).

[188] D. A. Lidar and T. A. Brun, eds., *Quantum Error Correction* (Cambridge University Press, Cambridge, 2013).
[189] E. H. Lieb, "Convex trace functions and the Wigner–Yanase–Dyson conjecture", Advances in Mathematics **11**, 267–288 (1973).
[190] E. H. Lieb and M. B. Ruskai, "Proof of the strong subadditivity of quantum-mechanical entropy", Journal of Mathematical Physics **14**, 1938–1941 (1973).
[191] G. Lindblad, "Completely positive maps and entropy inequalities", Communications in Mathematical Physics **40**, 147–151 (1975).
[192] S. Lloyd, "Capacity of the noisy quantum channel", Physical Review A **55**, 1613 (1997).
[193] H.-K. Lo and H. F. Chau, Unconditional security of quantum key distribution over arbitrarily long distances. Science **283**, 2050–2056 (1999).
[194] G. Ludwig, *Foundations of Quantum Mechanics I* (Springer, New York, 1983).
[195] E. Lutz and S. Ciliberto, "Information: from Maxwell's demon to Landauer's eraser", Physics Today **68**, 30–35 (2015).
[196] H. Maassen and J. B. M. Uffink, "Generalized entropic uncertainty relations", Physical Review Letters **60**, 1103 (1988).
[197] L. Mach, "Über einen Interferenzrefraktor", Zeitschrift für Instrumentenkunde **12**, 89–93 (1892).
[198] D. J. C. MacKay, *Information Theory, Inference & Learning Algorithms* (Cambridge University Press, Cambridge, 2002).
[199] O. Maroney, "Information Processing and Thermodynamic Entropy", *The Stanford Encyclopedia of Philosophy*, edited by E. N. Zalta (Metaphysics Research Lab, Stanford University, 2009).
[200] K. Maruyama, F. Nori, and V. Vedral, "Colloquium: the physics of Maxwell's demon and information", Reviews of Modern Physics **81**, 1–23 (2009).
[201] U. Maurer and R. Renner, "Abstract Cryptography", Proceedings of Innovations in Computer Science (ICS 2011) (2011), pp. 1–21.
[202] J. C. Maxwell, *Theory of Heat* (Longmans, Green and Co., London, 1871). Reprinted by Cambridge University Press (Cambridge, 2011).
[203] D. G. Mayo, *Error and the Growth of Experimental Knowledge*, Science and Its Conceptual Foundations (University of Chicago Press, Chicago, 1996).
[204] J. Mehra and H. Rechenberg, *The Completion of Quantum Mechanics, 1926–1941*, Vol. 6, The Historical Development of Quantum Theory, 1 (Springer, New York, 2000).
[205] A. R. Michaelis, *From Semaphore to Satellite* (International Telecommunication Union, Geneva, 1965).
[206] F. Miller, *Telegraphic Code to Insure Privacy and Secrecy in the Transmission of Telegrams* (C. M. Cornwell, New York, 1882).
[207] L. Mlodinow, *The Drunkard's Walk: How Randomness Rules Our Lives* (Pantheon Books, New York, 2008).
[208] H. Nagaoka, "Strong Converse Theorems in Quantum Information Theory", Proceedings of the ERATO Conference on Quantum Information Science (EQIS) (2001).
[209] M. A. Naimark, "Spectral functions of a symmetric operator", Izvestiya Akademii Nauk SSSR. Seriya Matematicheskaya **4**, 277–318 (1940). In Russian.
[210] J. Neyman and E. S. Pearson, "On the problem of the most efficient tests of statistical hypotheses", Philosophical Transactions of the Royal Society A: Mathematical, Physical and Engineering Sciences **231**, 289–337 (1933).
[211] M. A. Nielsen and I. L. Chuang, *Quantum Computation and Quantum Information*, Tenth anniversary edition (Cambridge University Press, Cambridge, 2010).
[212] M. Nussbaum and A. Szkoła, "The Chernoff lower bound for symmetric quantum hypothesis testing", The Annals of Statistics **37**, 1040–1057 (2009).

[213] T. Ogawa and M. Hayashi, A new proof of the direct part of Stein's lemma in quantum hypothesis testing, arXiv:quant-ph/0110125 (2001).
[214] T. Ogawa and H. Nagaoka, "Strong Converse and Stein's lemma in quantum hypothesis testing", IEEE Transactions on Information Theory **46**, 2428–2433 (2000).
[215] J. Ortigoso, "Twelve years before the quantum no-cloning theorem", American Journal of Physics **86**, 201–205 (2018).
[216] J. L. Park, "The concept of transition in quantum mechanics", Foundations of Physics **1**, 23–33 (1970).
[217] V. Paulsen, *Completely Bounded Maps and Operator Algebras*, Vol. 78, Cambridge Studies in Advanced Mathematics (Cambridge University Press, Cambridge, 2003).
[218] O. Penrose, *Foundations of Statistical Mechanics: A Deductive Treatment*, Vol. 22, International Series of Monographs in Natural Philosophy (Pergamon Press, Oxford, 1970).
[219] A. Peres, "Unperformed experiments have no results", American Journal of Physics **46**, 745 (1978).
[220] A. Peres, *Quantum Theory: Concepts and Methods*, Vol. 72, Fundamental Theories of Physics (Kluwer Academic Publishers, New York, 2002).
[221] A. Peres, "How the no-vloning theorem got its name", Fortschritte der Physik **51**, 458–461 (2003).
[222] J. Pierce, "The early days of information theory", IEEE Transactions on Information Theory **19**, 3–8 (1973).
[223] M. S. Pinsker, *Information and Information Stability of Random Variables and Processes* (Holden-Day, San Francisco, 1964).
[224] Y. Polyanskiy, H. Poor, and S. Verdú, "Channel coding rate in the finite blocklength regime", IEEE Transactions on Information Theory **56**, 2307–2359 (2010).
[225] Y. Polyanskiy and Y. Wu, "Lecture notes on information theory" (2019).
[226] S. Popescu and D. Rohrlich, "Quantum nonlocality as an axiom", Foundations of Physics **24**, 379–385 (1994).
[227] C. Portmann and R. Renner, "Cryptographic security of quantum key distribution", arXiv:1409.3525 [quant-ph] (2014).
[228] C. Portmann and R. Renner, "Security in quantum cryptography", arXiv:2102.00021 [quant-ph] (2021).
[229] R. T. Powers and E. Størmer, "Free states of the canonical anticommutation relations", Communications in Mathematical Physics **16**, 1–33 (1970).
[230] J. Preskill, "Lecture Notes for Ph219/CS219: Quantum Computation" (2004).
[231] W. Pusz and S. L. Woronowicz, "Functional calculus for sesquilinear forms and the purification map", Reports on Mathematical Physics **8**, 159–170 (1975).
[232] E. M. Rains, "Bound on distillable entanglement", Physical Review A **60**, 179 (1999).
[233] E. M. Rains, "A semidefinite program for distillable entanglement", IEEE Transactions on Information Theory **47**, 2921–2933 (2001).
[234] F. P. Ramsey and R. B. Braithwaite, "Truth and Probability", *The Foundations of Mathematics and other Logical Essays*, edited by B. Braithwaite (Kegan Paul, Trench, Trubner & Co, London, 1931), pp. 156–198. Reprinted in *Readings in Formal Epistemology: Sourcebook*, edited by H. Arló-Costa, V. F. Hendricks, and J. van Benthem, Springer Graduate Texts in Philosophy (Springer, Cham, Switzerland, 2016), pp. 21–45.
[235] M. Reimpell and R. F. Werner, "Iterative optimization of quantum error correcting codes", Physical Review Letters **94**, 080501 (2005).
[236] A. Reiserer, S. Ritter, and G. Rempe, "Nondestructive detection of an optical photon", Science **342**, 1349–1351 (2013).
[237] J. M. Renes, "Uncertainty relations and approximate quantum error correction", Physical Review A **94**, 032314 (2016).

[238] J. M. Renes and J.-C. Boileau, “Physical underpinnings of privacy”, Physical Review A **78**, 032335 (2008).

[239] J. M. Renes and J.-C. Boileau, “Conjectured strong complementary information tradeoff”, Physical Review Letters **103**, 020402 (2009).

[240] J. M. Renes and R. Renner, One-shot classical data compression with quantum side information and the distillation of common randomness or secret keys. IEEE Transactions on Information Theory **58**, 1985–1991 (2012).

[241] R. Renner, “Security of quantum key distribution”, PhD thesis (ETH Zürich, 2005).

[242] R. Renner and R. König, “Universally Composable Privacy Amplification Against Quantum Adversaries”, Proceedings of the Second Theory of Cryptography Conference (2005), pp. 407–425.

[243] H. P. Robertson, “The uncertainty principle”, Physical Review **34**, 163 (1929).

[244] R. T. Rockafellar, *Convex Analysis* (Princeton University Press, Princeton, 1970).

[245] J. S. Rosenthal, “Monty Hall, Monty Fall, Monty Crawl”, Math Horizons **16**, 5–7 (2008).

[246] S. M. Ross, *Introduction to Probability and Statistics for Engineers and Scientists*, Fifth edition (Academic Press, London, 2014).

[247] A. Rényi, *Foundations of Probability* (Holden-Day, San Francisco, 1970). Reprinted by Dover Publications (Mineola, NY, 2007).

[248] E. Schmidt, “Zur Theorie der linearen und nichtlinearen Integralgleichungen”, Mathematische Annalen **63**, 433–476 (1907).

[249] E. Schrödinger, “Discussion of probability relations between separated systems”, Mathematical Proceedings of the Cambridge Philosophical Society **31**, 555–563 (1935).

[250] E. Schrödinger, “Probability relations between separated systems”, Mathematical Proceedings of the Cambridge Philosophical Society **32**, 446–452 (1936).

[251] B. Schumacher, “Quantum coding”, Physical Review A **51**, 2738 (1995).

[252] B. Schumacher, “Sending entanglement through noisy quantum channels”, Physical Review A **54**, 2614 (1996).

[253] B. Schumacher and M. A. Nielsen, “Quantum data processing and error correction”, Physical Review A **54**, 2629 (1996).

[254] B. Schumacher and M. D. Westmoreland, “Sending classical information via noisy quantum channels”, Physical Review A **56**, 131 (1997).

[255] B. Schumacher and M. D. Westmoreland, “Approximate quantum error correction”, Quantum Information Processing **1**, 5–12 (2002).

[256] M. O. Scully and K. Drühl, Quantum eraser: a proposed photon correlation experiment concerning observation and ‘delayed choice’ in quantum mechanics. Physical Review A **25**, 2208–2213 (1982).

[257] S. Selvin, “A problem in probability (letter to the editor)”, The American Statistician **29**, 67 (1975).

[258] C. E. Shannon, “A mathematical theory of communication”, Bell System Technical Journal **27**, 379–423 (1948).

[259] C. E. Shannon, “Communication theory of secrecy systems”, Bell System Technical Journal **28**, 656 (1949).

[260] F. R. Shapiro, ed., *The Yale Book of Quotations* (Yale University Press, New Haven, 2006).

[261] P. W. Shor, “Algorithms for Quantum Computation: Discrete Logarithms and Factoring”, Proceedings of the 35th Annual Symposium on Foundations of Computer Science (1994), pp. 124–134.

[262] P. W. Shor, “Scheme for reducing decoherence in quantum computer memory”, Physical Review A **52**, R2493 (1995).

[263] P. W. Shor, “The quantum channel capacity and coherent information”, MSRI Workshop on Quantum Computation (2002).

[264] P. W. Shor and J. Preskill, "Simple proof of security of the BB84 quantum key distribution protocol", Physical Review Letters **85**, 441 (2000).
[265] P. W. Shor and J. A. Smolin, "Quantum error-correcting codes need not completely reveal the error syndrome", arXiv:quant-ph/9604006 (1996).
[266] B. Skyrms, *Choice and Chance: An Introduction to Inductive Logic*, Fourth edition (Wadsworth/Thomson Learning, Belmont, CA, 2000).
[267] M. L. Slater, "Lagrange multipliers revisted", Cowles Commission Discussion Paper, Math 403 (1950).
[268] D. Slepian and J. K. Wolf, "Noiseless coding of correlated information sources", IEEE Transactions on Information Theory **19**, 471–480 (1973).
[269] G. Smith and J. Yard, "Quantum communication with zero-capacity channels", Science **321**, 1812–1815 (2008).
[270] R. W. Spekkens, "Evidence for the epistemic view of quantum states: a toy theory", Physical Review A **75**, 032110 (2007).
[271] R. W. Spekkens, "Reassessing claims of nonclassicality for quantum interference phenomena", Perimeter Institute Seminar (2016).
[272] A. Steane, "Multiple-particle interference and quantum error correction", Proceedings of the Royal Society A **452**, 2551–2577 (1996).
[273] A. M. Steane, "Error correcting codes in quantum theory", Physical Review Letters **77**, 793 (1996).
[274] S. M. Stigler, *The History of Statistics: The Measurement of Uncertainty before 1900* (Belknap Press of Harvard University Press, Cambridge, 2003).
[275] W. F. Stinespring, "Positive functions on C*-algebras", Proceedings of the American Mathematical Society **6**, 211–216 (1955).
[276] A. D. Stone, *Einstein and the Quantum: The Quest of the Valiant Swabian* (Princeton University Press, Princeton, 2013).
[277] G. Strang, *Linear Algebra and Its Applications*, Fourth edition (Thomson, Brooks/Cole, Belmont, CA, 2006).
[278] E. C. G. Sudarshan, P. M. Mathews, and J. Rau, "Stochastic dynamics of quantum-mechanical systems", Physical Review **121**, 920–924 (1961).
[279] L. Szilárd, "Über die Entropieverminderung in einem thermodynamischen System bei Eingriffen intelligenter Wesen", Zeitschrift für Physik **53**, 840–856 (1929). Trans. by and C. Eckart, M. Knoller, and A. Rapoport as "On the decrease of entropy in a thermodynamic system by the intervention of intelligent beings", Behavioral Science **9**, 301–310 (1964).
[280] M. Tomamichel, *Quantum Information Processing with Finite Resources*, Vol. 5, SpringerBriefs in Mathematical Physics (Springer, Cham, Switzerland, 2016).
[281] M. Tomamichel and M. Hayashi, A hierarchy of information quantities for finite block length analysis of quantum tasks. IEEE Transactions on Information Theory **59**, 7693–7710 (2013).
[282] M. Tomamichel and A. Leverrier, "A largely self-contained and complete security proof for quantum key distribution", Quantum **1**, 14 (2017).
[283] M. Tribus and E. C. McIrvine, "Energy and information", Scientific American **224**, 178–184 (1971).
[284] T. Tsurumaru and M. Hayashi, Dual universality of hash functions and its applications to quantum cryptography. IEEE Transactions on Information Theory **59**, 4700–4717 (2013).
[285] A. Uhlmann, "The 'transition probability' in the state space of a *-algebra", Reports on Mathematical Physics **9**, 273–279 (1976).
[286] A. Uhlmann, "Relative entropy and the Wigner–Yanase–Dyson–Lieb concavity in an interpolation theory", Communications in Mathematical Physics **54**, 21–32 (1977).
[287] H. Umegaki, "Conditional expectation in an operator algebra. IV, Entropy and information", Kodai Mathematical Seminar Reports **14**, 59–85 (1962).

[288] J. van Tiel, *Convex Analysis: An Introductory Text* (Wiley, Chichester, UK, 1984).
[289] L. Vandenberghe and S. Boyd, “Semidefinite programming”, SIAM Review **38**, 49–95 (1996).
[290] V. S. Varadarajan, “Probability in physics and a theorem on simultaneous observability”, Communications on Pure and Applied Mathematics **15**, 189–217 (1962).
[291] G. S. Vernam, Cipher printing telegraph systems for secret wire and radio telegraphic communications. Transactions of the American Institute of Electrical Engineers **45**, 295–301 (1926).
[292] J. von Neumann, “Wahrscheinlichkeitstheoretischer Aufbau der Quantenmechanik”, Nachrichten von der Gesellschaft der Wissenschaften zu Göttingen, Mathematisch-Physikalische Klasse **1927**, 245–272 (1927).
[293] J. von Neumann, *Mathematische Grundlagen der Quantenmechanik* (Springer, Berlin, 1932). Trans. by R. T. Beyer as *Mathematical Foundations of Quantum Mechanics* (Princeton University Press, Princeton, 2018).
[294] J. von Neumann, “Discussion of a maximum problem” (1947). Reprinted in A. H. Taub, ed., *John von Neumann: Collected Works*, Vol. 6 (Pergamon Press, Oxford, 1963).
[295] J. von Neumann, “Various techniques used in connection with random digits”, Applied Mathematics Series (US National Bureau of Standards **12**, 36–38 (1951).
[296] L. Wang and R. Renner, “One-shot classical-quantum capacity and hypothesis testing”, Physical Review Letters **108**, 200501 (2012).
[297] J. Watrous, “Semidefinite programs for completely bounded norms”, Theory of Computing **5**, 217–238 (2009).
[298] J. Watrous, “Simpler semidefinite programs for completely bounded norms”, Chicago Journal of Theoretical Computer Science **2013**, 8 (2013).
[299] R. F. Werner, Quantum Information Theory—An Invitation, *Quantum Information*, Vol. 173, Springer Tracts in Modern Physics (Springer, Berlin, 2001), pp. 14–57.
[300] R. F. Werner and T. Farrelly, “Uncertainty from Heisenberg to today”, Foundations of Physics **49**, 460–491 (2019).
[301] J. E. Whitesitt, *Boolean Algebra and Its Applications* (Addison-Wesley, Reading, MA, 1961). Reprinted by Dover Publications (Mineola, NY, 1995).
[302] N. Wiener, “A new concept of communication engineering”, Electronics: The International Magazine of Electronics Technology **22**, 74–77 (1949).
[303] S. Wiesner, “Conjugate coding”, SIGACT News **15**, 78–88 (1983).
[304] E. P. Wigner, “The problem of measurement”, American Journal of Physics **31**, 6–15 (1963).
[305] A. Winter, “Coding theorem and strong converse for quantum channels”, Information Theory, IEEE Transactions on **45**, 2481–2485 (1999).
[306] A. Winter, “Tight uniform continuity bounds for quantum entropies: conditional entropy, relative entropy distance and energy constraints”, Communications in Mathematical Physics 347, 291–313 (2016).
[307] M. M. Wolf, *Quantum Channels and Operations: A Guided Tour*. Lecture Notes (2012).
[308] R. Wolf, *Quantum Key Distribution: An Introduction with Exercises*, Vol. 988, Lecture Notes in Physics (Springer, Cham, Switzerland, 2021).
[309] W. K. Wootters and W. H. Zurek, “A single quantum cannot be cloned”, Nature **299**, 802–803 (1982).
[310] A. D. Wyner, “The wire-tap channel”, Bell System Technical Journal **54**, 1355–1387 (1975).
[311] H. Yuen, R. Kennedy, and M. Lax, “Optimum testing of multiple hypotheses in quantum detection theory”, IEEE Transactions on Information Theory **21**, 125–134 (1975).
[312] H. P. Yuen, R. S. Kennedy, and M. Lax, “On optimal quantum receivers for digital signal detection”, Proceedings of the IEEE **58**, 1770–1773 (1970).
[313] G. U. Yule, “Critical notice”, British Journal of Psychology. General Section **12**, 100–107 (1921).
[314] L. Zehnder, “Ein neuer Interferenzrefraktor”, Zeitschrift für Instrumentenkunde **11**, 275–285 (1891).
[315] F. Zhang, ed., *The Schur Complement and Its Applications*, Vol. 4, Numerical Methods and Algorithms (Springer, Boston, 2005).

Index of symbols

Many symbols used only in single chapters are not listed here.

https://doi.org/10.1515/9783110570250-025

Index

https://doi.org/10.1515/9783110570250-026

www.ingramcontent.com/pod-product-compliance
Lightning Source LLC
Chambersburg PA
CBHW061159240926
51846CB00006B/1058
9783110570243